Revise AS

AQA Physics

Graham Booth & David Brodie

Contents

Chapter 3 Waves, imaging and information

Chapter 4 Waves, particles and the Universe

The AS/A2 Level Physics course

AS and A2

All Physics A Level courses are in two parts, with three separate units or modules in each part. Most students will start by studying the AS (Advanced Subsidiary) course. Some will then go on to study the second part of the A Level course, called the A2. It is also possible to study the full A Level course, both AS and A2, in any order.

How will you be tested?

Assessment units

For AS Physics, you will be tested by three assessment units. For the full A Level in Physics, you will take a further three units. AS Physics forms 50% of the assessment weighting for the full A Level.

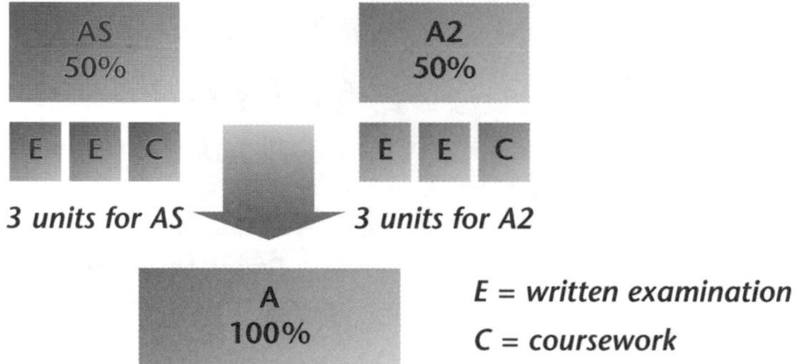

Units that are assessed by written examination can normally be taken in either January or June. Coursework units can be completed for assessment only in June. It is also possible to study the whole course before taking any of the unit tests. There is a lot of flexibility about when exams can be taken and the diagram below shows just some of the ways that the assessment units may be taken for AS and A Level Physics.

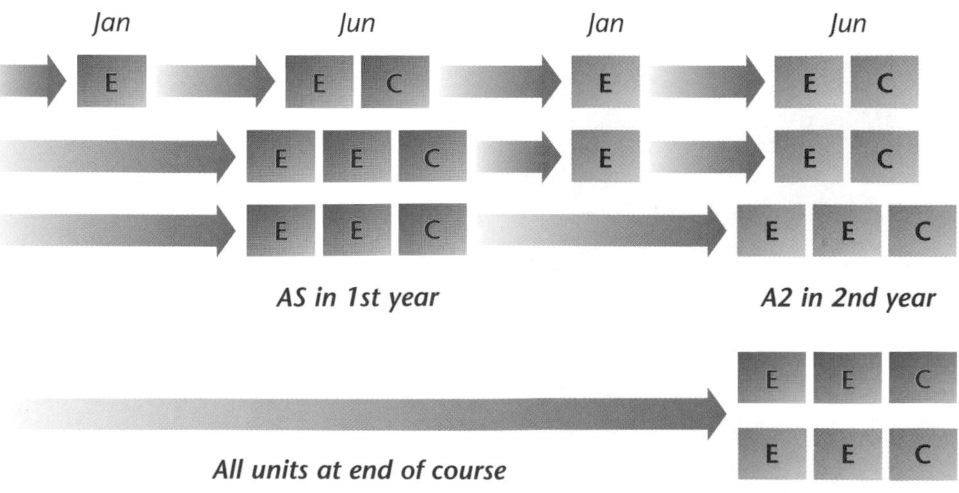

If you are disappointed with a unit result then you can re-sit the unit examination. The higher mark counts.

External assessment

For external assessment, written examinations are set and marked by teachers and lecturers from across the country, and overseen by the awarding bodies.

Internal assessment

Internal assessment is based on work that you do over a period of time, and is marked by your own teachers or lecturers. Some or all of your internally assessed work is based on practical and investigative skills and communication of your work.

What will you need to be able to do?

You will be tested by 'assessment objectives'. These are the skills and abilities that you should have acquired by studying the course. The assessment objectives for AS Physics are shown below.

Knowledge and understanding of science and how science works

- recall and understanding of scientific knowledge
- selection, organisation and communication of information in different forms

Application of knowledge and understanding of science and how science works

- analysis and evaluation of knowledge and processes
- applying knowledge and processes to unfamiliar situations
- assessing the validity, reliability and credibility of scientific information

How science works

- use ethical, safe and skilful practical methods
- select appropriate methods and make valid observations and measurements with appropriate precision and accuracy
- record and communicate observations and conclusions
- evaluate methods, results and conclusions

Specifications

AQA A Physics

ASSESSMENT UNIT	TOPIC, SECTION OR MODULE	CHAPTER REFERENCE	STUDIED	REVISED	PRACTICE QUESTIONS
1. Particles, quantum phenomena and electricity	*Particles and radiation*	4.1, 4.2, 4.3			
	Electromagnetic radiation and quantum phenomena	3.1, 4.3			
	Current electricity	2.1, 2.2, 2.3			
2. Mechanics, materials and waves	*Mechanics*	1.1, 1.2, 1.3, 1.4, 1.5, 1.6			
	Materials	1.7			
	Waves	3.1, 3.2, 3.3, 3.4			
3. Investigative and practical skills, internally assessed					

AS assessment analysis

Unit 1	*1 h 15 min test*	*40%*
Unit 2	*1 h 15 min test*	*40%*
Unit 3	*Internally assessed practical skills and investigative skills*	*20%*

AQA B Physics

ASSESSMENT UNIT	TOPIC, SECTION OR MODULE	CHAPTER REFERENCE	STUDIED	REVISED	PRACTICE QUESTIONS
1. Harmony and structure in the Universe	*The world of music*	*3.1, 3.2, 3.3, 3.4*			
	From quarks to quasars	*4.1, 4.2, 4.3, 4.4*			
2. Physics keeps us going	*Moving people, people moving*	*1.1, 1.2, 1.3, 1.4, 1.5, 1.6*			
	Energy and the environment	*1.4, 2.1, 2.2, 2.3*			
3. Investigative and practical skills, internally assessed					

A2 assessment analysis

Unit 1	*1 h 15 min test*	*40%*
Unit 2	*1 h 15 min test*	*40%*
Unit 3	*Internally assessed practical skills and investigative skills*	*20%*

Different types of questions in AS examinations

In AS Level Physics unit examinations, there may be combinations of short-answer questions and structured questions requiring both short answers and more extended answers. There may be some free-response and open-ended questions. There may also be some questions with a multiple-choice or 'objective' format.

Short-answer questions

A short-answer question may test recall or it may test understanding by requiring you to undertake a short, one-stage calculation. Short-answer questions normally have space for the answers printed on the question paper. Here are some examples (the answers are shown in *blue*):

What is the relationship between electric current and charge flow?

Current = rate of flow of charge.

The current passing in a heater is 6 A when it operates from 240 V mains. Calculate the power of the heating element.

$P = I \times V = 6\,A \times 240\,V = 1440\,W$

Which of the following is the correct unit of acceleration?

A $m\,s^{-1}$
B $s\,m^{-1}$
C $m\,s^{-2}$
D $m^2\,s^{-1}$

C $m\,s^{-2}$

Structured questions

Structured questions are in several parts. The parts are usually about a common context and they often become progressively more difficult and more demanding as you work your way through the question. They may start with simple recall, then test understanding of a familiar or an unfamiliar situation. The most difficult part of a structured question is usually at the end, where the candidate is sometimes asked to suggest a reason for a particular phenomenon or social implication.

When answering structured questions, do not feel that you have to complete one question before starting the next. The further you are into a question, the more difficult the marks are to obtain. If you run out of ideas, go on to the next question. Five minutes spent on the beginning of that question are likely to be much more fruitful than the same time spent racking your brains trying to think of an explanation for an unfamiliar phenomenon.

Here is an example of a structured question that becomes progressively more demanding.

(a) A car speeds up from $20\,m\,s^{-1}$ to $50\,m\,s^{-1}$ in 15 s.

Calculate the acceleration of the car.

acceleration = increase in velocity ÷ time taken
$= 30\,m\,s^{-1} \div 15\,s = 2\,m\,s^{-1}$

(b) The total mass of the car and contents is 950 kg.

Calculate the size of the unbalanced force required to cause this acceleration.

force = mass × acceleration
= 950 kg × 2 m s^{-2} = 1900 N

(c) Suggest why the size of the driving force acting on the car needs to be greater than the answer to (b).

The driving force also has to do work to overcome the resistive forces, e.g. air resistance and rolling resistance.

Extended answers

In AS Level Physics, questions requiring more **extended answers** will usually form part of structured questions. They will normally appear at the end of structured questions and be characterised by having at least three marks (and often more, typically five) allocated to the answers as well as several lines (up to ten) of answer space. These questions are also used to assess your abilities to communicate ideas and put together a logical argument.

The correct answers to extended questions are less well-defined than those for short-answer questions. Examiners may have a list of points for which credit is awarded up to the maximum for the question, or they may first of all judge the quality of your response as poor, satisfactory or good before allocating it a mark within a range that corresponds to that quality.

As an example of a question that requires an extended answer, a structured question on the use of solar energy could end with the following:

Suggest why very few buildings make use of solar energy in this country compared to countries in southern Europe. [5]

Points that the examiners might look for include:

- the energy from the Sun is unreliable due to cloud cover
- the intensity of the Sun's radiation is less in this country than in southern Europe due to the Earth's curvature
- more energy is absorbed by the atmosphere as the radiation has a greater depth of atmosphere to travel through
- fossil fuels are in abundant supply and relatively cheap
- the capital cost is high, giving a long payback time
- photo-voltaic cells have a low efficiency
- the energy is difficult to store for the times when it is needed the most

Full marks would be awarded for an argument that puts forward three or four of these points in a clear and logical way.

Free-response questions

AS Level Physics papers may or may not make use of free-response and open-ended questions. These types of questions allow you to choose the context and to develop your own ideas. Examples could include 'Describe a laboratory method of determining g, the value of free-fall acceleration' and 'Outline the evidence that suggests that light has a wave-like behaviour'. When answering these types of questions it is important to plan your response and present your answer in a logical order.

Exam technique

Links from GCSE

Advanced Subsidiary (AS) Physics builds from grade C in GCSE Science and GCSE Additional Science (combined) or GCSE Physics. This Study Guide has been written so that you will be able to tackle AS Physics from a GCSE Science background.

You should not need to search for important Physics from GCSE Science because this has been included where needed in each chapter. If you have not studied Science for some time, you should still be able to learn AS Physics using this text alone.

What are examiners looking for?

Examiners use instructions to help you to decide the length and depth of your answer. If a question does not seem to make sense, you may have misread it – read it again!

State, define or list

This requires a short, concise answer, often recall of material that can be learnt by rote.

Explain, describe or discuss

Some reasoning or reference to theory is required, depending on the context.

Outline

This implies a short response, almost a list of sentences or bullet points.

Predict or deduce

You are not expected to answer by recall but by making a connection between pieces of information.

Suggest

You are expected to apply your general knowledge to a 'novel' situation, one which you have not directly studied during the AS Physics course.

Calculate

This is used when a numerical answer is required. You should always use units in quantities and use significant figures with care. Look to see how many significant figures have been used for quantities in the question and give your answer to this degree of precision. If the question uses 3 (sig figs), then give your answer to 3 (sig figs) also.

Some dos and don'ts

Dos

Do answer the question

- No credit can be given for good Physics that is irrelevant to the question.

Do use the mark allocation to guide how much you write

- Two marks are awarded for two valid points – writing more will rarely gain more credit and could mean wasted time or even contradicting earlier valid points.

Do use diagrams, equations and tables in your responses

- Even in 'essay-type' questions, these offer an excellent way of communicating Physics. It is worth your while to practise drawing good clear and labelled sketches. Normally, you should do this without use of rulers or other drawing instruments.

Do write legibly

- An examiner cannot give marks if the answer cannot be read.

Do write using correct spelling and grammar. Structure longer essays carefully

- Marks are now awarded for the quality of your language in exams.

Don'ts

Don't fill up any blank space on a paper

- In structured questions, the number of dotted lines should guide the length of your answer.
- If you write too much, you waste time and may not finish the exam paper. You also risk contradicting yourself.

Don't write out the question again

- This wastes time. The marks are for the answer!

Don't contradict yourself

- The examiner cannot be expected to choose which answer is intended. You could lose a hard-earned mark.

Don't spend too much time on a part that you find difficult

- You may not have enough time to complete the exam. You can always return to a difficult calculation if you have time at the end of the exam.

What grade do you want?

Everyone would like to improve their grades but you will only manage this with a lot of hard work and determination. You should have a fair idea of your natural ability and likely grade in Physics and the hints below offer advice on improving that grade.

For a Grade A

You will need to be a very good all-rounder.

- You must go into every exam knowing the work extremely well.
- You must be able to apply your knowledge to new, unfamiliar situations.
- You need to have practised many, many exam questions so that you are ready for the type of question that will appear.

The exams test all areas of the specification and any weaknesses in your Physics will be found out. There must be no holes in your knowledge and understanding. For a Grade A, you must be competent in all areas.

For a Grade C

You must have a reasonable grasp of Physics but you may have weaknesses in several areas and you will be unsure of some of the reasons for the Physics.

- Many Grade C candidates are just as good at answering questions as the Grade A students but holes and weaknesses often show up in just some topics.
- To improve, you will need to master your weaknesses and you must prepare thoroughly for the exam. You must become a better all-rounder.

For a Grade E

You cannot afford to miss the easy marks. Even if you find Physics difficult to understand and would be happy with a Grade E, there are plenty of questions in which you can gain marks.

- You must memorise all definitions.
- You must practise exam questions to give yourself confidence that you do know some Physics. In exams, answer the parts of questions that you know first. You must not waste time on the difficult parts. You can always go back to these later.
- The areas of Physics that you find most difficult are going to be hard to score on in exams. Even in the difficult questions, there are still marks to be gained. Show your working in calculations because credit is given for a sound method. You can always gain some marks if you get part of the way towards the solution.

What marks do you need?

As a rough guide, you will need to score an average of 40% for a Grade E, 60% for a Grade C and 80% for a Grade A:

average	*80%*	*70%*	*60%*	*50%*	*40%*
grade	*A*	*B*	*C*	*D*	*E*

A* grades are awarded for the A level qualification only and not for the AS qualification or individual units.

To achieve an A* grade, you need to achieve a

- grade A overall (80% or more on uniform mark scale) for the whole A level qualification
- grade A* (90% or more on the uniform mark scale) across your A2 units.

Essential mathematics

This section describes some of the mathematical techniques that are needed in studying AS Level Physics.

Quantities and units

Physical quantities are described by the appropriate words or symbols, for example the symbol R is used as shorthand for the value of a *resistance*. The quantity that the word or symbol represents has both a numerical value and a unit, e.g. 10.5 Ω. When writing data in a table or plotting a graph, only the numerical values are entered or plotted. For this reason headings used in tables and labels on graph axes are always written as (physical quantity)/(unit), where the slash represents division. When a physical quantity is divided by its unit, the result is the numerical value of the quantity.

Resistance is an example of a **derived** quantity and the ohm is a derived unit. This means that they are defined in terms of other quantities and units. All derived quantities and units can be expressed in terms of the seven **base** quantities and units of the SI, or International System of Units.

The quantities, their units and symbols are shown in the table. The candela is not used in AS or A2 Level Physics.

Quantity	Unit	Symbol
length	metre	m
mass	kilogram	kg
time	second	s
electric current	ampere	A
temperature difference	kelvin	K
amount of substance	mole	mol
luminous intensity	candela	cd

Equations

Physical quantities and homogeneous equations

The equations that you use in Physics are relationships between physical quantities. The value of a physical quantity includes both the numerical value and the unit it is measured in.

An equation must be **homogeneous**. That is, the units on each side of the equation must be the same.

For example, the equation:

4 cats + 5 dogs = 9 camels

is nonsense because it is not homogeneous.

Likewise:

4A + 5V = 9 Ω

is not homogeneous, and makes no sense.

However, the equations:

4 cats + 5 cats = 9 cats

and:

4A + 5A = 9A

are homogeneous and correct.

Checking homogeneity in an equation is useful for:

- finding the units of a constant such as resistivity

- checking the possible correctness of an equation; if the units on each side are the same, the equation may be correct, but if they are different it is definitely wrong.

When to include units after values

A physicist will often write the equation:

$$4\,A \quad + \quad 5\,A \quad = \quad 9\,A$$

as:

$$4 \quad + \quad 5 \quad = \quad 9$$

$$\textbf{total current} \quad = \quad 9\,A$$

That is, it is permissible to leave out the unit in working, providing that the unit is given with the answer.

The equals sign

The = sign is at the heart of Mathematics, and it is a good habit never to use it incorrectly. In Physics especially, **the = sign is telling us that the two quantities either side are physically identical**. They may not look the same in the equation, but in the observable world we would not be able to tell the difference between one and the other. We cannot, for example, tell the difference between (4 + 5) cats and 9 cats however long we stare at them. This is a rule that is so obvious that people often forget it.

We can tell the difference between (4 + 5 + 1) cats and 9 cats (if they keep still for long enough).

(4 + 5 + 1) cats ≠ 9 cats

The ≠ sign means **not** equal. An equals sign in this 'equation' would not be telling the truth.

We cannot tell the difference between (4 + 5 + 1) cats and (9 + 1) cats.

(4 + 5 + 1) cats = (9 + 1) cats

The equals sign is telling the truth.

This reveals another rule that is also quite simple, but people often find it hard because they forget that = signs can only ever be used to tell the truth about indistinguishability. The new rule is that **you cannot change one side of an equation without changing the other in the same way.**

Rearranging equations

Often equations need to be used in a different form from that in which they are given or remembered. The equation needs to be rearranged. The rules for rearranging are:

- both sides of the equation must be changed in exactly the same way
- add and subtract before multiplying and dividing, and finally deal with roots and powers.

For example, suppose that you know the equation $v^2 = u^2 + 2as$ (and you know that the = sign here is telling the truth), and you want to work out u.

First, subtract 2as from both sides.

$$v^2 - 2as = u^2$$

(Check that you feel happy that this = sign is being honest. It will be if you have done the same thing to both sides of the original equation.)

Second, 'square root' both sides of the equation.

$$\sqrt{(v^2 - 2as)} = u$$

The hard part is not changing the equation (provided you remember the rules) but knowing what changes to make to get from the original form to the one you want.

It's not rocket science, but it does take practice.

For simple equations such as V = IR there is an alternative method, using the 'magic triangle'.

Write the equation in to the triangle. Cover up the quantity you want to work out. The pattern in the triangle tells you the equation you want.

The alternative method is to apply the rules as above, and change both sides of the equation in the same way. So if you divide both sides by R, you get V/R = I.

For equations such as these, it makes sense to use whichever method works best for you.

Drawing graphs

Graphs have a number of uses in Physics:

- they give an immediate, visual display of the relationship between physical quantities
- they enable the values of quantities to be determined
- they can be used to support or disprove a hypothesis about the relationship between variables.

When plotting a graph, it is important to remember that values determined by experiment are not exact. Every measurement has a certain level of accuracy and a certain level of precision.

An accurate measurement is one that agrees with the 'true' value. (Of course, the only way to find out a value of a physical quantity is to measure it, or to calculate it from other measured values. So the 'true' value is an ideal, and may be impossible to know.)

A precise measurement is one that agrees closely with repeated measurements. Precise measurements generally allow values to be given with confidence to more decimal places.

Perfect accuracy and perfect precision are fundamentally impossible. For these reasons, having plotted experimental values on a grid, the graph line is drawn as the best straight line or smooth curve that represents the points. Where there are 'anomalous' results, i.e. points that do not fit the straight line or curve, these should always be checked. If in doubt, ignore them, but do add a note in your experimental work to explain why you have ignored them and suggest how any anomalous results could have arisen.

Graphs, gradients and rates of change

Quantities such as velocity, acceleration and power are defined in terms of a **rate of change** of another quantity with time. This rate of change can be determined by calculating the gradient of an appropriate graph. For example, *velocity* is the *rate of change of displacement with time*. Its value is represented by the gradient of a displacement–time graph. Different techniques are used to determine the gradient of a straight line and a smooth curve. For a straight line:

- determine the value of Δy, the change in the value of the quantity plotted on the y-axis, using the whole of the straight line part of the graph

A common error when determining the gradient of a graph is to work it out using the gridlines only, without reference to the scales on each axis.

- determine the corresponding value of Δx
- calculate the gradient as $\Delta y \div \Delta x$

For a smooth curve, the gradient is calculated by first drawing a tangent to the curve and then using the above method to determine the gradient of the tangent. To draw a tangent to a curve:

- mark the point on the curve where the gradient is to be determined
- use a pair of compasses to mark in two points on the curve, close to and equidistant from the point where the gradient is to be determined
- join these points with a ruler and extend the line beyond each point.

These techniques are illustrated in the diagrams below.

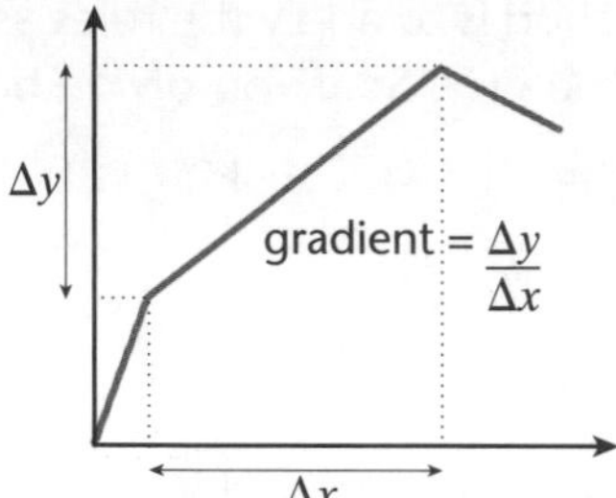

This calculation provides the value of the gradient of the middle section of the graph only. The other two sections have different gradients.

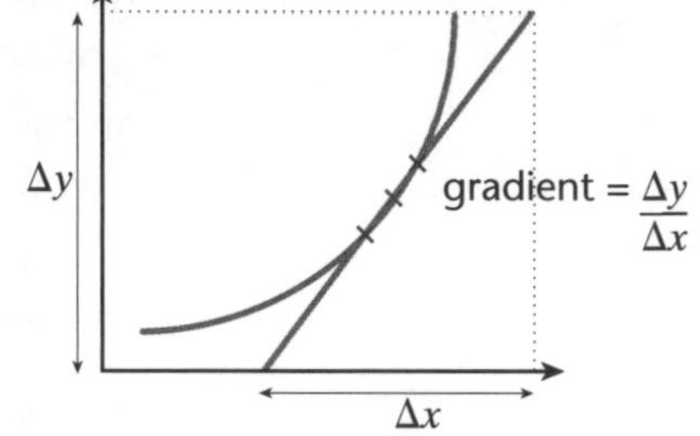

This method provides the value of the gradient at the middle of the three red marks. The graph has a continuously changing gradient that starts small and increases.

The area under a curve

The area between a graph line and the horizontal axis, often referred to as the 'area under the graph' can also yield useful information. This is the case when the product of the quantities plotted on the axes represents another physical quantity.

For example, on a *speed–time* graph this area represents the distance travelled. In the case of a straight line, the area can be calculated as that of the appropriate geometric figure. Where the graph line is curved, then the method of 'counting squares' is used.

- Count the number of complete squares between the graph line and the horizontal axis.
- Fractions of squares are counted as '1' if half the square or more is under the line, otherwise '0'.
- To work out the physical significance of each square, multiply together the quantities represented by one grid division on each axis.

These techniques are illustrated in the diagrams below:

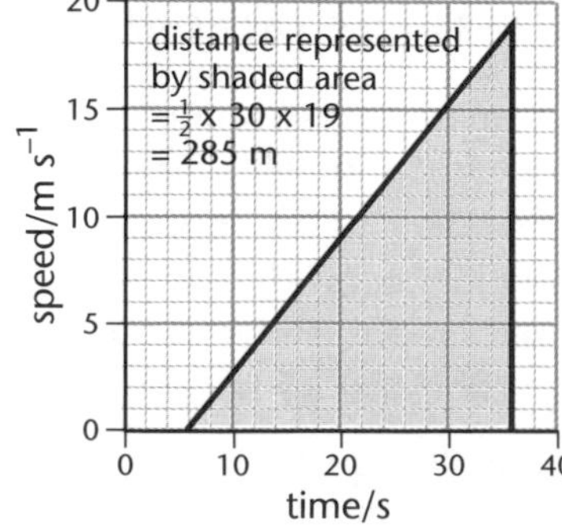

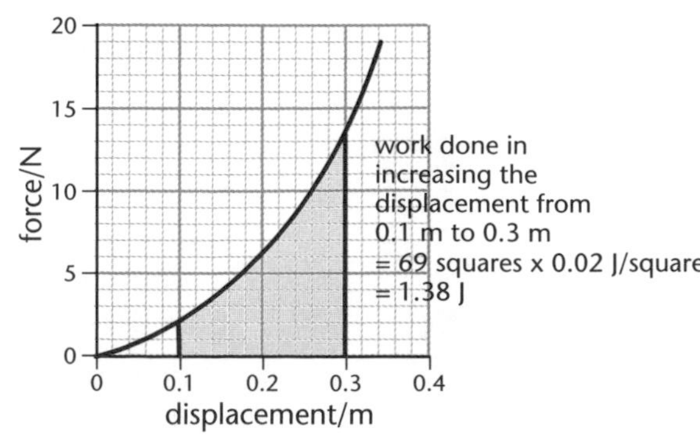

Equations from graphs

By plotting the values of two variable quantities on a suitable graph, it may be possible to determine the relationship between the variables. This is straightforward when the graph is a straight line, since all straight line graphs have an equation of the form $y = mx + c$, where m is the gradient of the graph and c is the value of y when x is zero, i.e. the intercept on the y-axis. The relationship between the variables is determined by finding the values of m and c.

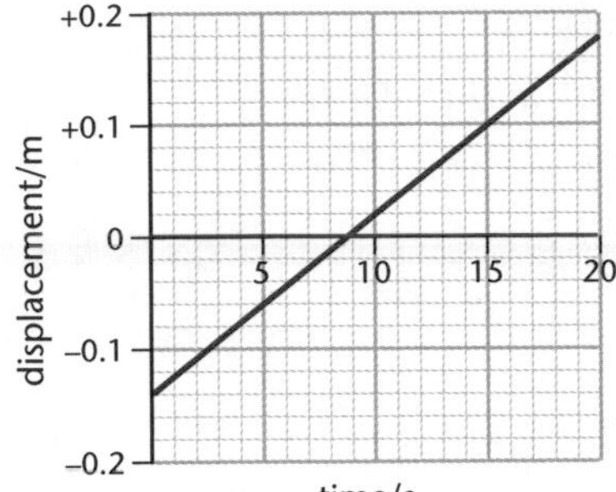

gradient = 0.32 ÷ 20
= $0.016\,\text{m s}^{-1}$
intercept on y-axis = −0.14

equation is s = 0.016t − 0.14

The straight line graph on the left shows how the displacement, s, of an object varies with time, t.

The equation that describes this motion is:

s = 0.016t − 0.14

(Note that no units are shown in the equation, for simplicity. The unit for displacement, s, is the metre and the unit for time, t, is the second. The physical value 0.016 has unit metre per second, and the physical value 0.14 is measured in metres.)

There is an important and special case of the equation $y = mx + c$. This is the case for which c is zero, so that $y = mx$. The straight line of this graph now passes through the graph's origin.

m is the gradient of the graph and is constant (since the graph is a single straight line). Note that different symbols can be used for the constant gradient in place of m. It is quite common, for example, to use k. That gives $y = kx$.

In Physics, we are often seeking simple patterns, and this is about as simple as patterns can be. If $y = mx$ or $y = kx$, whatever changes happen to x then y always changes by the same proportion.

Trigonometry and Pythagoras

In the right-angled triangle shown here, the sides are labelled o (opposite), a (adjacent) and h (hypotenuse). The relationships between the size of the angle θ and the lengths of these sides are shown on the left.

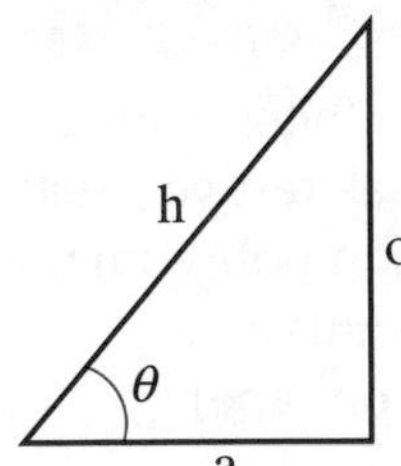

- sin θ = o/h
- cos θ = a/h
- tan θ = o/a

Pythagoras' theorem gives the relationship between the sides of a right-angled triangle: $h^2 = a^2 + o^2$

Multiples

For AS Level Physics, you are expected to be familiar with the following multiples of units:

Name	Multiple	Symbol
pico-	10^{-12}	p
nano-	10^{-9}	n
micro-	10^{-6}	μ
milli-	10^{-3}	m
kilo-	10^{3}	k
mega-	10^{6}	M
giga-	10^{9}	G

For example, the symbol MHz means 1×10^{6} Hz and mN means 1×10^{-3} N.

Four steps to successful revision

Step 1: Understand

- Mark up the text if necessary – underline, highlight and make notes.
- Re-read each paragraph slowly.

GO TO STEP 2

Step 2: Summarise

- Now make your own revision note summary:
  What is the main idea, theme or concept to be learnt?
  What are the main points? How does the logic develop?
  Ask questions: Why? How? What next?
- Use bullet points, mind maps, patterned notes.
- Link ideas with mnemonics, mind maps, crazy stories.
- Note the title and date of the revision notes
  (e.g. Physics: Electricity, 3rd March).
- Organise your notes carefully and keep them in a file.

This is now in **short-term memory**. You will forget 80% of it if you do not go to Step 3.
GO TO STEP 3, but first take a 10 minute break.

Step 3: Memorise

- Take 25 minute learning 'bites' with 5 minute breaks.
- After each 5 minute break test yourself:
  Cover the original revision note summary
  Write down the main points
  Speak out loud (record on tape)
  Tell someone else
  Repeat many times.

The material is well on its way to **long-term memory**.
You will forget 40% if you do not do Step 4. **GO TO STEP 4**

Step 4: Track/Review

- Create a Revision Diary (one A4 page per day).
- Make a revision plan for the topic, e.g. 1 day later, 1 week later, 1 month later.
- Record your revision in your Revision Diary, e.g.
  Physics: Electricity, 3rd March 25 minutes
  Physics: Electricity, 5th March 15 minutes
  Physics: Electricity, 3rd April 15 minutes
  ... revisit each topic at monthly intervals.

Chapter 1

Force, motion and energy

The following topics are covered in this chapter:

- Vectors
- Representing and predicting motion
- Force and acceleration
- Energy and work
- Vehicles in motion
- More effects of forces
- Force and materials

1.1 Vectors

LEARNING SUMMARY

After studying this section you should be able to:

- recognise and describe forces
- distinguish between a vector and a scalar quantity
- add together two vectors and subtract one vector from another
- split a vector into two parts at right angles to each other

Force as a vector quantity

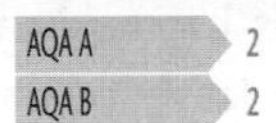

Forces can be of different types: **gravitational** forces are caused by and affect objects with mass, **friction** forces oppose relative motion and the force pushing up on you at the moment, the **normal contact** force, is due to compression of the material that you are sat on.

In this context 'normal' means 'at right angles to the surface'.

Forces also have things in common:

- all forces can be described as **object A pulls/pushes object B**
- all forces can be represented in both size and direction by an arrow on a diagram.

The first bullet point here is emphasising that all forces are caused by objects and they act on other objects.

Here are some examples.

A gravitational force

the Earth pulls the Moon

A friction force

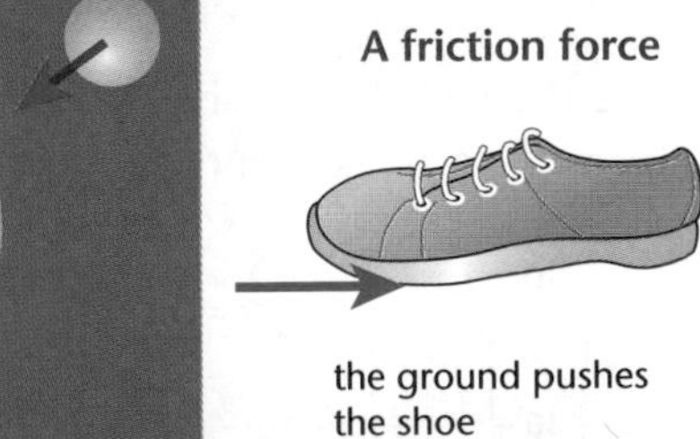

the ground pushes the shoe

A normal contact force

the chair pushes the person

Physical quantities that have direction as well as size are called **vectors**. Quantities with size only are **scalars**. Some examples are given in the table.

When representing a vector quantity on a diagram, an arrow is always used to show its direction.

Vectors	Scalars
force	mass
velocity	speed
acceleration	length
displacement	distance
field strength	energy

Weight as a vector acting at the centre of gravity

AQA A 2
AQA B 2

Gravitational forces act at a distance with no contact being necessary. These forces are always drawn as if they act at the **centre of gravity** of the object.

The centre of gravity of a traffic cone, a rubber ring and a person

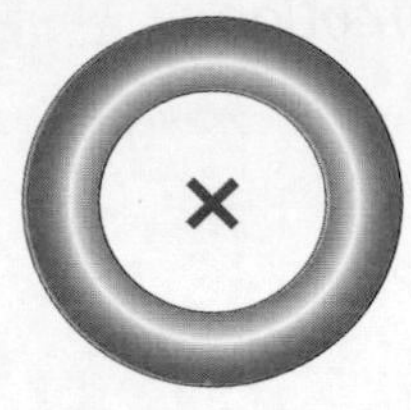

The gravitational force acting on an object due to the planet or moon whose surface it is on is known as its weight. The centre of gravity of the object is the point at which its weight can be considered to act.

More vectors

AQA A 2
AQA B 2

Information about the **speed** of an object only states how fast it is moving; the **velocity** also gives the direction. Speed is a scalar quantity while velocity is a vector quantity. For an object moving along a straight line, positive (+) and negative (–) are usually used to indicate movement in opposite directions.

As long as it is moving, the **distance** travelled by the object is increasing, but its **displacement** can increase or decrease and, like velocity, can have both positive and negative values.

The fixed point is usually the starting point of the motion or the object's rest position.

Displacement is a vector quantity; it specifies both the distance and direction of an object measured from a fixed point.

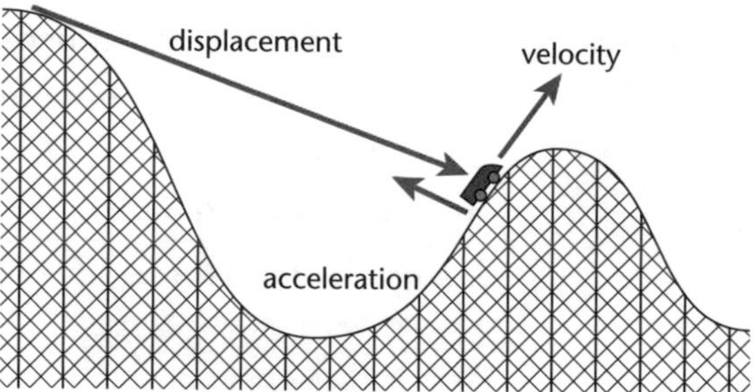

The diagram below shows a distance–time graph and a displacement–time graph for one particular journey.

Check that you understand why distance has only positive values, but displacement has both positive and negative values.

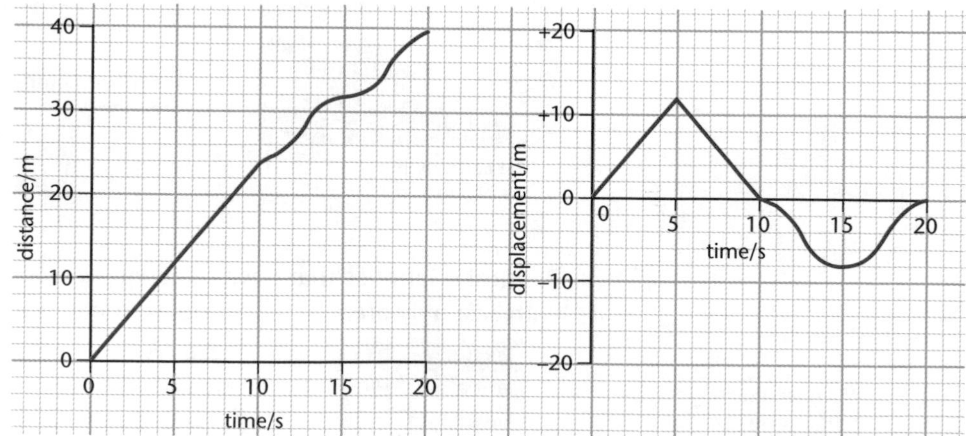

The same symbols are used for speed and velocity. When an object is changing speed/velocity, u is often used for the initial speed/velocity, with v being used for the final speed/velocity. The symbol c is usually reserved for the speed of light.

Information from the graphs can be used to calculate the **average speed** and **average velocity** over any time period, using the relationships:

KEY POINT

average speed	= distance travelled	÷ time taken	$v = \Delta d \div \Delta t$
average velocity	= displacement	÷ time taken	$v = \Delta s \div \Delta t$

These show that over the 20 s time period, the average speed is 2 m s^{-1} and the average velocity is 0.

Adding scalars and vectors

AQA A	2
AQA B	2

To add together two scalar quantities the normal rules of arithmetic apply, for example, 2 kg + 3 kg = 5 kg and no other answers are possible. When adding vector quantities, both the size and direction have to be taken into account.

What is the sum of a 2 N force and a 3 N force acting on the same object? The answer could be any value between 1 N and 5 N, depending on the directions involved.

The sum, or resultant, of two vectors such as two forces acting on a single object is the single vector that could replace the two and have the same effect.

There are three steps to finding the sum of two vectors. These are illustrated by working out the sum of a 2 N force acting up the page and a 3 N force acting from left to right, both forces acting on the same object.

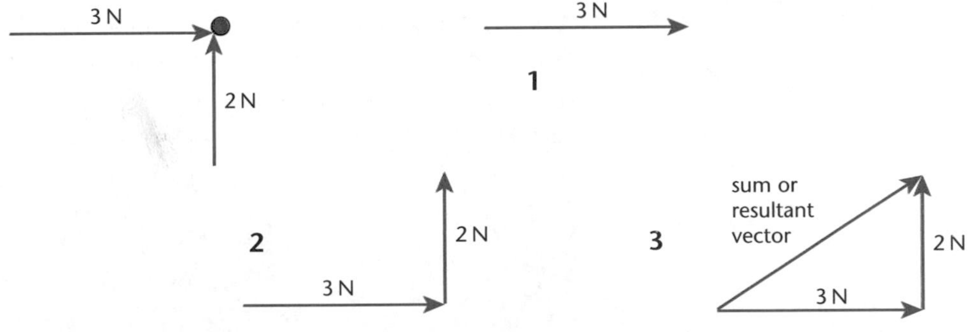

The order in which you draw the vectors does not matter; you should arrive at the same answer whichever one you start with.

- Draw an arrow that represents one of the vectors in both size and direction.
- Starting where this arrow finishes, draw an arrow that represents the second vector in size and direction.
- The sum, or resultant of the two vectors is represented (in both size and direction) by the single arrow drawn from the **start** of the first arrow to the **finish** of the second arrow.

In the example given, the size of the resultant force is 3.6 N and the direction is at an angle of 34° to the 3 N force. These figures were obtained by scale drawing.

Although a scale drawing is often the quickest way of working out the resultant of two vectors, the size and direction can also be calculated. This is straightforward when the vectors act at right angles, but needs more complex mathematics in other cases.

In the worked example:

The notation 'tan^{-1}' means 'the angle whose tangent is'.

- the size of the resultant force can be calculated using Pythagoras' theorem as $\sqrt{(2^2 + 3^2)}$
- the angle between the resultant force and the 3 N force can be calculated using the definition of tangent as $\tan^{-1}(2 \div 3)$.

Checking that the resultant is zero

AQA A 2
AQA B 2

The drawing method described above can be used to find the sum of any number of vectors by drawing an arrow for each vector, starting each new arrow where the previous one finished. The resultant of the vectors is then represented by the single arrow that starts at the beginning of the first vector and ends where the last one finishes. If the vectors being added together form a closed figure, i.e. the last one finishes where the first one starts, it follows that the sum is zero. This is what you would expect to find when working out the resultant force at a point in a stable structure, for example.

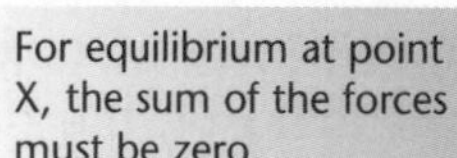

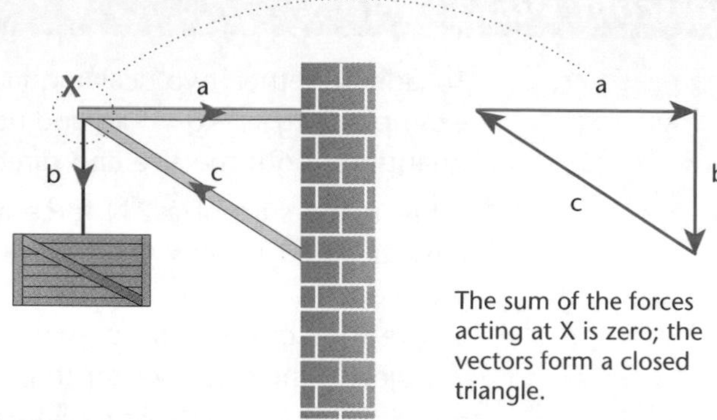

The sum of the forces acting at X is zero; the vectors form a closed triangle.

Splitting a vector in two

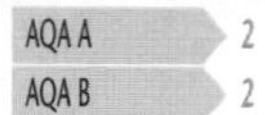

A vector has an effect in any direction except the one which is at right angles to it. Sometimes a vector has two independent effects which need to be isolated.

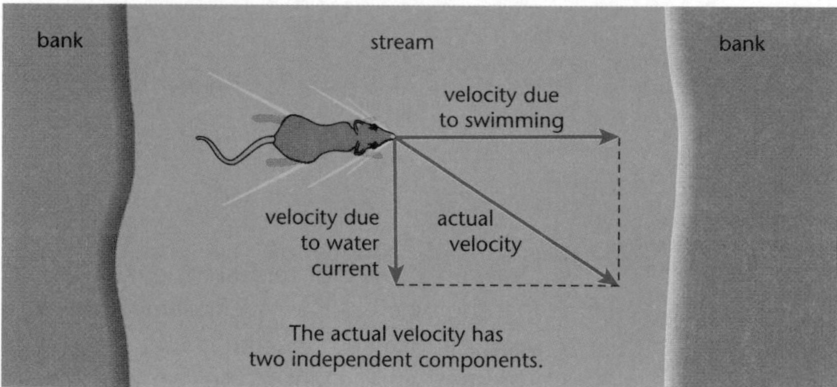

The actual velocity has two independent components.

Just as the combined effect of two vectors acting on a single object can be calculated, two separate effects of a single vector can be found by splitting the vector into two **components**. Provided that the directions of the two components are chosen to be at right angles, each one has no effect in the direction of the other so they are considered to act independently.

The process of splitting a vector into two components is known as **resolving** or **resolution of** the vector.

The diagram below shows the tension (T) in a cable holding a radio mast in place. The force is pulling the mast both vertically downwards and horizontally to the left.

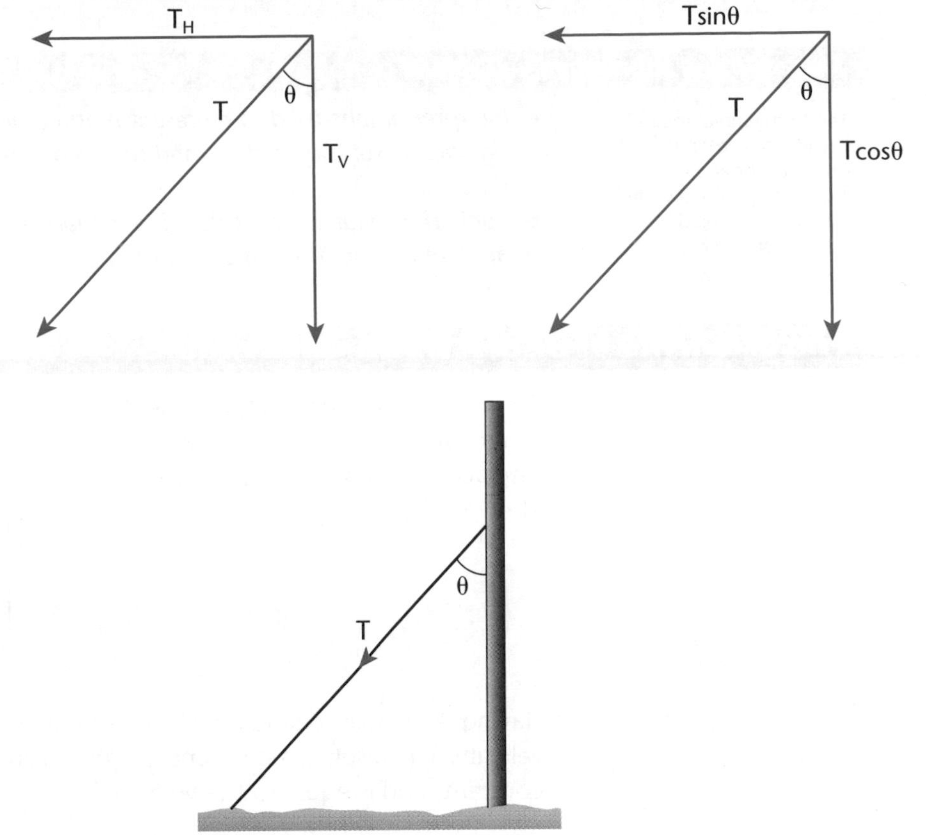

To find the effect of the tension in horizontal and vertical directions, **T** can be split into two components, shown as $\mathbf{T_H}$ (the horizontal component) and $\mathbf{T_V}$ (the vertical component) in the diagram. You should check that according to the rules of vector addition, $\mathbf{T_H} + \mathbf{T_V} = \mathbf{T}$.

> To find the magnitude of $\mathbf{T_H}$ and $\mathbf{T_V}$ the following rules apply:
> - the component of a vector **a** at an angle θ to its own angle is acosθ
> - the component of a vector **a** at an angle (90° – θ) to its own angle is asinθ
>
> KEY POINT

The application of these rules is shown in the diagram above.

Progress check

1 Work out the resultant of the two forces shown in the diagram.

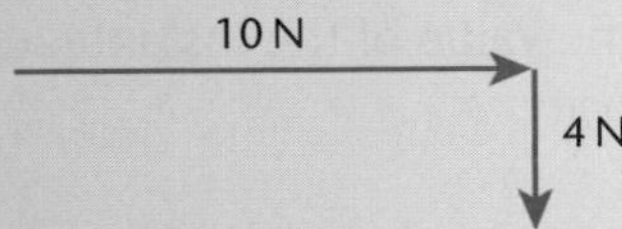

2 A car is being towed with a rope inclined at 20° to the horizontal. The tension in the rope is 350 N.
Work out the horizontal and vertical components of the tension in the rope.

2 Horizontal component = 329 N Vertical component = 120 N
1 10.8 N at an angle of 22° to the 10 N force

1.2 Representing and predicting motion

After studying this section you should be able to:

The term 'curve' means the line drawn on the graph to show the relationship between the quantities. It could be straight or curved.

- *interpret graphs used to represent motion, and understand the physical significance of the gradient and the area between the curve and the time axis*
- *apply the equations that describe motion with uniform acceleration*
- *analyse the motion of a projectile*

LEARNING SUMMARY

The gradients of displacement–time graphs

AQA A 2
AQA B 2

For a stationary body, displacement is unchanging. The gradient of a displacement–time graph is zero. The velocity is zero.

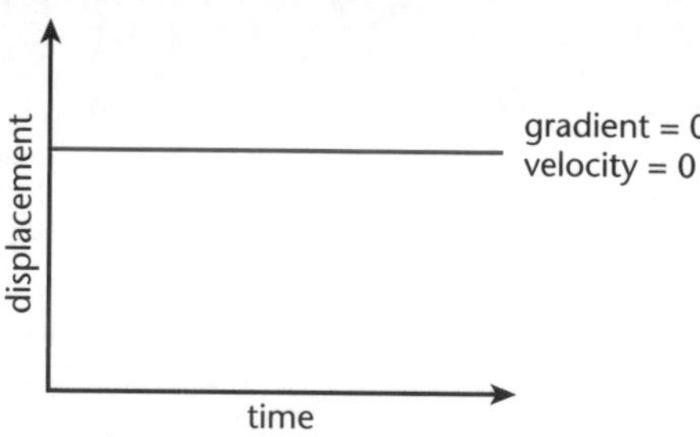

Having a steadily changing displacement is the same thing as having a steady velocity. The displacement–time graph is a straight line. The gradient of the line is constant, and is equal to the velocity.

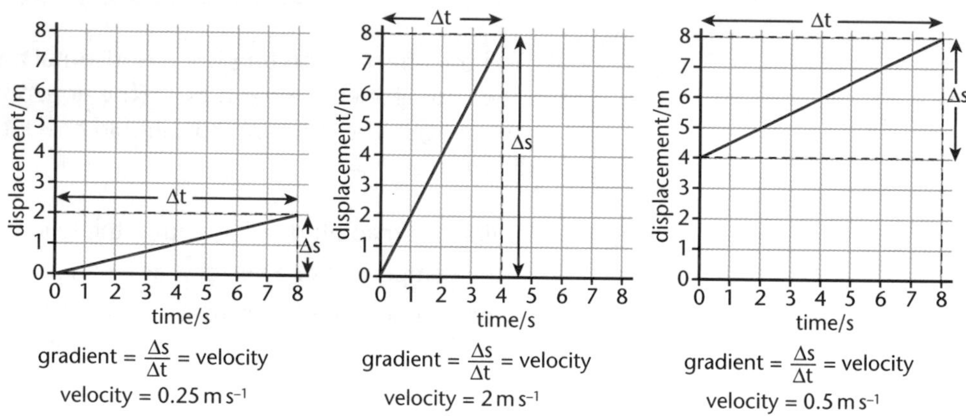

The motion of an accelerating car is a little more complicated. The displacement changes slowly to start with, and then changes faster and faster. For any point on the graph, the value of the gradient is the same as the value of the instantaneous velocity.

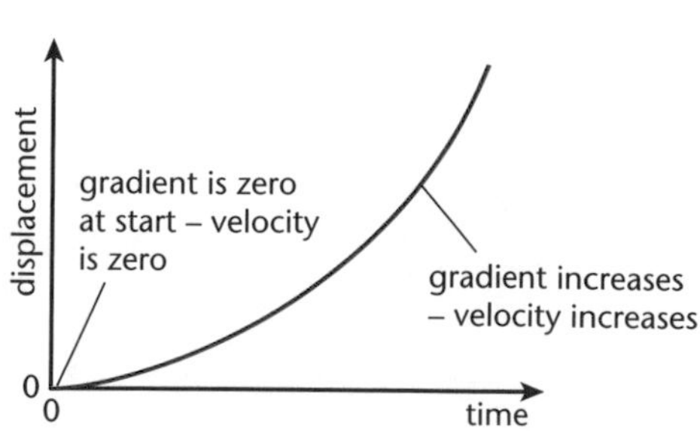

Average and instantaneous values

Calculations of the average velocity over a period of time do not show all of the changes that have taken place. **Instantaneous** values of velocity are defined in terms of **the rate of change** of displacement and are represented by the gradient of the displacement–time graph at a particular instant.

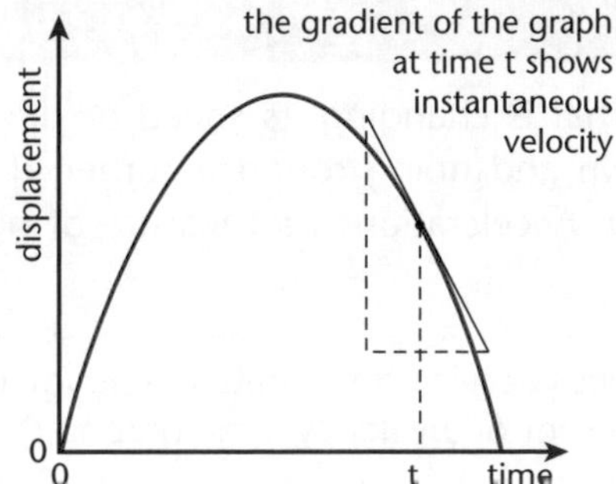

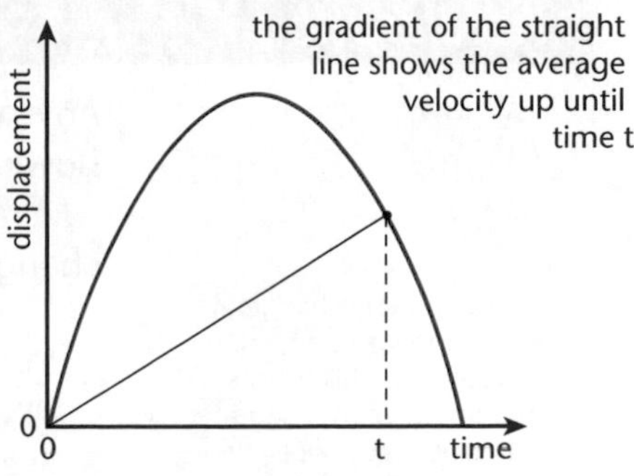

Note that at time t, the instantaneous velocity is negative but the average velocity for the whole journey so far is positive.

> Instantaneous velocity = rate of change of displacement
>
> It is represented by the gradient of a displacement–time graph at a particular instant.
>
> KEY POINT

The gradient of a displacement–time graph can have either a positive or a negative value, so it shows the direction of motion as well as the speed.

Displacement–time and velocity–time graphs

AQA A 2
AQA B 2

It is possible to represent the same motion in different ways. Think of the journey of a stone that is flicked upwards.

Firstly, think in terms of velocity. The initial flick gives it a large upwards velocity but this decreases as the stone rises. For one instant at the top of the journey, the velocity is zero, and then it becomes a downwards velocity. The upwards velocity is positive, and the downwards velocity is negative. The downwards velocity increases in size until the stone has fallen back to where it started.

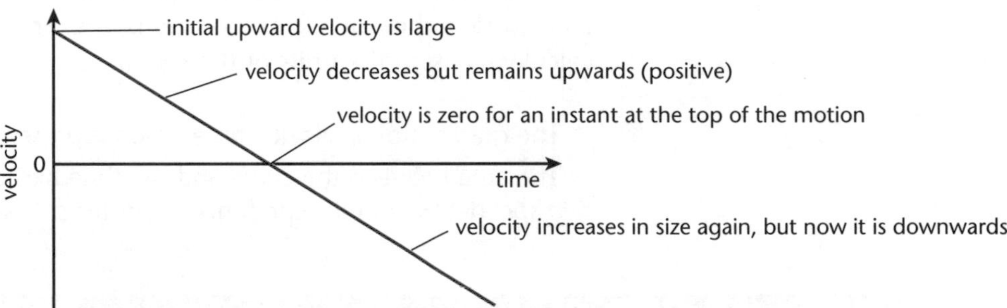

Now think about the displacement. It starts at zero, and is zero again when the stone gets back to where it started. In between, the displacement is always positive. But the displacement does not increase steadily when the stone is first flicked. It increases most rapidly at the start, when the stone is moving fast. Around the top of the journey, displacement changes slowly. The result is a curved line.

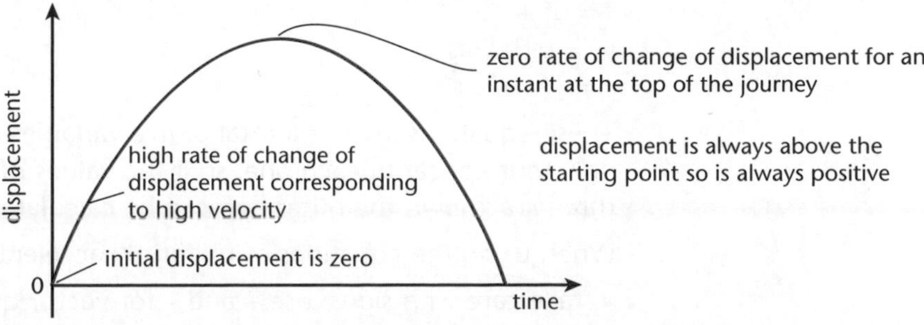

The graphs represent the same motion in different ways. They do not look the same.

Acceleration

AQA A 2
AQA B 2

Any object that is changing its speed or direction is **accelerating**. Speeding up, slowing down and going round a corner at constant speed are all examples of **acceleration**. Acceleration is a measure of how quickly the velocity of an object changes.

In Physics, any change in velocity involves an *acceleration*. This means that an object moving at constant speed, but changing direction, is accelerating.

> **KEY POINT**
> Instantaneous acceleration = rate of change of velocity. It is represented by the gradient of a velocity–time graph at a particular instant.
> average acceleration = change in velocity ÷ time $a = \Delta v \div \Delta t$
> Acceleration is a vector quantity and is measured in $m\,s^{-2}$.

More about velocity–time graphs

AQA A 2
AQA B 2

Since acceleration is equal to rate of change of velocity, it is also equal to the gradient of a velocity–time graph.

Remember, the area of a triangle = $\frac{1}{2}$ × base × height.

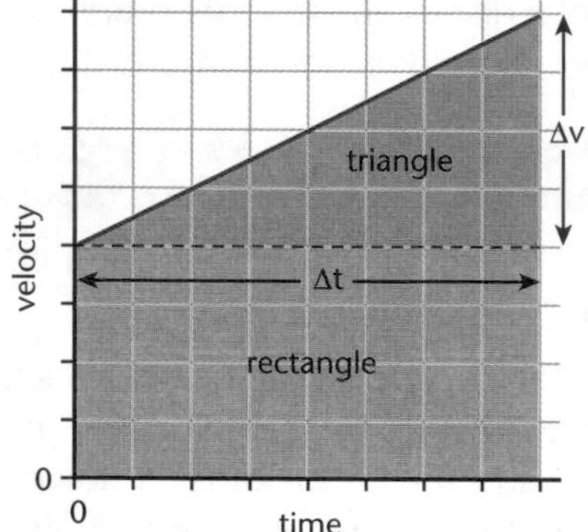

It is also possible to work out displacement from a velocity–time graph. It is the area 'under the curve'. That is, it is the area between the line and the time, calculated using the units of the two axes.

> **KEY POINT**
> The gradient of a velocity–time graph represents the acceleration.
> The area between the curve and the time axis of a velocity–time or a speed–time graph represents the distance travelled.

Uniform acceleration

AQA A 2
AQA B 2

The gradient of the velocity–time graph (see previous section) has a constant value; it represents a constant or **uniform** acceleration. A number of equations link the variable quantities when an object is moving with uniform acceleration. They are:

$v = u + at$

$s = ut + \frac{1}{2}at^2$

$v^2 = u^2 + 2as$

$s = \frac{1}{2}(u + v)t$

The symbols have the meanings already defined:
u = initial velocity
v = final velocity
s = displacement
a = acceleration
t = time

These equations involve a total of five variables but only four appear in each one, so if the values of three are known the other two can be calculated.

When using the equations of uniformly accelerated motion:

- take care with signs: use + and – for vector quantities such as velocity and acceleration that are in opposite directions
- remember that s represents displacement; this is not the same as the distance travelled if the object has changed direction during the motion.

Motion in two dimensions

AQA A 2
AQA B 2

People who play sports such as tennis and squash know that, no matter how hard they hit the ball, they cannot make it follow a horizontal path. The motion of a ball through the air is always affected by the Earth's pull.

There is a simple relationship between the total distances travelled after t, 2t etc. by an object falling vertically. Can you spot it?

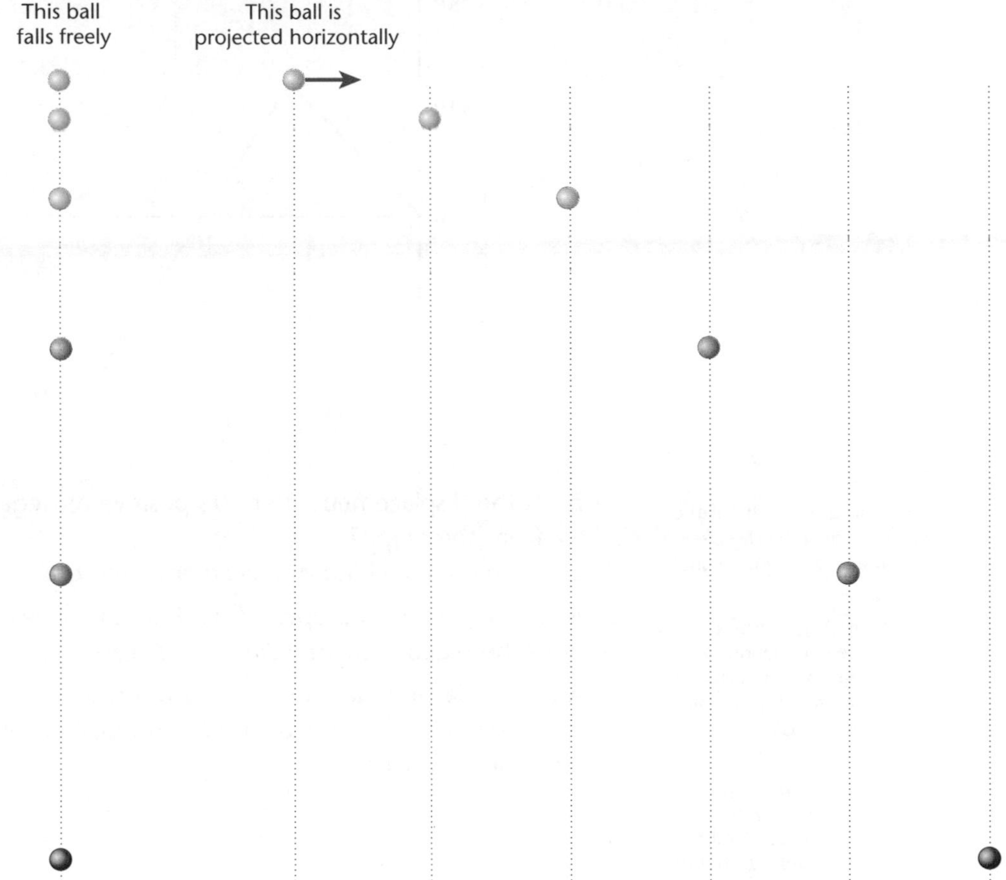

The diagram above shows the results of photographing the motion of a ball projected horizontally alongside one released so that it falls freely. The photographs are taken at equal time intervals.

The free-falling ball travels increasing distances in successive time intervals. This is because the vertical motion is accelerated motion, so the average speed of the ball over each successive time interval increases.

The ball projected horizontally travels equal distances horizontally in successive time intervals, showing that its horizontal motion is at **constant speed**. Vertically, its motion matches that of the free-falling ball, showing that **its vertical motion is not affected by its horizontal motion**.

The motion of any object can be resolved in two directions at right angles to each other. The two motions can then be treated separately.

KEY POINT: When an object has both horizontal and vertical motion, these are independent of each other.

This important result means that the horizontal and vertical motions can be analysed separately:

- for the horizontal motion at constant speed, the equation $\mathbf{v = s \div t}$ applies
- for the vertical, accelerated, motion the equations of motion with uniform acceleration apply.

Progress check

1 a The diagram below shows a velocity–time graph. Calculate the displacement and the acceleration for each labelled section of the graph.

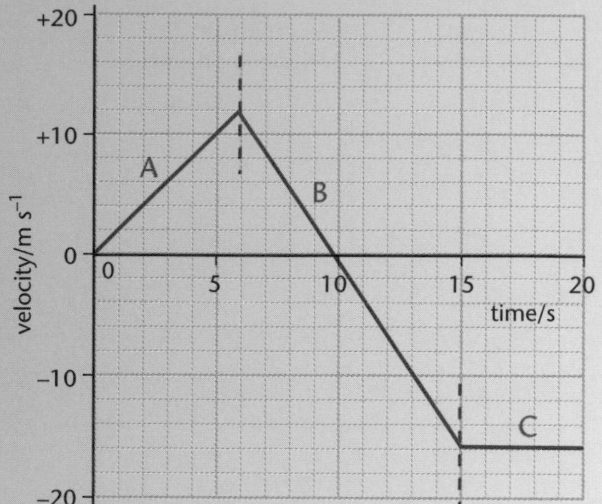

Hint for Q2: what is the velocity of the clay pigeon at its maximum height?

b Is the displacement after 20 s positive or negative? How can you tell this from the graph?

c Sketch a matching displacement–time graph for the whole journey.

- Use the constant speed equation to calculate the distance travelled horizontally by the ball in this time.
- Apply the equations of motion with uniform acceleration to the vertical motion to find the time that the ball is in the air. Note that vertically u = 0.

2 A clay pigeon is fired upwards with an initial velocity of 23 m s^{-1}. What height does the pigeon reach? Take g = 10 m s^{-2}.

3 A tennis player is standing 5.5 m from the net. She hits the ball horizontally at a speed of 32 m s^{-1}. How far has the ball dropped when it reaches the net? Take g = 10 m s^{-2}.

1 a A: 36 m; 2 m s^{-2};
B: 64 m; -3.1 m s^{-2}
C: 80 m; 0 m s^{-2}
b Negative. The positive area for first 10 s period (sections A and B of the graph) is smaller than the negative area for the second 10 s period.
c Graph
2 26.45 m
3 0.15 m

1.3 Force and acceleration

After studying this section you should be able to:

LEARNING SUMMARY

- *predict the effect of resistance to motion on acceleration*
- *distinguish the effects of balanced and unbalanced forces*
- *recall and use the relationship between resultant force, mass and acceleration*
- *distinguish between mass and weight*
- *understand and apply Newton's first and third laws of motion*

Getting going

Changing motion requires a force. Starting, stopping, getting faster or slower and changing direction all involve forces. One way to start an object moving is to release it from a height and allow the Earth to pull it down. The Earth's pull causes the object to accelerate downwards at the rate of $10\,m\,s^{-2}$.

The force that pushes a cycle or other vehicle forwards is called the driving force, or motive force.

To set off on a cycle, you push down on the pedal. The chain then transmits this force to the rear wheel. The wheel pushes on the road and, provided that there is enough friction to prevent the wheel from slipping, the push of the road on the wheel accelerates the cycle forwards.

However, as every cyclist knows, this acceleration is not maintained. The cyclist eventually reaches a speed at which the driving force no longer causes the cycle to accelerate. This is due to the **resistive** forces acting on the cycle. The main one is **air resistance**, although other resistive forces act on the bearings and the tyres.

Air resistance or **drag** also affects the motion of an object released and allowed to fall vertically; in fact it opposes the motion of anything that moves through the air. This is due to the air having to be pushed out of the way. The faster the object moves, the greater the volume of air that has to be displaced each second and so the greater the resistive force.

The diagram shows the directions of the forces acting on a cyclist travelling horizontally and a ball falling vertically.

The single arrow acting on the front of the cyclist represents all the resistive forces acting against the motion.

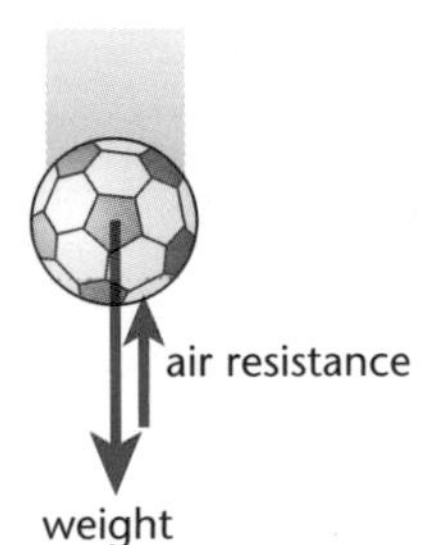

The effect of the resistive forces is to reduce the size of the resultant, or unbalanced, force that causes the acceleration. How this affects the speed of the cycle and ball is shown by the graph.

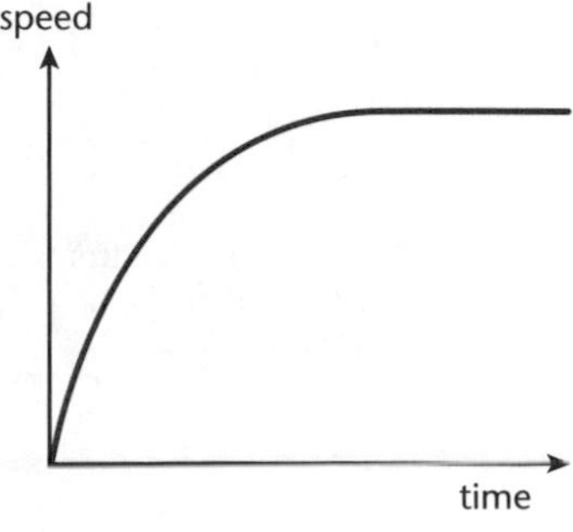

The decreasing gradient of the speed–time graph shows that the acceleration decreases as the resultant force becomes smaller. At a speed where the resistive force is equal to the driving force, the forces are balanced and resultant force is zero, so the object travels at a constant speed.

When the forces acting on a moving object are balanced, the constant speed is called its terminal speed or terminal velocity.

To summarise:

- an object accelerates if there is a resultant, or unbalanced, force acting on it
- the acceleration is proportional to the resultant force (provided that the mass of the object stays the same) and in the direction of the resultant force.

What else affects the acceleration?

AQA A 2
AQA B 2

As the number of passengers on a bus increases, its ability to accelerate away from the bus stop decreases. Acceleration depends on mass as well as force.

Results of experiments using a constant force to accelerate different masses show that:

the acceleration of an object is inversely proportional to its mass (when force stays the same).

If the mass of an object is doubled, its acceleration is halved for the same pulling force.

The dependence of acceleration on both the resultant force and the mass is summarised by the relationship:

KEY POINT

resultant force = mass × acceleration

$\Sigma F = ma$

Where the unit of force, the newton, is defined as the force required to cause a mass of 1 kg to accelerate at 1 m s^{-2}.

The symbol Σ, meaning 'sum of' is used here to emphasise that the relationship applies to the resultant force on an object, and not to individual forces.

More about balanced and unbalanced forces

AQA A 2
AQA B 2

Unbalanced forces always produce acceleration. Balanced forces never produce acceleration.

A car experiences resistive force, in the opposite direction to its motion, but these forces are invisible and all too easy to forget about. The car needs a driving force if it is to keep going at steady speed, so that there is no resultant force. The driving force and the resistive forces are then balanced.

For a spacecraft or any other object in space, where there is no resistance to motion, it is relatively easy to imagine that unbalanced force always produces acceleration and balanced forces never produce acceleration. Exactly the same rules apply here on Earth, we just have to remember to take account of resistive forces.

It reminds us that space is normal. Where we live is an unusual place in the Universe. Physics has to deal with the normal and also the unusual.

The force of gravity

AQA A 2
AQA B 2

In Physics, we use the word weight to mean force of gravity. The weight of a body depends on its location in the Universe. We measure weight in newtons.

We measure mass in kilograms. Mass is not a force. A body has the same mass wherever it goes, as long as it does not lose or gain material.

There is a clash here with everyday language. When people talk about dieting to 'lose weight' then they want there to be less of them. To a physicist, it is mass that they are trying to change.

The problem is that everyday language does not have to distinguish carefully between mass and weight because they are closely related here on Earth. Physics must try to think big, and be universal.

The force of gravity a body experiences depends on its mass, and there is a relatively simple 'conversion factor', which we can write in an equation:

$$g = W/m$$

The conversion factor tells us the force experienced for each kilogram of weight. It is measured in N kg^{-1}. It is called the **gravitational field strength**. On the Earth's surface, the gravitational field strength is just over 9.8 N kg^{-1}, which in calculations we often round to 10 N kg^{-1}.

In deep space, far from any gravitational pull, g is zero.

The acceleration due to gravity, at the Earth's surface, is just over 9.8 m s^{-2}, which in calculations we often round to 10 m s^{-2}. The two quantities have identical values. We use g as an abbreviation for acceleration due to gravity as well as for gravitational field strength.

About Newton's laws of motion

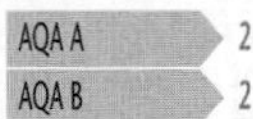

First law

Newton's first law agrees with the statements:

Unbalanced forces always produce acceleration. Balanced forces never produce acceleration.

The formal statement is:

An object stays at rest or in uniform motion in a straight line unless there is a resultant (or unbalanced) force acting on it.

Second law

Newton's second law is slightly more technical. It states that resultant force is proportional to rate of change of **momentum**. Momentum is the mass of a body multiplied by its velocity. A planet has a lot of momentum, a truck accelerating onto a motorway has less, and a drip of water from a tap has less again. All of these bodies are experiencing resultant force, however, and for each one this is proportional to the rate of change of momentum. Using p for momentum, p = mv. Newton's second law says:

$$F \propto \Delta(mv)/\Delta t$$

If the mass of a body isn't changing:

$$F \propto m\Delta v/\Delta t$$

Can you see the similarity between this relationship and the equation F = ma?

The quantity 'change in momentum', $\Delta(mv)$, has its own name – **impulse**.

Third law

Newton's third law says that it is impossible for one body (which includes you) to exert a force on another without experiencing a force of the same size in the opposite direction. If you push on a wall, it is not just the wall that experiences force – so do you. The two forces are in opposite directions, and are the same size.

If you push backwards on a floor with your foot, again you experience a force that is the same size as your push. It is how you walk.

A fish flicks water backwards, and experiences a forwards force. A plane blasts air backwards, and experiences a forwards thrust. Newton's third law is involved in all propulsions, whether of animal or machine.

It works on a big scale as well. The force of the Earth on the Moon is the same size as the force of the Moon on the Earth. (The Moon experiences a larger acceleration, because it has less mass.)

The formal statement of Newton's third law is:

To every action there is an equal and opposite reaction.

Putting it slightly differently:

If object A exerts a force on object B, then B exerts a force equal in size and opposite in direction on A.

Newton's third law and not Newton's third law

AQA A 2

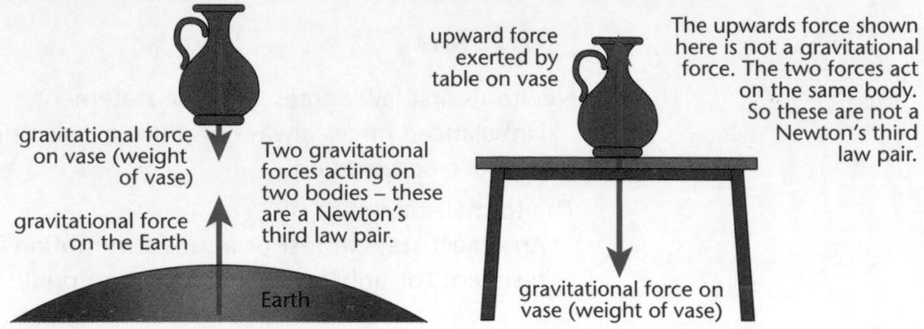

Newton's third law

The Earth exerts a gravitational force on a vase. The vase exerts a gravitational force on the Earth. The vase accelerates (much) more than the Earth because of its (much) smaller mass.

- The two forces are: of the same type – both forces of gravity.
- The two forces act on different objects.
- The two forces are inevitable consequences of each other – there is no escape from Newton's third law.

NOT Newton's third law

A table under the vase exerts an upwards force on it. But these are not a 'Newton's third law pair' of forces. The gravitational pair are still there, and the new upwards force doesn't affect them.

- Now, the two forces acting on the vase are not both of the same type. The upwards force due to the table is not a gravitational force. This is enough to show that they are not a 'Newton's third law pair'.
- The weight of the vase and the upwards force of the table act on the same body, the vase. Again, these forces can't be Newton's third law pairs.
- The upwards force of the table is not an absolutely inevitable consequence of the force of gravity pulling the vase down.

Progress check

1 **a** The total mass of a cyclist and her cycle is 85 kg.
Calculate the acceleration as she sets off from rest if the driving force is 130 N.
b The driving force remains constant at 130 N. What is the size of the resistive force when she is travelling at terminal velocity?

2 A fully laden aircraft weighs 300 000 kg. It accelerates from rest to a take-off speed of 70 m s^{-1} in 20 s.
a Calculate the acceleration of the aircraft.
b Calculate the size of the resultant force needed to cause this acceleration.
c Explain why the force from the engines must be greater than the answer to **b**.

3 A tightrope walker stands in the middle of a rope.
a What exerts the upward force on the tightrope walker?
b Explain why it is not possible for the rope to be perfectly horizontal.

4 Explain how you can tell from a vector diagram that an object is in a state of rest or uniform motion.

1 **a** 1.53 m s^{-2} **b** 130 N
2 **a** 3.5 m s^{-2} **b** 1.05 × 10^{6} N **c** The force also has to act against air resistance and friction.
3 **a** The tension in the rope.
b A rope can only pull in its own direction. For the tension to have a vertical component, the rope must be inclined to the horizontal.
4 The arrows that represent the vectors form a closed figure; i.e. the last arrow finishes where the first one starts.

1.4 Energy and work

After studying this section you should be able to:

LEARNING SUMMARY

- *identify situations where a force is working*
- *understand that energy can transfer from system to system and recognise that most energy transfers involve some dissipation of energy*
- *explain that some systems act as energy stores*
- *explain that during some energy transfers, physical work can be done, and in real energy transfers heating takes place*
- *use Sankey diagrams to illustrate energy transfer processes*
- *apply the principle of conservation of energy*
- *predict rate of transfer of energy by heating (thermal transfers)*
- *classify energy resources*
- *explain global energy balance and equilibrium temperature, and the greenhouse effect*

Working

AQA A 2
AQA B 2

Every event requires **work** to make it happen. Any force that causes movement is doing work. Pushing a supermarket trolley is working, as is throwing or kicking a ball. However, holding some weights above your head is not working; it may cause your arms to ache, but the force on the weights is not causing any movement!

How much work a force does depends on:

- the size of the force
- the direction of movement
- the distance that an object moves.

The phrase, 'distance moved in the direction of the force', is used here instead of 'displacement' to emphasise that the force and displacement it causes are measured in the same direction.

KEY POINT

The work done by a force that causes movement is defined as:

work = average force × distance moved in the direction of the force

$$W = F \times s$$

Work is measured in joules (J) where 1 joule is the work done when a force of 1 N moves its point of application 1 m in the direction of the force.

Here are some examples of forces that are working and forces that are not working.

A pylon that supports electricity transmission cables is not working, but a wind that causes the cables to move is!

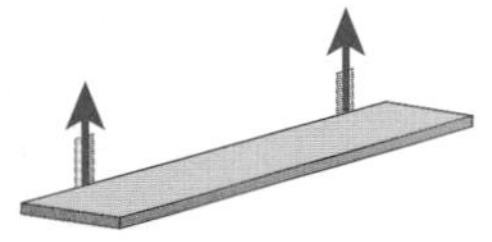

The shelf is not moving, so no work is being done.

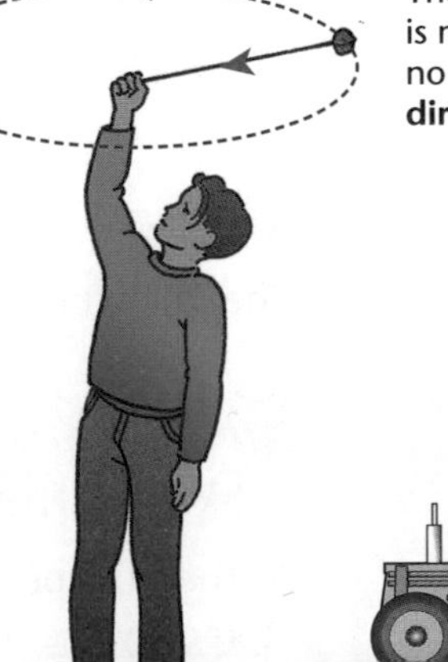

The tension in the string is not working as there is no movement **in the direction of the force.**

This force causes movement **in the direction of the force.**

The horizontal component of the force on the log is doing the work here.

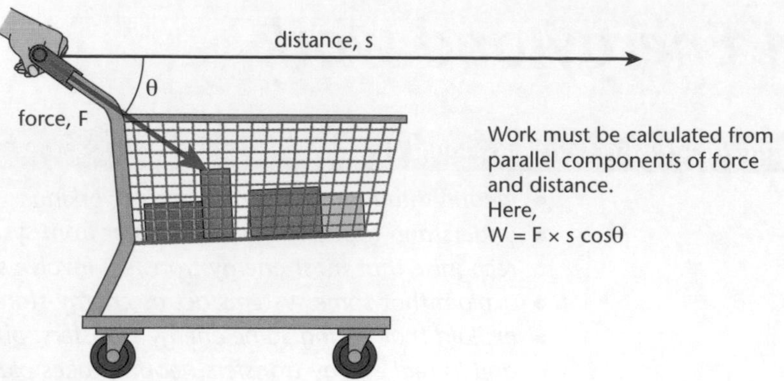

In circular motion, the direction of the velocity is along a tangent to the circle.

In the case of the stone being whirled in a horizontal circle, the stone's velocity is always at **right angles** to the force that maintains the motion, so the force is not working.

Although the force pulling the log and the movement it causes are not in the same direction, the force on the log has a component in the same direction as the log's movement. In calculating the work done the horizontal component of the force is multiplied by the horizontal distance moved by the log.

Energy stores and transfers

AQA A 2
AQA B 2

'Systems' store energy. A reservoir full of water in the mountains acts as en energy store, as does a hammer raised up above a nail. A flying bullet stores energy, as does a flywheel. A fuel, together with an oxygen supply, acts as an energy store. So does an electrical battery. A cylinder full of hot gas is an energy store.

In all cases, energy can transfer out from the store into other systems. Sometimes the energy transfers to another storage system, and sometimes it is dissipated in the surroundings. In that case, the surroundings are heated.

Energy transfers always involve working or heating or both of these. Working and heating are both energy transfer processes. They can be called mechanical and thermal processes.

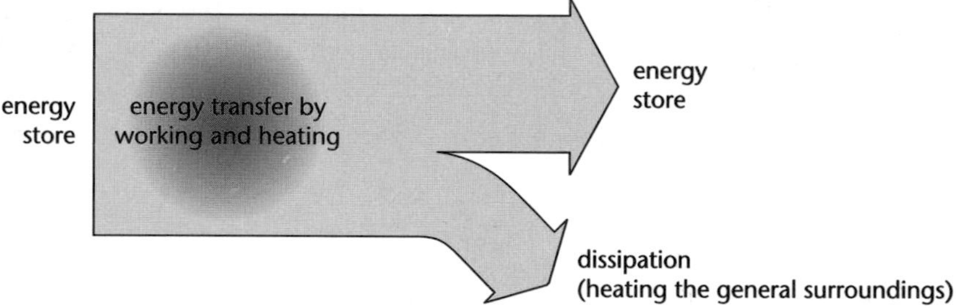

A representation of energy flow like this is sometimes called a **Sankey diagram**.

The following are some examples of everyday energy transfers.

A bus accelerates away from a bus stop and then maintains a steady speed

As the bus speeds up, energy stored in the fuel–oxygen system transfers to energy stored by the motion of the bus. This store is called the kinetic energy of the bus. Other energy from the fuel–oxygen transfers to the surroundings, heating them up just a little. Some energy is transferred by the hot exhaust gases, and some is transferred by the action of forces of resistance. This energy becomes thinly spread, and we say it has dissipated.

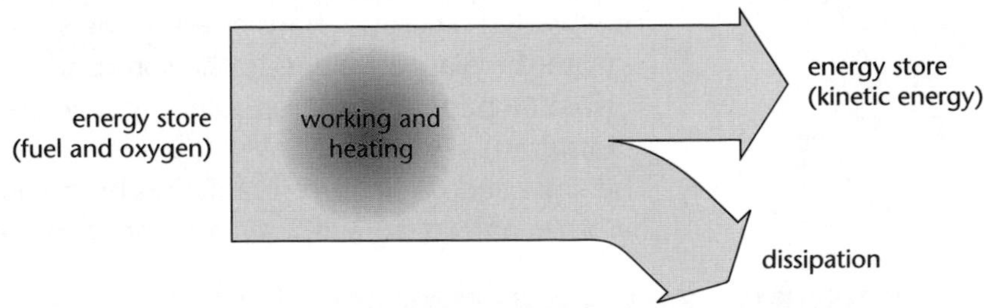

Once the bus is moving at steady speed, there is no more increase in its kinetic energy, so all of the energy transferred from the fuel and oxygen is being dissipated in the surroundings.

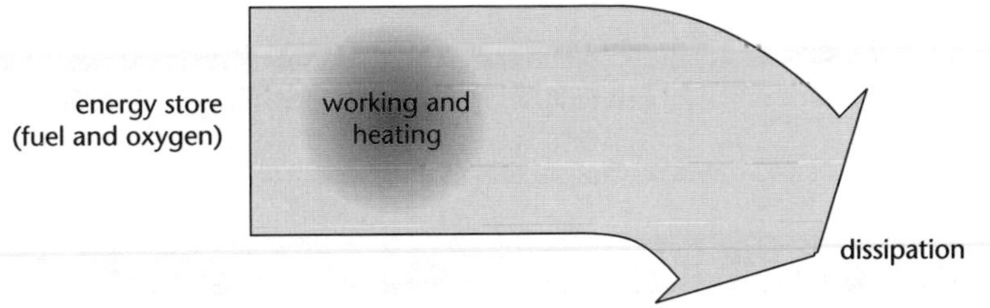

An electric motor lifts a load at a steady speed

An electric motor is an energy transfer device.

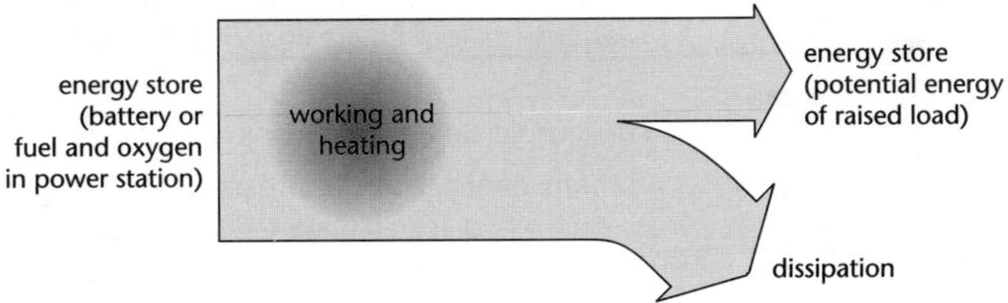

As the motor lifts a load, the load gains gravitational potential energy, which is one way in which a mechanical system can store energy. Its gravitational potential energy changes because its height above the Earth's surface changes. Except right at the start and finish, there is no gain or loss of kinetic energy of the rising load, since its speed is constant. There is, though, some dissipation due to resistive forces (which include frictional forces).

A filament lamp lights a room

In the steady state, the parts of the lamp are emitting energy and absorbing energy at the same rate, so there is no change in temperature.

When a filament lamp is switched on it takes a fraction of a second to reach its operating temperature.

Once the lamp has reached its steady state, of the 60 J of energy passing into the lamp each second from the electricity supply, typically 3 J passes out as light. This is shown in the diagram. The Sankey diagram gives a visual indication of the relative proportions of energy transferred into a desirable output and wasted.

Sankey diagrams are a useful way of showing the energy flow through a process where there are several stages, for example the energy flow through a power station.

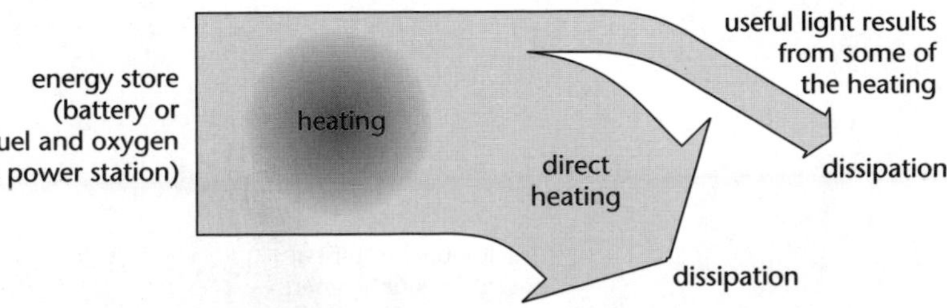

The 'wasted' energy is transferred to the surroundings in two main ways:

- from the glass envelope to the surrounding air by conduction and convection
- from all parts of the lamp as non-visible electromagnetic radiation (mainly infra-red).

The light and the infra-red radiation both cause heating when they are absorbed by other objects, so all the energy input to the lamp ends up as heat!

Work and energy

AQA A 2
AQA B 2

In a purely mechanical system where there is no energy transfer by heating, the energy transfer from a store or source of energy is equal to the amount of work done.

> **KEY POINT**
> Energy is sometimes described as the ability, in an ideal mechanical system, to do work. Like work, energy is measured in joules, J.

Conservation of energy

AQA A 2
AQA B 2

Note that the total width of the arrows in the Sankey diagrams is always the same before and after the transfer, to show that the total amount of energy stays the same. Like mass and charge, energy is a conserved quantity.

> **KEY POINT**
> The principle of conservation of energy states that:
> energy cannot be created or destroyed.

Take care not to confuse 'conservation of energy' with 'conservation of energy resources'. Energy resources such as coal can be used up, and conservation in this context means preserving them as long as possible.

This simple statement means that energy is never **used up**.

In many energy transfer processes the amount of energy at each stage cannot easily be quantified; it is difficult to put a figure on the total kinetic and potential energy of the moving parts in an engine for example. There are simple formulae for calculating the kinetic energy and gravitational potential energy of individual objects.

> **KEY POINT**
> kinetic energy = $\frac{1}{2}$ × mass × (speed)^2
> $$E_k = \frac{1}{2}mv^2$$
> change in gravitational potential energy = weight × change in height
> $$\Delta E_p = mg\Delta h$$

The formula for change in gravitational potential energy is only valid for changes in height close to the Earth's surface, where g has a constant value.

The principle of conservation of energy applies to all energy transfers.

When a streamlined object is falling freely in the atmosphere at a low speed, the resistive forces are small and so little energy is transferred to the surroundings. In this case the principle of conservation of energy can be applied to the transfer of gravitational potential energy to kinetic energy. When a parachutist is falling at a constant speed conservation of energy still applies, but the energy transfer is from gravitational potential energy to heat in the atmosphere.

Energy transfer of falling objects

Potential energy, kinetic energy, and predicting the speed of a falling body

AQA A 2
AQA B 2

For a body that falls freely, without resistance to motion:

potential energy lost = kinetic energy gained

After falling height Δh and acquiring velocity v:

$$mg\Delta h = \tfrac{1}{2}mv^2$$
$$2g\Delta h = v^2$$
$$v = \sqrt{2g\Delta h}$$

This equation provides a way of predicting velocity of fall from height fallen.

Power

AQA A 2
AQA B 2

Power is a measure of the work done or energy transferred each second.

Typically, a car engine has an output power of 70 kW, compared to 1.0 kW for a hairdryer and 2 W for a clock.

KEY POINT

Power is the rate of working or energy transfer.

Average power can be calculated as work done ÷ time taken, $P = \Delta W \div \Delta t$

Power is measured in watts (W), where 1 watt is a rate of working of 1 joule per second ($J\ s^{-1}$).

Power and efficiency

AQA A 2
AQA B 2

In choosing or designing a machine to do a particular job, important factors to consider include the **power input** and the **power output**. You would expect a hairdryer with a high power output to dry your hair faster than one with a low power output, but one with a high power input may not do the job any faster than one with a lower power input, it may just be less efficient!

Do not confuse power with speed. A double-decker bus may be slower than many cars, but it can transport 60 passengers a given distance much faster than any car!

Steam trains have a very high power input, but a very low efficiency at transferring this to power output

A comparison of the power output and power input of a device tells us the efficiency. A ratio is a mathematical comparison, so we make a ratio of useful power output and total power input, and we call that the efficiency of the device.

That is:

efficiency = useful power output ÷ total power input

It is quite common to multiply the answer here by 100 to give a percentage efficiency.

A fluorescent lamp, for example, can be described as having an efficiency of 20%.

Energy and efficiency

AQA A 2
AQA B 2

For a particular period of time, we can measure the useful energy output and the total energy input of a device, and perform the division to find the efficiency.

efficiency = useful energy output in a specified time ÷ total energy output in the same time

So, for the filament lamp that has an energy input of 60 J in 1 second, and a useful energy output of 3 J, also in 1 second:

efficiency = 3 ÷ 60 = 0.05

We can multiply this by 100, and describe the lamp as being 5% efficient.

Thermal energy transfers

AQA B 2

Newton's law of cooling

If two bodies have different temperatures then energy normally flows from the one at the higher temperature to the one at the lower temperature. So a body that has a higher temperature than its surroundings cools down. If the surroundings are significantly larger than the body, the energy will spread out and the temperature rise of the surroundings will be small.

Newton's law of cooling states that the rate of change of temperature of a cooling body is proportional to its temperature difference with the surroundings.

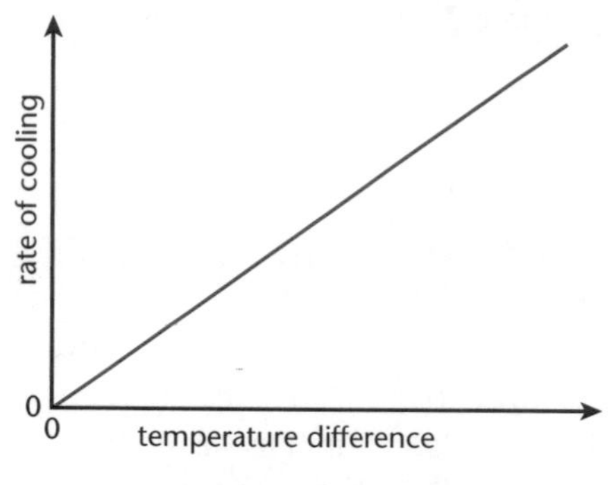

This graph expresses Newton's law of cooling.

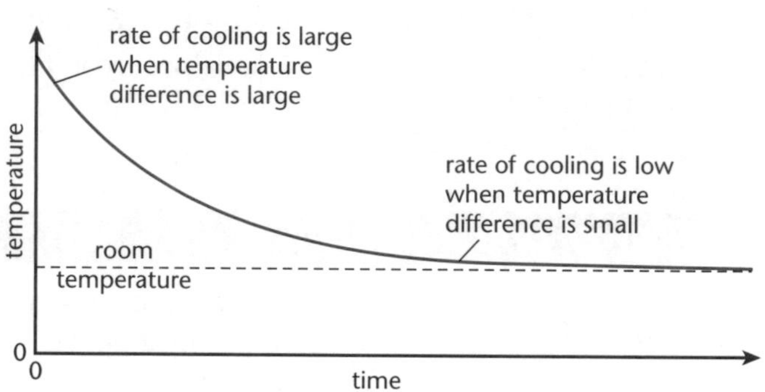

This graph, for a heated body cooling to room temperature, is a consequence of Newton's law of cooling.

Factors affecting rate of energy transfer through a material

Suppose that a solid material separates two bodies that are at different temperatures. There will be a tendency for energy to flow through the material, from the hotter to the cooler body.

Factors that affect the rate of energy transfer are:

- the **thermal conductivity** of the material, which is a measure of its ability to transfer energy by thermal conduction, k
- the area through which the energy can flow, A
- the temperature gradient across the material, which is the total temperature difference divided by the thickness of the material, $\Delta T/\Delta x$

Rate of energy transfer by thermal conduction $= k A \Delta T/\Delta x$

The quantity $k/\Delta x$ is called the **U value** of the insulation.

Energy resources

AQA B 2

The Sun is almost, but not quite, the only source of energy for life on Earth. Energy from the Sun not only keeps us alive, but drives the temperature differences that give rise to ocean currents and winds. Energy from the Sun evaporates water, some of which returns as rain or snow to the land to create rivers. Dammed rivers provide hydroelectric generation systems, or HEP.

Wind turbines:

Maximum power of turbine motion from wind in ideal conditions

$= \frac{1}{2}\pi r^2 \rho v^3$

where r is the radius of the turbine, ρ is the density of the air and v is the air speed.

There are some sources of energy other than the Sun, however. The most used of these are nuclear resources, based on fission of large nuclei that were created in previous generations of stars. The mass of these nuclei, a small proportion of the total that they have, ceases to exist during the fission process, and the mass loss is matched by energy release.

Geothermal energy resources take advantage of the heat of the inner Earth that is due to the natural radioactivity of our planet.

Tides slowly drain energy away from the relative motions of Earth and Moon, and we can use the flowing water to generate electricity.

Our distance from the Sun is important. The power of sunlight decreases rapidly with distance from the Sun, obeying the **inverse square law**.

Supposing the Sun acts as a single point, the intensity of energy received at an area at distance r from the Sun, measured as the energy arriving per unit time per unit area, is related to the Sun's total power output (P) by the equation:

Intensity, $I = P / 4\pi r^2$

Planetary energy balance

AQA B 2

No body is totally isolated from any other, and no body is so cold that it could never become colder. All bodies emit some thermal radiation, and they also receive it from other bodies.

For a body to have a constant temperature it must emit and receive energy at the same rate. A body reaches a natural **equilibrium temperature**.

If it were to warm up, it would emit energy faster, and this opposes the warming.

If it cooled, the rate of emission would be less, so again there would be a tendency for balance or equilibrium to be restored.

Thus the planet Mercury has a natural equilibrium average surface temperature, as does the planet Venus.

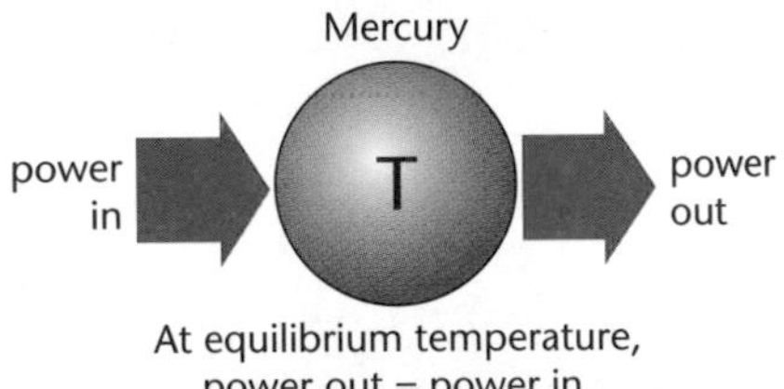

At equilibrium temperature, power out = power in.

Venus, however, has an atmosphere that resists the outward flow of radiant energy more than it resists the inward flow. Equilibrium is reached when outward flow and inward flow are the same and the planet's temperature has risen so that outward flow can be large enough. This what is called the greenhouse effect.

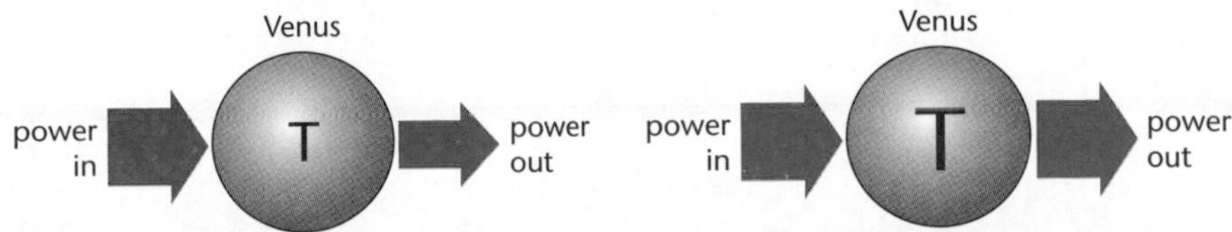

If energy escapes more slowly than it arrives, then temperature rises.

A hotter object tends to emit energy more rapidly. Temperature rises until power out = power in. This produces a new (hotter) equilibrium temperature.

There is a natural greenhouse effect on Earth, as well. The problem is that human activity has increased the concentration of greenhouse gases in the atmosphere – particularly carbon dioxide. It now seems very likely that human activity is producing an additional greenhouse effect, which is in turn very likely to produce climate changes that are hard to predict and that will escalate for as long as humans add to the greenhouse gases.

Progress check

1 A 240 N force is used to drag a 50 kg mass a distance of 5.0 m up a slope (see diagram). The mass moves through a vertical height of 1.8 m. Take $g = 10\ \text{m s}^{-2}$.
Calculate:

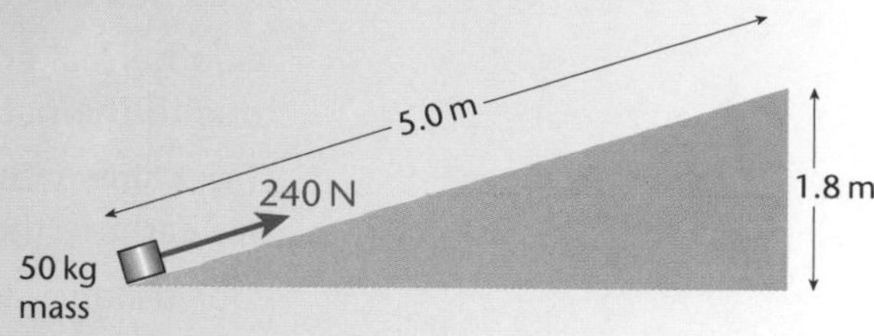

- **a** The work done on the mass.
- **b** The gravitational potential energy gained by the mass.
- **c** The efficiency of this way of lifting the mass through a vertical height of 1.8 m.
- **d** The time taken to drag the mass up the slope by a motor that has an output power of 150 W.
- **e** Draw a Sankey diagram for the process.

2 A car pulls a trailer with a force of 750 N, transmitted through the tow bar. The tow bar is inclined at 20° to the horizontal.
Calculate the work done by the car in pulling the trailer through a horizontal distance of 200 m.

3 a Explain why the planet Venus has a constant average temperature despite continuous arrival of energy from the Sun.

b Explain why the temperature is much higher than would be expected for a similar planet in similar orbit but with no atmosphere.

1 **a** 1200 J
b 900 J
c 0.75
d 8.0 s
e Diagram should show relative amounts of gravitational potential energy and energy dissipated.
2 1.4×10^5 J
3 **a** Planet radiates out to space at same rate as energy is received.
b There is a greenhouse effect / gases resist outward flow of energy / temperature must be high in order that rate of emission is high enough.

1.5 Vehicles in motion

After studying this section you should be able to:

- *explain the difference between laminar and turbulent flow*
- *calculate the drag force acting on a vehicle and explain how it varies with speed*
- *explain why objects reach a terminal velocity*
- *describe how balance of forces, horizontally and vertically, is achieved in flight*

LEARNING SUMMARY

Streamlines and turbulence

AQA B 2

In a wind tunnel the air moves over a stationary object such as a car body to model the movement of the car through the air. This is a valid model, as it is the relative speed of the air and the object that determines the pattern of flow.

Viscosity is a measure of the resistance of a fluid to flowing. Syrup and tar are viscous fluids; hydrogen has a very low viscosity.

To be as fuel-efficient as possible, a car, train or aircraft needs to be designed to minimise the resistive forces acting on it. To investigate the resistive or **drag** forces due to movement through the air, a car body is placed in a **wind tunnel**. Here the car body is held stationary and the air moves around it; vapour trails show the air flow over the car body. The diagram below shows the air flow at a low speed over a car body. The flow is said to be **laminar** or **streamlined**.

In laminar flow:

- the air moves in layers, with the layer of air next to the car body being stationary and the velocity of the layers increasing away from the car body
- particles passing the same point do so at the same velocity, so the flow is regular
- the drag force is caused by the resistance of the air to layers sliding past each other
- more **viscous** air has a greater resistance to relative motion and exerts a bigger drag force
- the drag force is proportional to the speed of the car relative to the air.

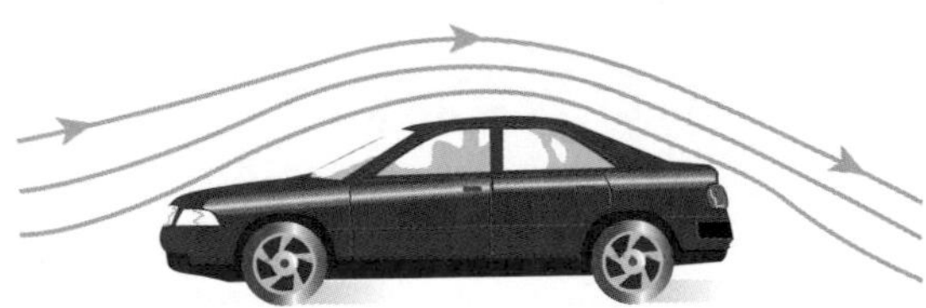

The air flow over a car travelling at low speed

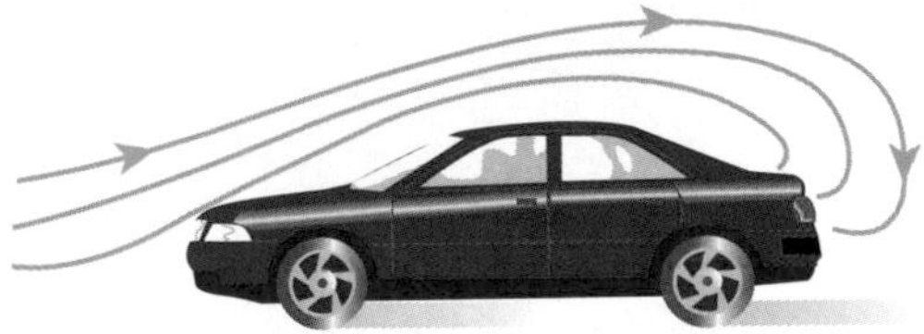

Turbulence occurs at higher speeds

As the speed of the air passing over the car body is increased, the flow pattern changes from laminar to **turbulent**. This is shown in the diagram above.

The changeover from laminar flow to turbulent occurs at a speed known as the **critical velocity**. Turbulent flow causes much more drag than laminar flow.

In turbulent flow:

- the air flow is disordered and irregular
- the drag force depends on the **density** of the air and not the viscosity
- the drag force is proportional to the $(\text{speed})^2$ where speed is relative to the air.

Viscous resistive force increases as velocity increases. For a vehicle at low speed, viscous force is small, and force can produce relatively high acceleration. At higher speeds the same force must overcome the increased viscous force. Eventually, for a given applied force, there is no more acceleration.

A fluid exerts a **viscous drag** on a body that is moving through it. The extent to which a fluid does this is quantified in terms of its viscosity. A more viscous fluid, like heavy oil or treacle, exerts a larger viscous drag than does a less viscous one, such as air. The size of the viscous drag is related to fluid viscosity, N, body size represented as effective sphere radius, r, and body velocity, v, by an equation that is known as **Stokes' Law:**

$F = 6\pi Nrv$

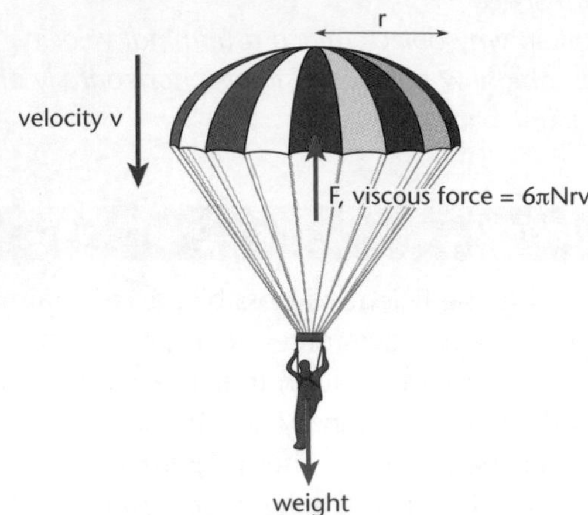

Terminal velocity

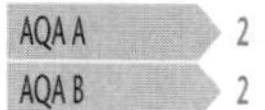

If an object is falling through a fluid, e.g. air or water, as its speed increases, the drag on it will also increase. Eventually a speed is reached where the upward force will equal the weight of the object. As there is no net force on the object the acceleration will be zero. The object will fall at a constant velocity. This velocity is called the **terminal velocity**.

Similarly for a car, or other vehicle, when it reaches a speed where the sum of all the forward forces, i.e. the thrust, equals the sum of all the resistive forces, the drag, the vehicle will move at a terminal velocity.

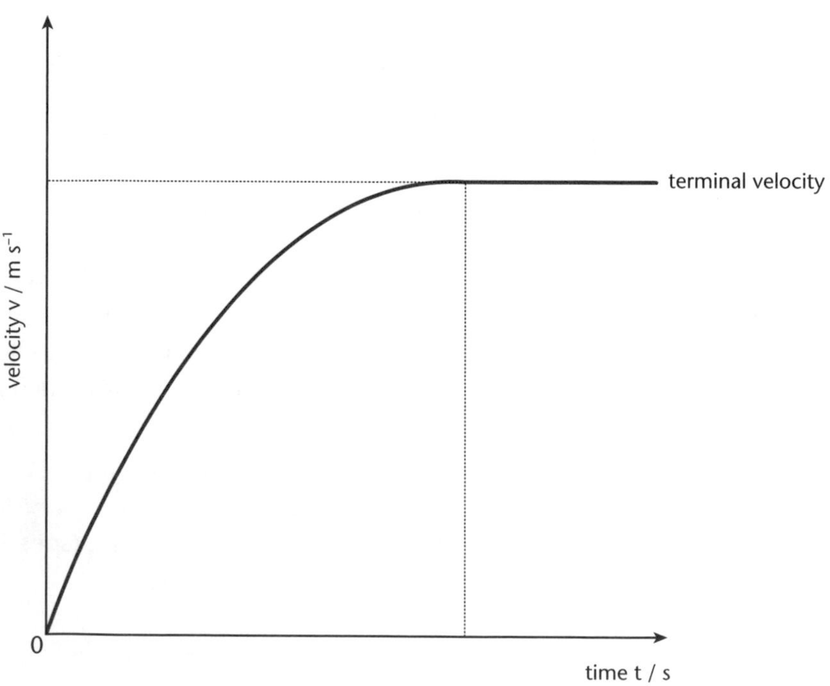

Forces in flight

AQA B 2

Drag is the force of air resistance that opposes motion. For motion at steady speed, drag must be balanced by a driving force or a thrust provided by an aircraft engine.

Lift is the upwards force on the wings of an aircraft, which starts with the flow of air across the wing surfaces. The wing shape is called an aerofoil. Its flow creates a lower pressure on the upper wing surface than on the lower. For steady flight, lift must be big enough to balance weight.

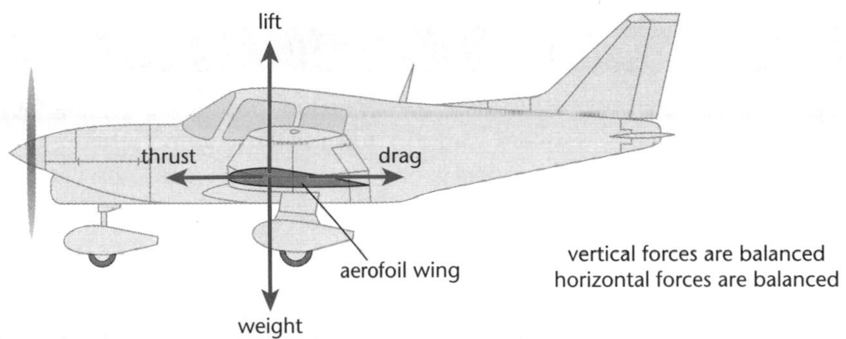

It is easier to consider the vertical forces and the horizontal forces independently, rather than as a complete set of four. Knowledge of whether the forces are balanced or unbalanced allows the flight of the aircraft to be predicted.

Progress check

1 A car is travelling at the legal maximum speed of $28\,m\,s^{-1}$ on a single carriageway. The driver notices a hazard in the road ahead.

a If the driver's reaction time is 0.5 s and the car decelerates at $7\,m\,s^{-2}$ during braking, calculate the stopping distance.

b Calculate the size of the force required to decelerate the car if its total mass is 850 kg.

2 State **two** differences between laminar flow and turbulent flow of air over an object such as a car body.

3 Suggest why it is dangerous for an aircraft to attempt to take off when there is a layer of ice on its wings.

1 **a** 70 m **b** 5950 N

2 Any two from: laminar flow is regular; in laminar flow layers of air slide over each other; in laminar flow the drag force is due to resistance of the layers of air to sliding over each other; in laminar flow the drag force is proportional to the relative speed of the object and the air; in laminar flow the drag is proportional to the viscosity of the air.

3 the ice increases the turbulence, reducing the lift.

1.6 More effects of forces

After studying this section you should be able to:

- *calculate the moment of a force and apply the principle of moments to a stable object*
- *calculate the density of a solid, liquid or gas*

The turning effect

AQA A 2

When forces are used to open doors, steer and pedal bicycles or turn on taps, they are causing turning. The effect that a force has in turning an object round depends on:

- the size of the force
- the perpendicular (shortest) distance between the **force line** and the **pivot** (the axis of rotation).

Both of these factors are taken into account when measuring the turning effect, or **moment**, of a force.

The 'line of action' of a force is the line drawn along the direction in which the force acts.

KEY POINT

Moment of a force = force × perpendicular distance from line of action of force to pivot.

The moment of a force, also known as torque, is measured in N m.

The diagram shows how the same force used to open a door can produce very different effects, according to the direction in which it is applied.

The moment of a force is a vector that can only have one of two directions; either clockwise or anticlockwise.

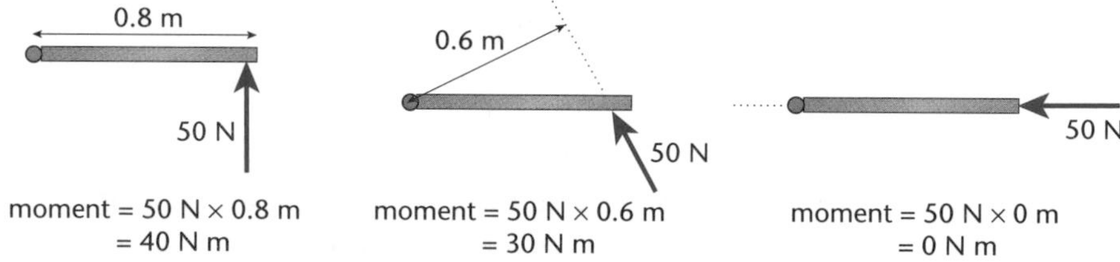

moment = 50 N × 0.8 m = 40 N m

moment = 50 N × 0.6 m = 30 N m

moment = 50 N × 0 m = 0 N m

A question of balance

AQA A 2

When turning on a tap or steering a bicycle, two forces are normally used. The forces act in opposite directions, but they each produce a moment in the same direction. A pair of forces acting like this is called a **couple**. The combined moment is equal to the sum of the moments of the individual forces.

If the forces that make the couple are equal in size, the moment of the couple = size of one force × shortest (perpendicular) distance between the force lines.

Forces in structures

AQA A 2

If a beam or rod is **uniform** its mass is evenly distributed so that it is the same as if all the mass was concentrated at a single point at the centre of the beam. The weight is the force acting vertically downwards from this point.

The resultant force and resultant moment at any point in a stable structure are both zero. Either of these principles can be used to find the values of forces F_1 and F_2 in the ropes that support the beam shown in the diagram overleaf.

The resultant force and resultant moment both have to be zero for the structure to be stable.

Resolving horizontal and vertical components will always give the same answers as taking moments.

- Both F_1 and F_2 could be found by resolving the vertical components of all the forces (which must balance) and the horizontal components of all the forces (which must also balance).
- Since F_1 does not have a moment about A, taking moments about this point gives a relationship between F_2 and the forces we know.
- Similarly, taking moments about B gives a relationship between F_1 and the known forces.

To take moments about A, F_2 needs to be resolved into horizontal and vertical components. The moment of the horizontal component is zero and that of the vertical component is in the anticlockwise direction. The equation is:

$$F_2 \cos 60° \times 4.0\,\text{m} = 200\,\text{N} \times 2.0\,\text{m} + 80\,\text{N} \times 1.0\,\text{m}$$

$$F_2 = 480\,\text{N m} \div (\cos 60° \times 4.0\,\text{m}) = 240\,\text{N}.$$

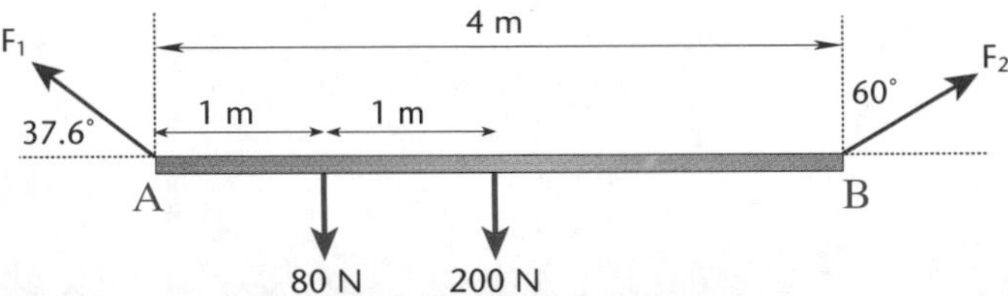

Density

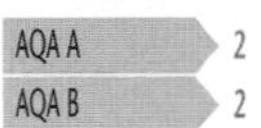

The density of a material is a measure of how close-packed the particles are. Gases at atmospheric pressure are much less dense than solids and liquids because the particles are more widespread.

Typical densities of some everyday materials in kg m^{-3} are:

air 1.2
water 1000
concrete 5000
iron 9000

KEY POINT

Density of a material = mass per unit volume.

$$\rho = m/V$$

Density is measured in kg m^{-3} or g cm^{-3}.

Density is measured by dividing the mass of a specimen of the material by its volume.

Forces in fluids

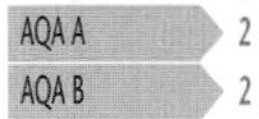

A ship or an iceberg experiences an upwards force exerted, in effect, by the water. This is called **upthrust**. Any fluid exerts an upthrust on a body immersed (or partly immersed) in it. The size of the upthrust is equal to the weight (in newtons, of course) of the fluid that the body displaces. This statement is called **Archimedes' principle**.

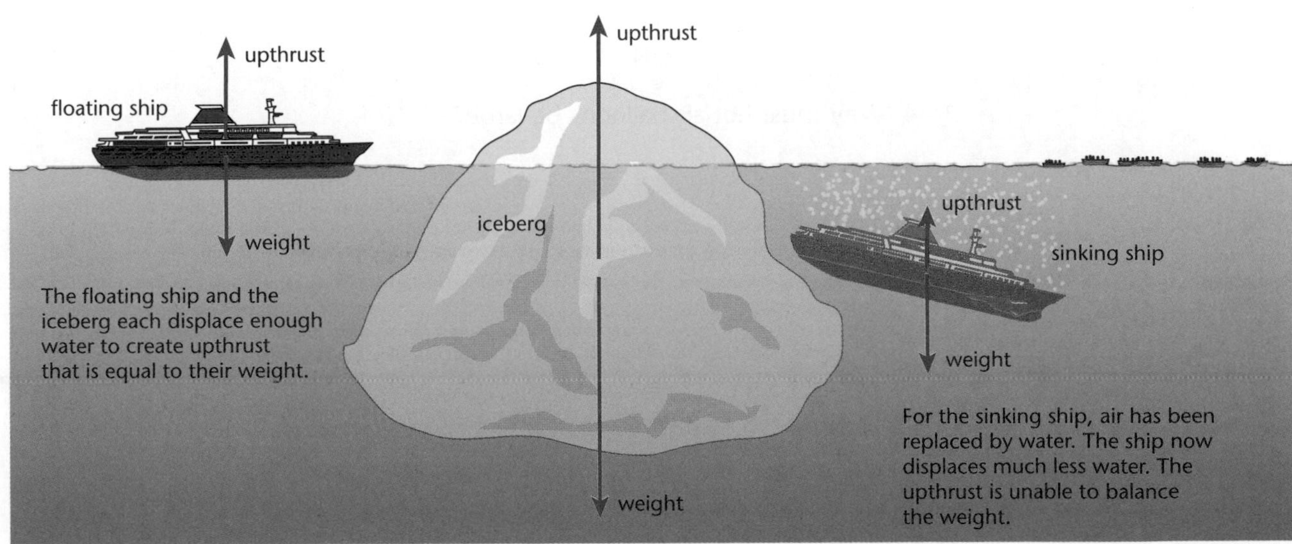

Note that when floating ice melts, its mass and weight do not change, so it takes up exactly the same volume as the volume of water it displaced when solid. Melting sea ice, therefore, has no effect on global sea levels.

Note also that the Archimedes' principle applies to bodies in gases as well as in liquids. For a body to float in air, it must displace a volume of air with a greater weight than its own. That is, it must be less dense than the air.

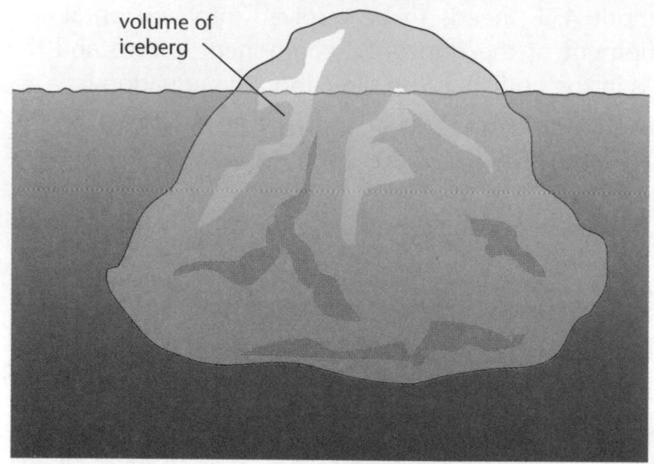

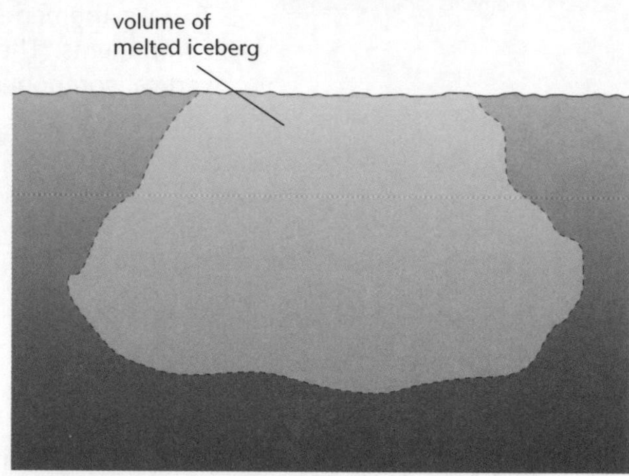

Progress check

1 Use the principle of moments to calculate the tension in the rope, T, in the diagram below.

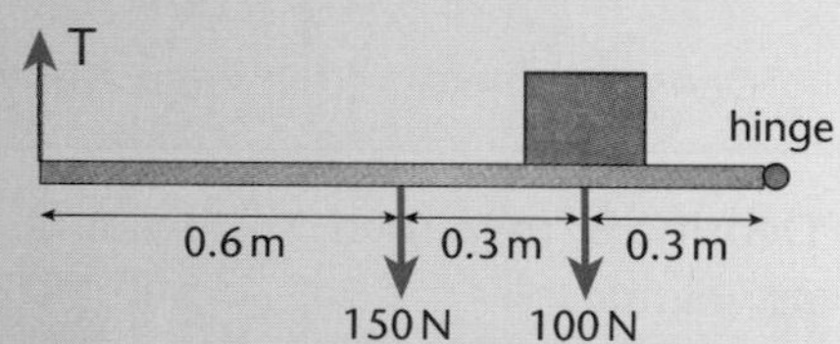

2 Describe how you would measure the density of air.

3 Use the principle of moments to calculate the value of force F_1 acting on the balanced beam in the diagram below.

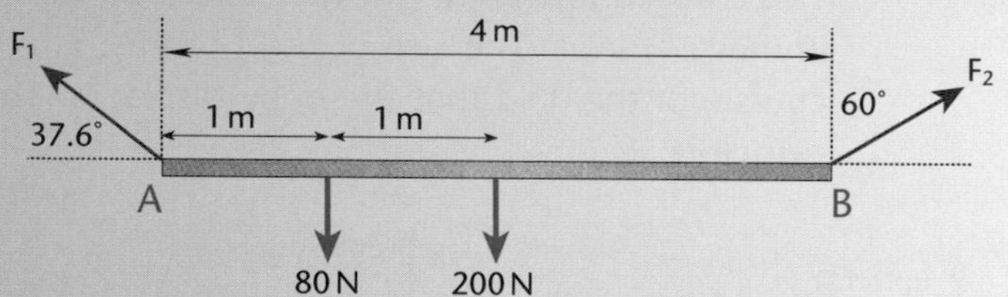

4 Why must hot air balloons be large?

1 100 N

2 Weigh an air-filled container of known volume. Remove the air using a vacuum pump and weigh again. Subtract the mass of the empty container from that of the full container to find the mass of the air and divide this by the volume of the container.

3 262 N

4 So that they displace a weight of air that is equal to their own average weight, including the load they carry. (The part of the balloon that is less dense than air must have a large volume so that the average density of the whole structure is the same as or less than the density of the air.)

1.7 Force and materials

LEARNING SUMMARY

After studying this section you should be able to:

- *describe the difference between elastic and plastic behaviour*
- *state Hooke's law and appreciate its limitations*
- *contrast the behaviour of a ductile, a brittle and a polymeric solid when stretched*
- *explain how stress and strain are used in measurements of the Young modulus*

Elastic or plastic

AQA A 2

No material is absolutely rigid. Even a concrete floor changes shape as you walk across it. The behaviour of a material subjected to a tensile (pulling) or compressive (pushing) force can be described as either **elastic** or **plastic**.

a material is **elastic** if it returns to its original shape and size when the force is removed

a material is **plastic** if it does not return to its original shape and size when the force is removed

Most materials are elastic for a certain range of forces, up to the **elastic limit**, beyond which they are plastic. Plasticine and playdough are plastic for all forces.

Hooke attempted to write a simple rule that describes the behaviour of all materials subjected to a tensile force.

If the extension is proportional to the stretching force, then doubling the force causes the extension to double.

KEY POINT

Hooke's law states that:

the extension, x, of a sample of material is proportional to the stretching force:

$x \propto F$

This can be reinterpreted as: $F = kx$

where k represents the stiffness of the sample and has units of $N\,m^{-1}$

A polymeric solid is one made up of long chain molecules.

Metals and springs 'obey' Hooke's law up to a certain limit, called the **limit of proportionality**. For small extensions, the extension is proportional to the stretching force. Rubber and other **polymeric** solids do not show this pattern of behaviour.

The graphs below contrast the behaviour of different materials subjected to an increasing stretching force.

Because the limits of proportionality and elasticity lie close together on the force–extension curve, the term 'elastic limit' is often used to refer to both points.

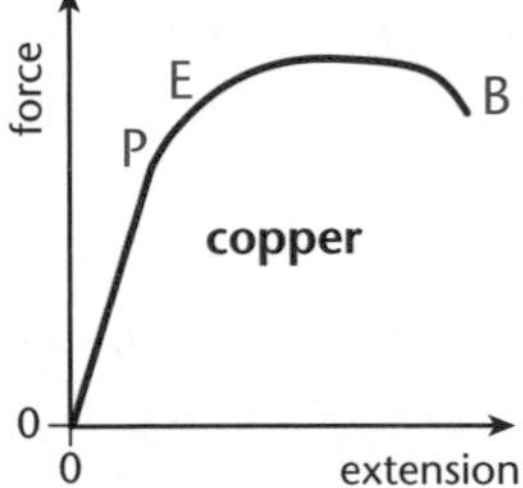

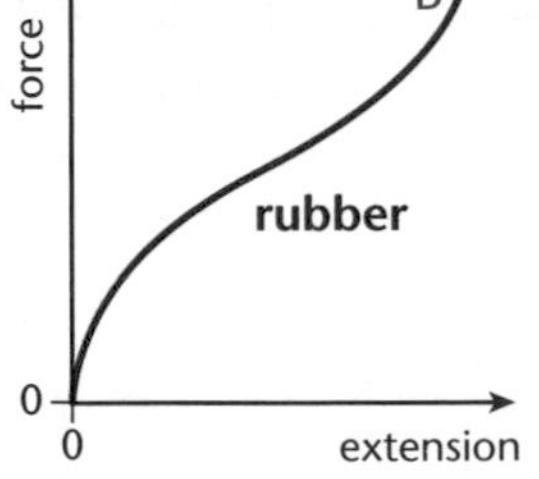

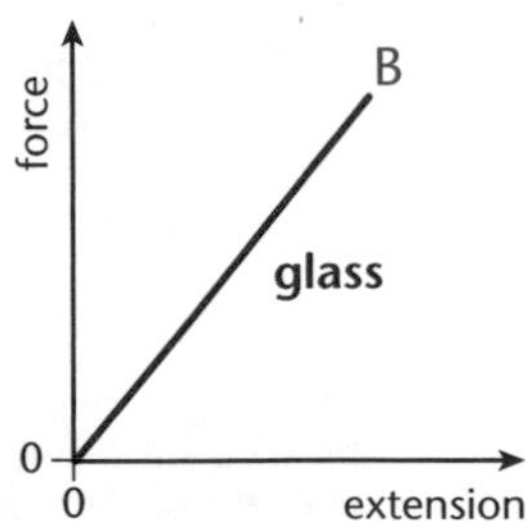

P = limit of proportionality E = elastic limit B = breaking point

- **Copper** is a **ductile** material, which means that it can be drawn into wires. It is also **malleable**, which means that it can be reshaped by hammering and bending without breaking. When stretched beyond the point E on the graph it retains its new shape.
- **Rubber** does not follow Hooke's law and it remains elastic until it breaks.
- **Glass** is **brittle**; it follows Hooke's law until it snaps.

Here are other properties of materials:

- **Kevlar** is **tough**; it can withstand shock and impact.
- **Mild steel** is **durable**; it can withstand repeated loading and unloading.
- **Diamond** is **hard**; it cannot be easily scratched.

Storing energy

AQA A 2
AQA B 2

The force–extension graph on the right shows that the more a material is stretched, the greater the force that is needed.

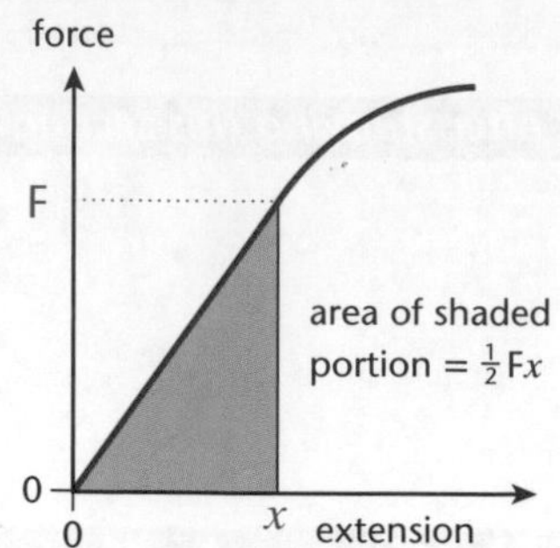

Section 1.4 (page 39) shows that work must be done when a force causes a displacement. So work must be done to stretch a material. Energy must be provided to the material to stretch it. You only have to pull on a stiff spring to experience this. That energy becomes available when the material returns to its original shape. You have to take care how you let go of a stretched stiff spring. So, a stretched material acts as an energy store. (Many non-electrical clocks, for example, use a spring as an energy store, without which they would stop quite quickly due to natural dissipation.)

The energy stored as a material is deformed and is represented by the area between the curve and the extension axis.

If a force F produces an extension x then the energy stored is equal to $\frac{1}{2}Fx$.

Since $F = kx$, energy stored = $\frac{1}{2}kx^2$.

Stress, strain and the Young Modulus

AQA A 2

The extension that a force produces depends on the dimensions of the sample as well as the material that it is made from. To compare the elastic properties of materials in a way that does not depend on the sample tested, measurements of **stress** and **strain** are used instead of force and extension.

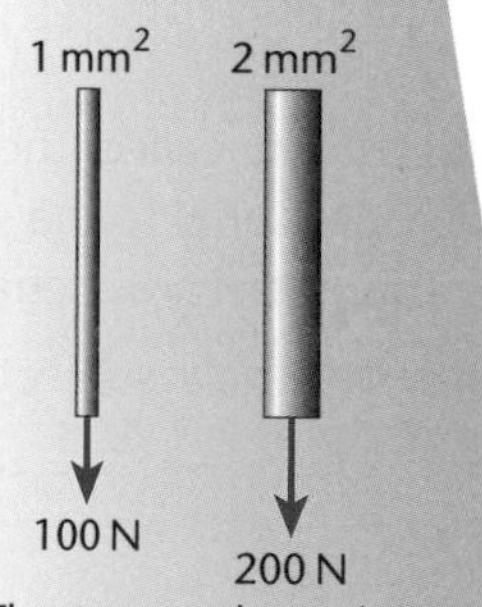

The stress on these wires is the same, even though the forces are different.

For some materials, e.g. copper and mild steel, the stress when the material breaks is less than the ultimate tensile stress.

KEY POINT

stress = normal force per unit area

$\sigma = F/A$

Like pressure, stress is measured in Pa, where 1 Pa = 1 N m^{-2}

strain = extension per unit length

$\varepsilon = x/l$

Strain has no units.

The strength of a material is measured by its **breaking stress** or **ultimate tensile stress**. This is the maximum stress the material can withstand before it fractures or breaks and may not be the actual stress when breaking occurs. Its value is independent of the dimensions of the sample used for the test.

Using measurements of stress and strain, different materials can be compared by the **Young modulus**, E.

The Young modulus for mild steel is 2.1×10^{11} Pa and that for copper is 0.7×10^{11} Pa. This means that for the same stretching force, a sample of copper stretches three times as much as one of mild steel with the same dimensions.

KEY POINT

Young modulus = stress ÷ strain

$E = \sigma/\varepsilon = F/A \div x/l = Fl/xA$

The Young modulus has the same units as stress, Pa.

The greater the value of the Young modulus the **stiffer** the material, i.e. the less it stretches for a given force.

Progress check

1 Describe the difference between a **ductile** material and one that is **brittle**.

2 A force of 150 N applied to a spring causes an extension of 25 cm.
   a Calculate the energy stored in the spring.
   b What assumption is necessary to answer a?

3 The Young modulus for copper is 1.3×10^{11} Pa.

   A tensile force of 150 N is applied to a sample of length 1.50 m and cross-sectional area 0.40 mm^2. (1.0 mm^2 = 1.0×10^{-6} m^2).

   Assuming that the limit of proportionality is not exceeded, calculate the extension of the copper.

1 A ductile material can be drawn into wires; a brittle material snaps easily.
2 a 18.75 J
b the limit of proportionality has not been exceeded
3 4.3 mm

Sample questions and model answers

The final velocity, v, is implicit in the question, although not clearly stated. As the car brakes to a halt, the final velocity must be 0.

Note that the acceleration is negative, as the speed in the direction of the velocity is decreasing.

1 In an emergency stop, a driver applies the brakes in a car travelling at 24 m s^{-1}. The car stops after travelling 19.2 m while braking.

(a) Calculate the average acceleration of the car during braking. [3]

The initial velocity, u, final velocity, v, and displacement, s, are known.
The unknown quantity is the acceleration, a.
The appropriate equation is $v^2 = u^2 + 2as$ — 1 mark
Rearrange to give $a = (v^2 - u^2)/2s$ — 1 mark
$= (0 - 24^2)/(2 \times 19.2)$
$= -15$ m s^{-2} — 1 mark 

(b) Sketch a graph that shows how the speed of the car changes during braking, assuming that the acceleration is uniform. Mark the value of the intercept on the time axis.

Show on your graph the feature that represents the distance travelled during braking. [3]

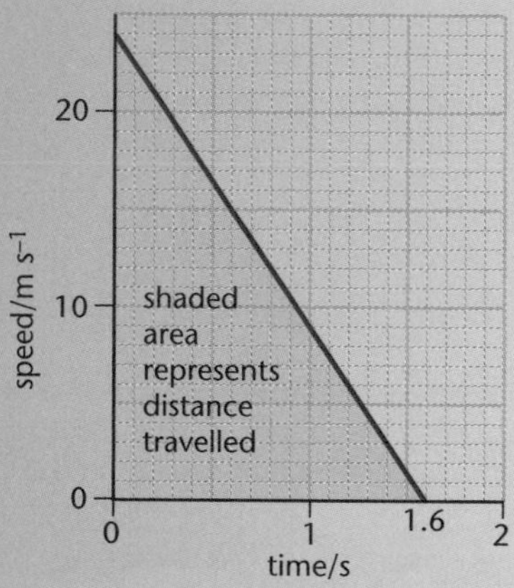

The graph shows a straight line from 24 m s^{-1} to 0 m s^{-1} — 1 mark
The intercept on the time axis is 1.6 s — 1 mark
The area between the graph line and the time axis represents the distance travelled. — 1 mark

Again, the negative sign shows that the direction of the force is opposite to the direction of motion.

(c) Calculate the size of the force required to decelerate the driver, of mass 75 kg, at the same rate as the car. [3]

force = mass × acceleration — 1 mark
= 75 kg × −15 m s^{-2} — 1 mark
= −1125 N — 1 mark

The cue word 'suggest' indicates that you are not meant to have studied this example as part of your course, but should be able to reason it from your understanding of forces and motion.

(d) Suggest what is likely to happen to a driver who is not wearing a seat belt. The car does not have air bags. [2]

The driver will carry on moving (1 mark) as there is nothing to stop her/him until (s)he hits the steering wheel and windscreen (1 mark).

Sample questions and model answers (continued)

2 A wooden pole is held in an upright position by two wires in tension. These forces are shown in the diagram below.

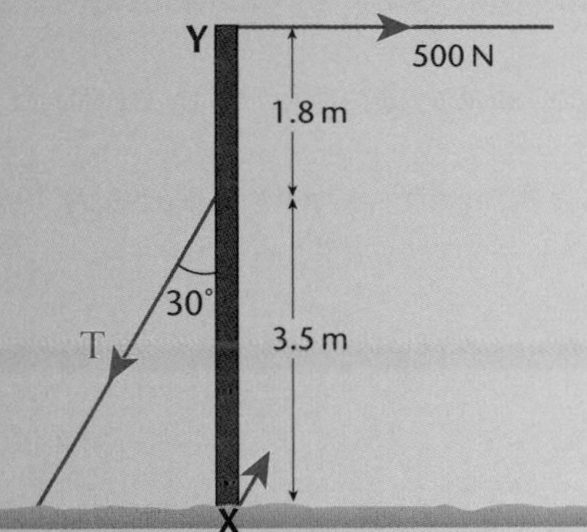

The pole is fixed at its base.

(a) Calculate the moment of the 500 N horizontal force about X. [2]

moment = force x perpendicular distance to pivot
= 500 N x 5.3 m — 1 mark
= 2650 N m — 1 mark

(b) Calculate the value of the force T. [3]

The horizontal component of the force T must also have a moment of 2650 N m: — 1 mark

$T\cos 60° \times 3.5\,m = 2650\,N\,m$ — 1 mark

$T = 2650\,N\,m \div (\cos 60° \times 3.5\,m) = 1514\,N$ — 1 mark

Always write down the formula that you intend to use, and each step in the working. It would be very easy in this case to use a wrong distance. Simply writing down a wrong answer always gets no marks, but showing how you arrive at that wrong answer allows credit to be awarded for the correct process.

Only the horizontal component of T has a turning effect; its moment must counterbalance that of the 500 N horizontal force.

Having written down the physical principle, the next step is to write it in the form of an equation. Tcos60° or Tsin30° is the horizontal component of T.

The final step is to calculate the value of T. Although it is not essential to show the units of physical quantities at each step in the calculation, it is good practice and you MUST include the correct unit with your answer to each part.

This problem cannot be solved by resolving T and the 500 N force as there is a force at X (you can see that this is not just a vertical reaction force by taking moments about Y – there must be a horizontal component with a moment that balances the horizontal component of T).

Practice examination questions

1 A microlight aircraft heads due north at a speed through the air of $18\,\text{m s}^{-1}$.
A wind from the west causes the aircraft to move east at $6\,\text{m s}^{-1}$.
Draw a vector diagram and use it to work out the resultant velocity of the aircraft. [3]
Take the value of free-fall acceleration, g, to be $10\,\text{m s}^{-2}$.

2 The diagram shows a light fitting held in place by two cables.

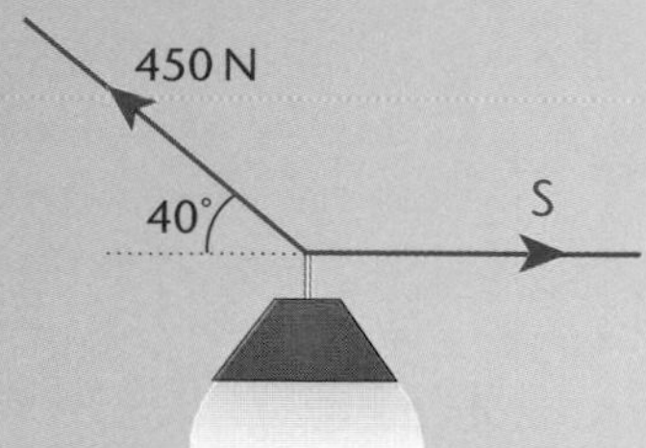

(a) Calculate the vertical component of the 450 N force. [2]

(b) Write down the weight of the light fitting. [1]

(c) Explain why a vector diagram that represents the forces on the light fitting is a closed triangle. [2]

(d) Draw the vector diagram and use it to find the value of the force S. [2]

3 The diagram is a velocity–time graph for a train travelling between two stations. A positive velocity represents motion in the forwards direction.

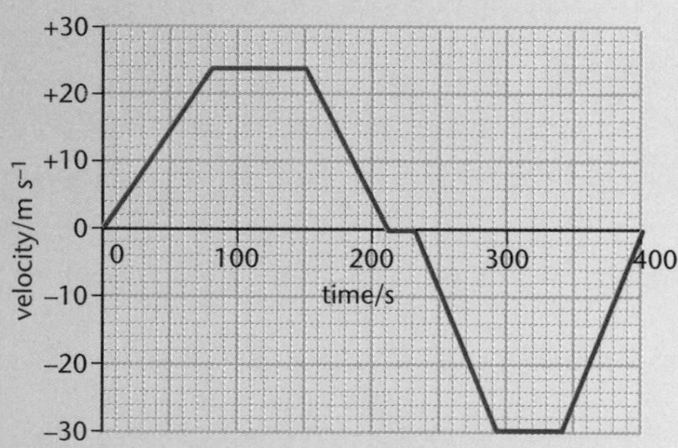

(a) Calculate the acceleration of the train during the first 80 s. [3]

(b) During which other times shown on the graph is the train increasing speed? [1]

(c) For what length of the time was the train:
(i) not moving? [1]
(ii) travelling at a constant non-zero speed? [1]

(d) Calculate the total distance travelled by the train. [2]

(e) At the end of the time shown on the graph, how far was the train from its starting point? [1]

4 In a game of cricket, a ball leaves a bat in a horizontal direction from a height of 0.39 m above the ground.

Practice examination questions (continued)

The speed of the ball is $35\ \text{m s}^{-1}$.

(a) Calculate the time interval between the ball leaving the bat and reaching the ground. [3]

(b) What horizontal distance does the ball travel in this time? [2]

(c) Fielders are not allowed within 15 m of the bat.
What is the maximum speed that the ball should leave the bat, travelling in a horizontal direction, for the player not to be caught out? [2]

5 A motorist travelling at the legal speed limit of $28\ \text{m s}^{-1}$ (60 mph) takes his foot off the accelerator as he passes a sign showing that the speed limit is reduced to $14\ \text{m s}^{-1}$ (30 mph). The car decelerates at $2.0\ \text{m s}^{-2}$.

(a) For what time interval is the motorist exceeding the speed limit? [2]

(b) How far does the car travel in that time? [2]

6 A fountain is designed so that the water leaves the nozzle and rises vertically to a height of 3.5 m.

(a) Calculate the speed of the water as it leaves the nozzle. [3]

(b) For how long is each drop of water in the air? [2]

Take the value of free-fall acceleration, g to be $10\ \text{m s}^{-2}$.

7 An aircraft has a total mass, including fuel and passengers, of 70 000 kg.
Its take-off speed is $60\ \text{m s}^{-1}$ and it needs to reach that speed before the end of the runway, which is 1500 m long.

(a) Calculate the minimum acceleration of the aircraft. [3]

(b) Calculate the average force needed to achieve this acceleration. [3]

(c) Explain how the resultant force on the aircraft is likely to change during take-off. [2]

8 A car pulls a trailer with a force of 150 N. This is shown in the diagram.

150 N

(a) According to Newton's third law, forces exist in pairs.
  (i) What is the other force that makes up the pair? [1]
  (ii) Write down the size and direction of this force. [1]

(b) The mass of the trailer is 190 kg. As it sets off from rest, there is a resistive force of 20 N.
  (i) Calculate the size and direction of the resultant force on the trailer. [1]
  (ii) Calculate the initial acceleration of the trailer. [3]

(c) The force from the car on the trailer is maintained at 150 N.
When the car and trailer are travelling at constant speed:
  (i) what is the size of the resultant force on the trailer? [1]
  (ii) write down the size and direction of the resistive force on the trailer. [1]

Practice examination questions (continued)

9 The diagram shows the forces acting on a child on a playground slide. Air resistance is negligible.

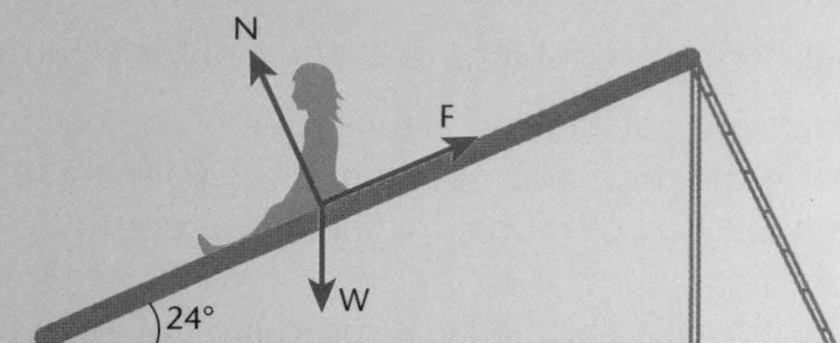

(a) The mass of the child is 40 kg.

(i) Calculate the size of the force W. [1]

(ii) Describe the force that, along with W, makes up the 'equal and opposite pair' of forces described by Newton's third law. [1]

(b) The size of the friction force is 90 N.

(i) Calculate the component of W parallel to the slide. [2]

(ii) Calculate the acceleration of the child. [3]

(c) The slide is 5.5 m long. If the child maintains this acceleration, calculate her speed at the bottom. [3]

10 An astronaut on a space walk pushes against the side of the spacecraft.

(a) Explain how this causes the velocity of both the astronaut and the spacecraft to change. [2]

(b) The astronaut can change his velocity by firing jets of nitrogen gas. Explain how this changes the velocity of the astronaut. [2]

11 A crane lifts a 3 tonne (3000 kg) load through a vertical height of 18 m in 24 s. Calculate:

(a) the work done on the load [3]

(b) the gain in gravitational potential energy of the load [1]

(c) the power output of the crane [2]

(d) the power input to the crane if the efficiency is 0.45. [2]

12 In a hydroelectric power station, water falls at the rate of $2.0 \times 10^5\ \mathrm{kg\,s^{-1}}$ through a vertical height of 215 m before driving a turbine.

(a) Calculate the loss in gravitational potential energy of the water each second. [3]

(b) Assuming that all this energy is transferred to kinetic energy of the water, calculate the speed of the water as it enters the turbines. [3]

(c) The water leaves the turbines at a speed of $8\ \mathrm{m\,s^{-1}}$. Calculate the maximum input power to the turbines. [2]

(d) The electrical output is 250 MW. Calculate the efficiency of this method of transferring gravitational potential energy to electricity. [2]

Practice examination questions (continued)

13 The diagram shows a child's toy. The spring is compressed a distance of 0.15 m using a force of 25 N. When the release mechanism is operated the ball rises into the air.

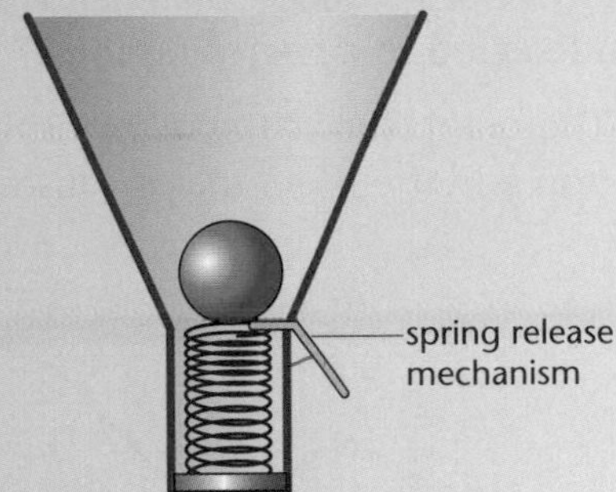

(a) Calculate the energy stored in the spring. [2]

(b) The mass of the ball is 0.020 kg. Calculate:

(i) the maximum speed of the ball [3]
(ii) the maximum vertical distance that the ball travels. [3]

14 The diagram shows a garden tool being used to cut through a branch.

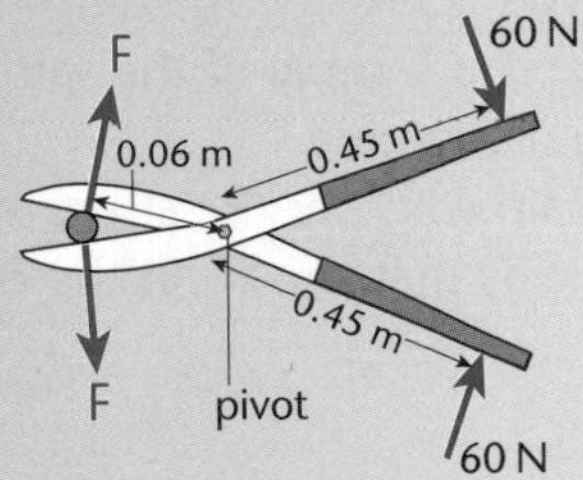

(a) Calculate the moment of the force on each handle. [2]

(b) Assuming equilibrium, calculate the value of the resistive force, F, that acts on each blade. [2]

(c) Suggest why a similar tool designed to cut through thicker branches has longer handles. [2]

15 The diagram shows the base of a steel girder that supports the roof of a building. The downward force in the girder is 4500 N and it is fixed to a square steel plate, each side of which is 0.30 m long.

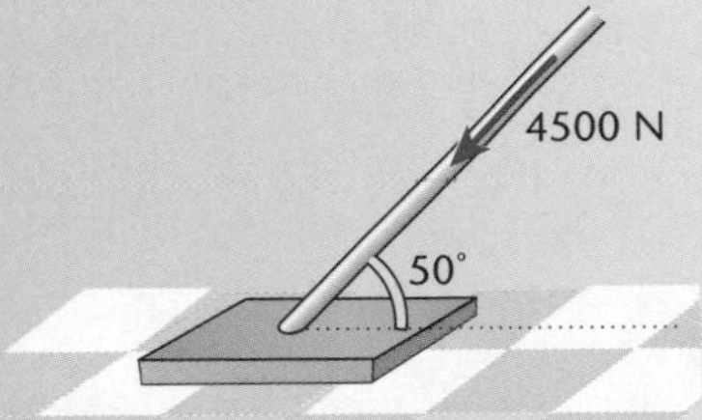

(a) Calculate the values of the horizontal and vertical components of the force in the girder. [2]

(b) Calculate the pressure that the steel plate exerts on the ground. [3]

Practice examination questions (continued)

(c) Explain why the girder is fixed to a steel plate rather than being fixed directly to the ground. [2]

16 (a) How does a ductile material behave differently from a brittle one when subjected to a stretching force? [2]

(b) The graph shows the relationship between the stress on a sample of cast iron and the strain that it causes. The curve ends when the sample breaks.

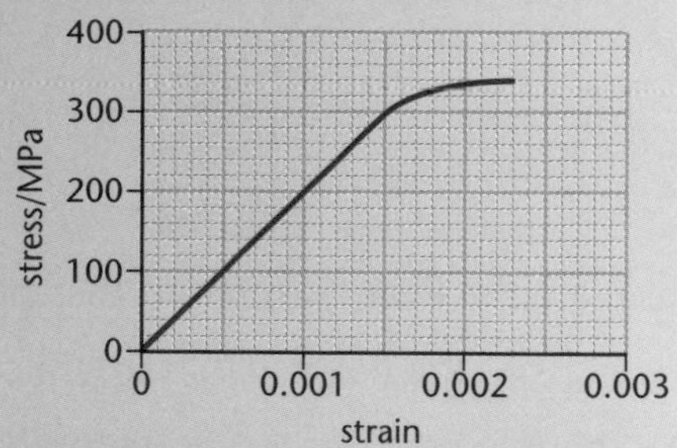

(i) Write down the value of the breaking stress of the cast iron. [1]

(ii) For what range of stresses does the cast iron follow Hooke's law? [1]

(iii) Calculate the Young modulus for the cast iron in the region where it follows Hooke's law. [2]

(iv) A second sample of the same material has twice the diameter of the original sample.
Explain whether the stress–strain graph for this sample would be the same as that shown in the diagram. [2]

Chapter 2

Electricity

The following topics are covered in this chapter:

- *Charge, current and energy transfer*
- *Resistance*
- *Circuits*

2.1 Charge, current and energy transfer

After studying this section you should be able to:

- *compare the phenomena of mass and charge of particles*
- *explain that free electrons and ions can be charge carriers*
- *recall and use the relationship between charge flow and current*
- *recall and use the relationship between potential difference, energy transfer and charge*
- *compare energy transfer to charge carriers by sources with energy transfer from charge carriers in circuit components*
- *explain the difference between electromotive force and potential difference*

LEARNING SUMMARY

Charge

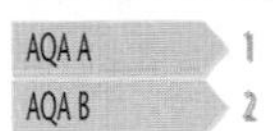

1
2

Charge is the name that we give to the ability of a body to exert and experience electrical force. The size of the force depends, among other things, on the amounts of charge involved.

It is useful to compare charge with mass. A body with mass can exert gravitational force on others that also have mass. The size of the force depends, among other things, on the amounts of mass involved.

There are differences as well as similarities. Gravitational forces act on huge scales and are most significant between large bodies. Electric forces act on smaller scales and they are most important when between small bodies. Also, gravitational forces are always attractive, while electric forces can be attractive or repulsive. So we have to suppose that while there is only one kind of mass, there are two kinds of charge. These need names – so we call them positive and negative. Charges that are the same always repel. Charges that are different always attract.

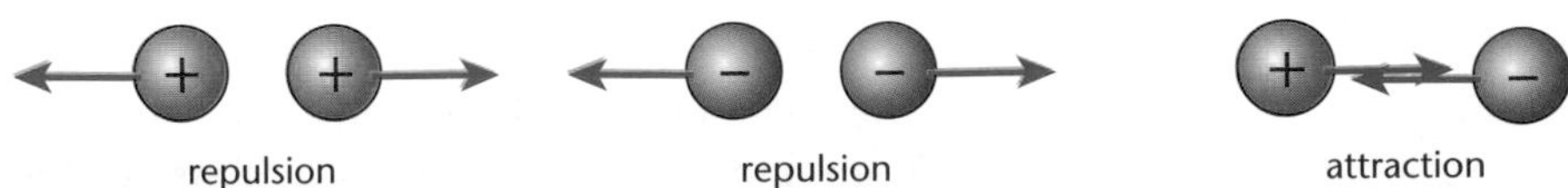

Electrons, protons and neutrons are particles which all have mass. Only two of them though, have electric charge. Protons have positive charge and electrons have negative charge. Neutrons get their name from their lack of electrical behaviour – they are electrically neutral.

Charge carriers

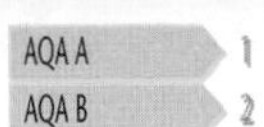

1
2

Within a metal, some of the electrons are free to move between individual atoms. The electrons act as **charge carriers** inside the metal. When they move together in large numbers there is a flow – there is an electric current.

An ordinary atom has a balance of electrons and protons, and is electrically neutral. However, one or more electrons can be removed from an atom, or extra electrons can be added. The atom then becomes an ion. Ions in solutions can be mobile. This happens in salt solutions, for example. The salt solution can conduct electricity thanks to the presence of these mobile charge carriers.

If a gas is ionised then it starts to conduct electricity. A flame, for example, contains ions, as does the gas inside a lighting tube.

Electric current

Electric current is 'rate of flow of charge'. Every word in the phrase counts – current is not just a flow of charge, but a 'rate of flow of charge'. The size of a current depends on how much charge is flowing and on how quickly. If ΔQ is amount of charge and Δt is the time it takes to flow past a point, then we can write an equation for current:

Current, $I = \Delta Q/\Delta t$

The charge carried by one electron, $e = -1.60 \times 10^{-19}$ C. 1 C of charge is equivalent to that carried by 6.25×10^{18} electrons.

Current is measured in amperes or amps, or A for short.

We define the unit of charge in terms of the unit of current. The unit of charge is called the coulomb.

One coulomb is the amount of charge that flows past a point when a steady current of 1 A passes for 1 second.

Current and drift velocity

The free electrons in a metal are in constant random motion. A typical speed in a metal at room temperature is of the order of 1×10^{6} m s^{-1}.

the random motion of a free electron, with collisions causing changes in direction

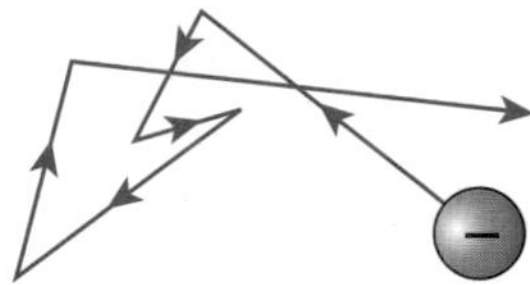

+ –

drift velocity is slow compared to the random motion of the electrons

Applying an electric force in the form of a voltage does not cause the free electrons to suddenly dash for the positive terminal. The random motion continues, but in addition the body of electrons moves at low speed in the direction negative to positive.

In a metal, a typical drift velocity is of the order of 1 mm s^{-1}, but in a semiconductor of similar dimensions carrying a similar current it is likely to be higher than this because the concentration of charge carriers is much less.

The overall speed of the body of electrons is called the **drift velocity (V)**.
There are three key variables that are related to the drift velocity:

- current (I)
- charge carrier concentration, n, defined as the number of charge carriers (free electrons in the case of a metal) per unit volume
- the cross-sectional area of the specimen, A.

These lead to the relationship for a metal $I = nAev$, where e is the electronic charge.

A similar expression applies to non-metallic conductors. As the charge on each ion can vary between non-metals, the expression becomes $I = nAqv$, where q is the ionic charge.

Charge and energy transfer

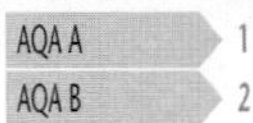

1
2

Charge is not used up or changed into anything else as it moves round a circuit. It **transfers energy**. It gains energy from the source of electricity (mains, battery or power supply) and transfers this energy out to the surroundings by way of the circuit components as it flows through them.

In the case of a filament lamp, the energy transfer from the charge is due to collisions of the electrons moving through the filament. With a motor or other electromagnetic device the charge loses energy as it has to work against repulsive forces in moving through the armature.

There are two separate physical processes:

- energy transfer **to** the charge from the source
- energy transfer **from** the charge through components such as lamps and motors to the world outside the circuit.

In each case, rather than measure the energy transfer for a single charge carrier, the transfer to and from **each coulomb** of charge is measured.

Electromotive force and potential difference

Electromotive force, e.m.f., measures the energy transfer per unit charge from the source. It can be written as:

$$E = W_{source}/q$$

A 6 V battery transfers 6 J of energy to each coulomb of charge that it moves around a circuit.

Electromotive force is a mis-named quantity. It describes energy transfer rather than force.

KEY POINT

Electromotive force, symbol E, is defined as being the **energy transfer from the source in driving unit charge round a complete circuit including through the source itself.** It is measured in volts (V), where 1 volt = 1 joule/coulomb (1 J C^{-1}).

Potential difference, p.d., between the terminals of a component measures the work done per unit charge as it flows through the component. It can be written as:

$$V = W_{component}/q$$

KEY POINT

The potential difference between two points, symbol V, is defined as being the **energy transfer per unit charge from the charge to the circuit between the two points.** Like e.m.f., it is measured in V.

- Electromotive force and potential difference are both defined in terms of the work done per unit charge, a key difference being whether the work is done **on** the charge or **by** the charge.
- The unit of e.m.f. and p.d. is the volt. *One volt is the potential difference between two points if 1 J of energy is transferred when 1 C of charge passes between the points.*

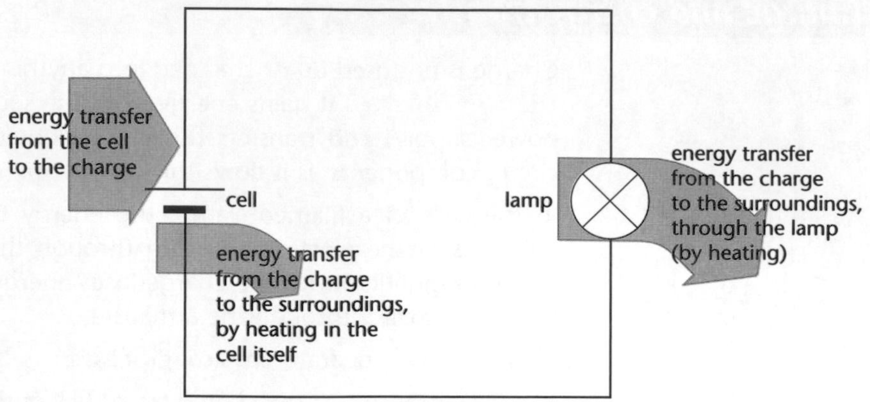

For this circuit:

energy in = energy out

energy transfer from the cell to the charge	=	energy transfer from the charge to the surroundings, through the lamp	+	energy transfer from the charge to the surroundings, by heating in the cell itself

The same charge flows through all parts of the circuit, so the whole equation can be divided by charge. That gives:

e.m.f. of cell, E	=	p.d. across the lamp, V	+	energy lost by cell per unit charge, also known as 'lost volts'

$$E = V + \text{lost volts}$$

Progress check

Use the definitions of current and potential difference.

1 Charge flows through a lamp filament at the rate of 15 C in 60 s. In the same time, 1500 J of energy is transferred to the filament.
Calculate:

a the current in the filament

b the potential difference across the filament.

2 A silicon diode has a cross-sectional area of 4.0 mm^2 (4.0×10^{-6} m^2).
The concentration of charge carriers is 2.5×10^{27} m^{-3} and each charge has a charge of 1.60×10^{-19} C.
Calculate the drift velocity when the current in the diode is 1.5 A.

1 a 0.25 A b 100 V
2 9.4×10^{-4} m s^{-1}

2.2 Resistance

After studying this section you should be able to:

LEARNING SUMMARY

- *recall and use the formula for resistance*
- *recall and use the formula for resistivity*
- *state Ohm's law and describe its limitations*
- *sketch the current–voltage characteristics of a filament lamp, a wire at constant temperature and a diode*
- *explain how resistance varies with temperature for metals and semiconductors, and how this can be used for environmental sensing*
- *distinguish between resistance and conductance, and between resistivity and conductivity*
- *describe superconductivity*
- *predict energy transfer from values of quantities selected from current, voltage resistance and time*
- *predict the heating power of a resistor from values of quantities selected from current, voltage and resistance*

Resistance

1
2

Resistance is a measure of the opposition to current. The greater the resistance of a component, the smaller the current that passes for a given voltage.

KEY POINT

Resistance is defined and calculated using the formula:

Resistance = voltage ÷ current or R = V ÷ I

The unit of resistance is the ohm (Ω).

Different components

1
2

It is possible to vary the potential difference between the terminals of a circuit component. To put it another way, it is possible to vary the voltage applied to the component. Then the current that results from different applied voltages can be measured. Different components behave differently, and the patterns of behaviour become easier to see, and then to explain, if the data is represented on graphs.

The graphs represent typical results obtained for a metal wire kept at a constant temperature, a filament lamp and a diode.

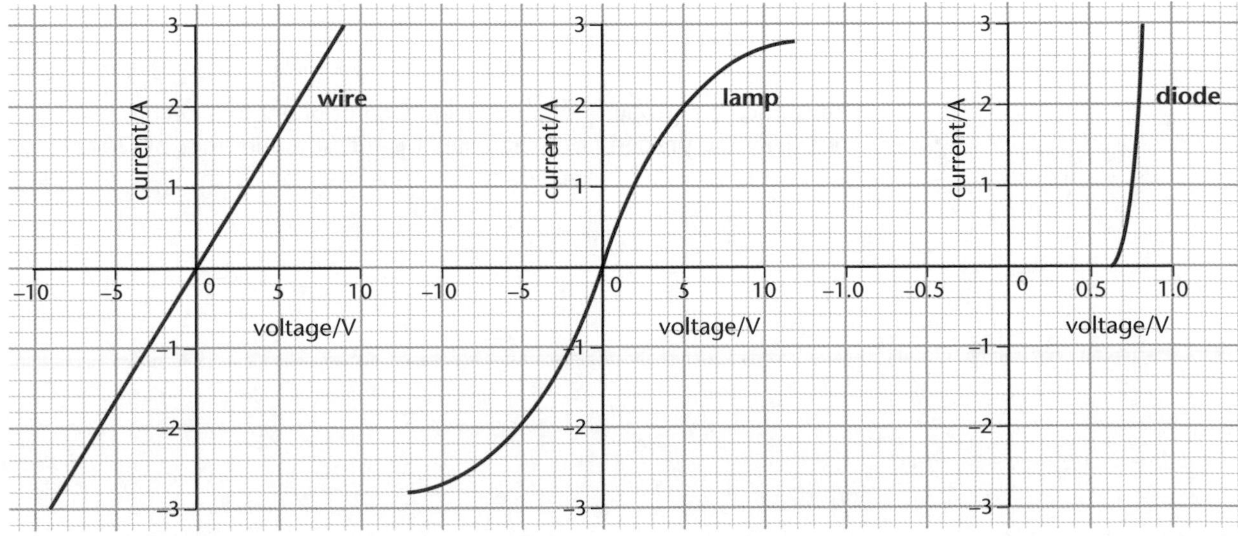

The text in *italics* is **Ohm's law**, one that he formulated based on his own experiments. Because it only applies to metallic conductors kept at a constant temperature, it is of only limited use in predicting the values of electric currents.

KEY POINT

The current–voltage graph for the **wire** is a straight line through the origin, showing that the *current is directly proportional to the voltage.*

A **filament lamp** is an everyday example of a metallic conductor that changes in temperature during use. In a short time after it is switched on, the temperature of the filament increases from room temperature to about 2000°C. The current–voltage graph shows how this change in temperature affects the relationship between these quantities. Like the one for the wire at constant temperature, this graph is symmetrical. Both the wire and the lamp filament show the same pattern of behaviour no matter which way round the voltage is applied.

Calculations based on this graph show that **as the current in the filament increases, so does its resistance**. This is due to the higher temperature causing an increase in the amplitude of the vibrations, which in turn increases the frequency of the collisions that impede the electron flow.

The arrow shows the direction in which the diode allows conventional current to pass.

The graph showing the characteristics of the **diode** also shows the property that makes diodes different from other components; they only allow current to pass in one direction.

Resistivity

AQA A 1
AQA B 2

What factors determine the resistance of a resistor?

- Length
  Long wires have more resistance than short wires, so the longer the sample of material, the greater the resistance.
- Cross-sectional area
  Doubling the cross-sectional area of a sample also doubles the number of charge-carriers available to carry the current, halving the resistance.

KEY POINT

The resistance of a sample of material is directly proportional to its length and inversely proportional to its cross-sectional area.
This statement can be written as $R \propto l/A$.

Good conductors have a low resistivity; that of copper being $1.7 \times 10^{-8}\ \Omega$ m. The resistivity of carbon is about one thousand times as great, $3.0 \times 10^{-5}\ \Omega$ m.

The other factor that determines the resistance is the material that the resistor is made of. The resistive properties of a material are measured by its **resistivity**, symbol ρ. When this is taken into account the formula becomes

$$R = \frac{\rho l}{A}.$$

This formula can be used to calculate the resistance when the dimensions and material of a resistor are known or to calculate the dimensions required to give a particular value of resistance.

KEY POINT

Resistance is a property of an individual component in a circuit, but resistivity is a property of a material. It is measured in units of Ω m.

Resistivity and temperature

AQA A 1
AQA B 2

The resistivity of a material is temperature dependent. Metals and semiconductors such as graphite (carbon) and silicon behave differently.

For metals, resistivity increases with temperature. At higher temperatures, a flow of electrons through a metal experiences more collisions due to atomic vibration, so that electrons tend to lose energy more quickly.

NTC stands for negative temperature coefficient, indicating that its resistance decreases as temperature increases, and that the gradient of the resistance–temperature graph is negative.

For semiconductors, the higher temperature sets more electrons (charge carriers) free within the material. So the resistivity of a semiconductor can decrease rapidly when its temperature increases.

An NTC thermistor is a sample of semiconductor, with suitable connections, that can be used in a circuit. The resistance of such a thermistor is very temperature-dependent, so it can allow an electric circuit to 'respond' to temperature changes.

Because of their large change in resistance with temperature, thermistors are useful in applications such as heat-sensitive cameras, which need to be able to detect small changes in temperature.

These graphs show the variation in resistance with environmental conditions for a metallic conductor, a thermistor and a light-dependent resistor.

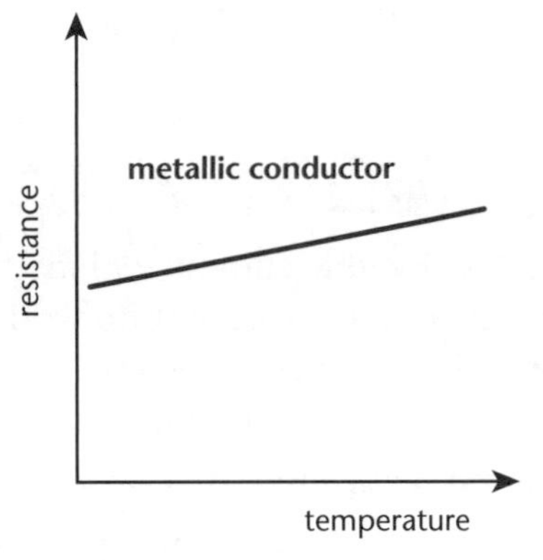

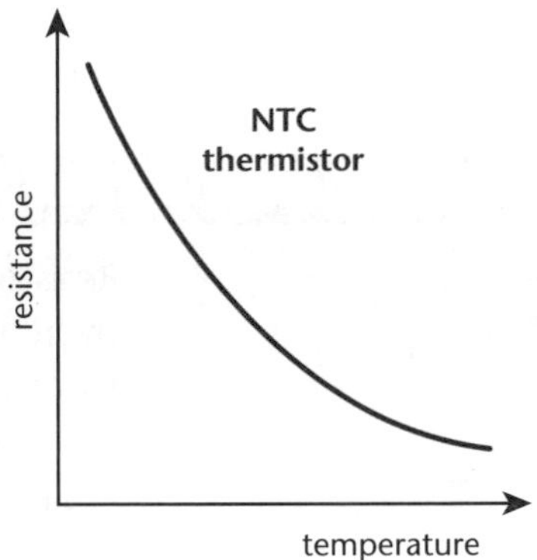

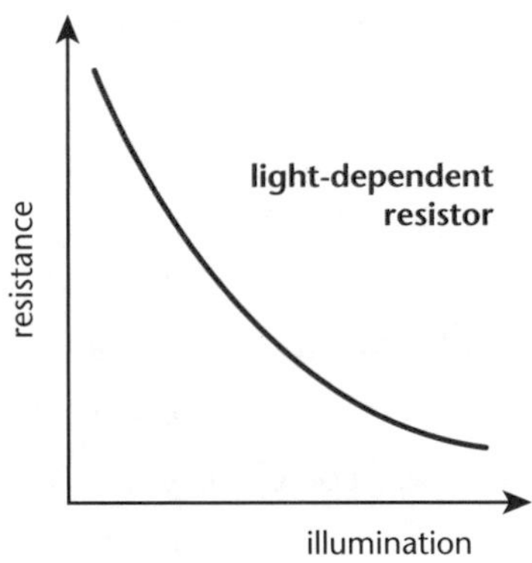

A light-dependent resistor, or LDR behaves in a similar way to a NTC thermistor, though it is light falling on a wafer of semiconductor that provides energy that sets electrons free. Again, the more free charge carriers there are per unit volume of a sample of material, the lower its resistance.

Conductance and conductivity

AQA A 1
AQA B 2

It is possible to consider the conductance of a component rather than its resistance. Conductance is the 'inverse' of resistance.

conductance = 1 / resistance

Conductance can be defined by the equation:

conductance = I/V and is measured in A V^{-1}.

In a similar way, conductivity is the inverse of resistivity. Conductivity is calculated from:

conductivity, $\sigma = l / R A$

Superconductivity

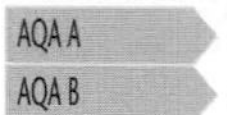
AQA A 1
AQA B 2

The resistivity of some (but not all) metals falls to zero (which means that conductivity becomes infinite) at extremely low temperatures. Superconducting cables can carry very large currents without any energy loss to the surroundings. There is no dissipation, and no potential difference is needed to keep the current going.

Unfortunately, keeping the temperature of a cable very low is difficult and expensive. Superconducting cables are only used in applications where a strong magnetic field is needed, and where people are prepared to pay. These include medical imaging equipment (MRI scanners) and scientific research (such as in particle accelerators).

Some more complex combinations of materials become superconducting at higher, but still very cold, temperatures. These are called high temperature superconductors, but the 'high' is relative. Existing materials still need to be well over 100 °C below the freezing point of water before they become superconducting.

Energy transfer by resistors

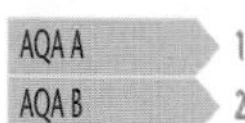
AQA A 1
AQA B 2

Resistors are heated by the current within them. Electrons must 'do work' to overcome resistance, just as a car must do work to overcome resistance of different kinds (friction and air resistance). The amount of work done is exactly equal to the amount of energy that is transferred by heating.

energy transferred by heating, ΔE = work done by electrons, $\Delta W = Vq$

where V is the potential difference between the ends of the resistor and q is the charge carried by the electrons. Note that this equation comes from the definition of potential difference, found on page 59.

Charge q is related to the current:

$$I = q/\Delta t$$

Which is the same as saying that:

$$q = I\,\Delta t$$

So, during a length of time Δt, energy transferred by heating:

$$\Delta E = V\,I\,\Delta t$$

In a resistor, $V = IR$, so we could write,

$$\Delta E = I^2\,R\,\Delta t$$

And also, $I = V/R$, so: $\Delta E = V^2\,\Delta t\,/\,R$

Power

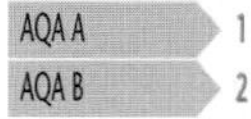
AQA A 1
AQA B 2

Power is the *rate of working or energy transfer.*

> **KEY POINT**
> The relationship between electrical power, current and potential difference is:
> *power = current × potential difference* or $P = IV$.
> Power is measured in watts (W), where 1 W = 1 J s^{-1}.

Unless stated otherwise, the term 'rate of' refers to 'rate of change with time', so power is the work done or energy transfer per second.

This relationship can be combined with the resistance equation to give:

- $P = I^2R$
- $P = V^2/R$

These three equivalent relationships can all be used to calculate the value of either a constant or a varying power.

A battery-operated lamp transfers energy at a constant rate. The power of a mains-operated lamp is constantly changing. The graphs compare the power of a battery-operated lamp with that of a mains lamp.

The smaller the value of Δt, the better the approximation that $E = IV\Delta t$ is to the area between the graph line and the time axis.

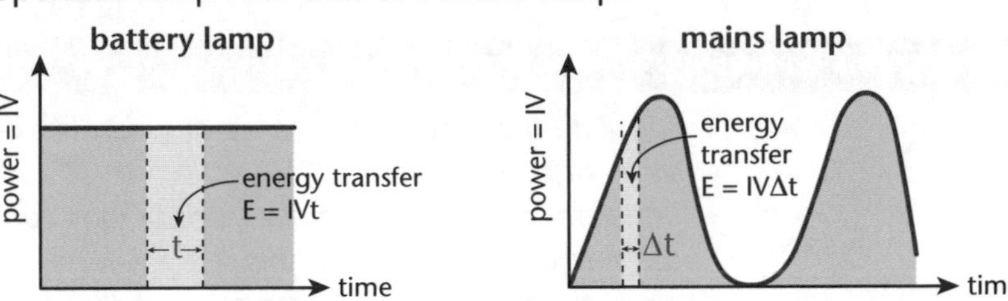

The SI unit of energy is the joule. For domestic purposes, energy is measured in kilowatt-hours (kW h), where 1 kW h is the energy transfer by a 1 kW appliance in 1 hour.

The total energy transfer in time Δt where the power is constant is equal to:

KEY POINT

$$\Delta E = \text{power} \times \text{time} = IV\Delta t.$$

For a varying power the energy transfer is represented by the area between the curve and the time axis of a graph of power against time. The shorter the value of Δt, the better the approximation of the above expression to this area.

Progress check

1 Use the resistance formula to work out the following:
   a The voltage across a lamp filament if the current passing is 2.4 A and the filament resistance is 7.3 Ω.
   b The resistance of an infra-red heater if the current is 9.5 A when it operates from the mains 240 V supply.
   c The current passing in a 6.8 Ω resistor when it is connected to a 9 V battery.

Note that 1 mm = 1×10^{-3} m.

First re-arrange the resistivity equation to make l the subject.

2 Use the resistivity formula to work out the following:
   a The resistivity of silicon. A 1 mm cube of silicon has a resistance of 2.20×10^{6} Ω when measured across opposite faces.
   b The length of carbon rod required to make a 3.9 Ω resistor. The resistivity of carbon is 3.00×10^{-5} Ω m and the rod has a cross-sectional area of 1.26×10^{-6} m².

3 An electrical heating element is to be made from nichrome wire. The current in the heating element is to be 5.00 A when the voltage across it is 12.0 V.
   a Calculate the resistance of the heating element. The resistivity of nichrome is 1.10×10^{-6} Ω m. The cross-sectional area of the wire used is 7.85×10^{-7} m².
   b Calculate the length of wire required.

4 What happens to the power of a filament lamp when the current is doubled?

5 When the current in a cell is 1.5 A it supplies energy at the rate of 18 W. Calculate the e.m.f. of the cell and the total resistance of the circuit.

6 a Show, using appropriate equations, what is meant by the statement 'conductivity is the inverse of resistivity'.
   b What are the units of conductivity?

1 a 17.5 V b 25.3 Ω c 1.32 A
2 a 2200 Ω m b 0.164 m
3 a 2.4 Ω b 1.71 m
4 It is quadrupled.
5 12 V and 8 Ω.
6 a $\sigma = 1/\rho = l/RA$.
   b $\Omega^{-1}\text{m}^{-1}$

2.3 Circuits

After studying this section you should be able to:

LEARNING SUMMARY

- *calculate the effective resistance of a number of resistors connected in series or parallel*
- *understand that a source of electricity may contribute to the resistance of a circuit and the effects of this*
- *apply the principles of conservation of charge and conservation of energy to circuit calculations*
- *describe how the resistance of a metallic conductor, a thermistor and a light-dependent resistor change with environmental conditions*

Series and parallel

AQA A 1
AQA B 1

Resistors can be combined in series, parallel or a combination of the two. There are simple formulae for calculating the effective resistance of two or more resistors connected together.

Series

The effective resistance of a number of resistors in series is always greater than the largest value resistor in the combination.

Adding another resistor in series in a circuit always decreases the current. This means that the effective resistance in the circuit has increased.

KEY POINT

The formula for calculating the effective resistance of a number of resistors in series is:

$$R = R_1 + R_2 + R_3.$$

To find the effective resistance of a 4.7 Ω, a 6.8 Ω and a 10 Ω resistor connected in parallel
$1/R = 1/4.7 + 1/6.8 + 1/10 = 0.460$
$R = 1/0.460 = 2.2\ \Omega$
A common error is to forget to carry out the final stage of the calculation.

This formula can be used for any number of resistors connected in series by adding on extra terms.

Parallel

When an additional resistor is added in parallel to a circuit, the current always increases. The additional resistor opens up another current path, without affecting the current in any of the existing paths. This results in less resistance in the circuit.

The effective resistance of a number of resistors in parallel is always smaller than the smallest value resistor in the combination.

KEY POINT

The formula for calculating the effective resistance of a number of resistors in parallel is:

$$1/R = 1/R_1 + 1/R_2 + 1/R_3.$$

As with the formula for resistors in series, any number of terms can be added to this expression.

Internal resistance of a cell

The internal resistance of a cell or power supply acts as an extra resistor in series with the rest of the circuit.

All parts of a circuit have resistance and energy is needed to move the charge carriers through them. A cell of e.m.f. 1.5 V, for example, does 1.5 J of work on each coulomb of charge that completes a circuit.
This work is done on moving the charge through:

- the connecting wires
- the circuit components
- the cell itself, due to the cell's own **internal resistance**, symbol r.

The greater the current in the cell, the more work is done against the internal resistance, so the less can be done in the external circuit.

How does internal resistance affect the circuit?

For example, a cell of e.m.f (E) 1.5 V and internal resistance (r) 0.2 Ω causes a current (I) of 0.5 A in an external resistor (R). The work done per coulomb of charge against the internal resistance = I × r = 0.5 A × 0.2 Ω = 0.1 V. This leaves 1.4 V for the external circuit.

The greater the current in the cell, the greater the reduction in potential difference.

The effect of internal resistance is to reduce the potential difference across the external circuit.

When the only component connected to a cell is a high-resistance voltmeter such as a digital voltmeter, the reading is the cell's e.m.f. This is because when it is connected to the voltmeter alone, virtually no current passes in the cell so no work is being done against internal resistance. However, when the cell is connected to a circuit, a voltmeter placed across its terminals reads less than the e.m.f.

A typical digital voltmeter has a resistance of 10 MΩ (1.0×10^7 Ω), so the current passing in it is very small, causing negligible reduction in the p.d. across the terminals.

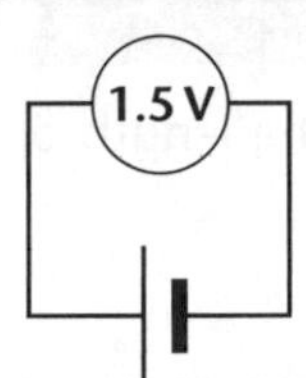

The high resistance voltmeter shows that the e.m.f. of the cell is 1.5 V. The cell supplies 1.5 J of energy to move each coulomb of charge around a complete circuit.

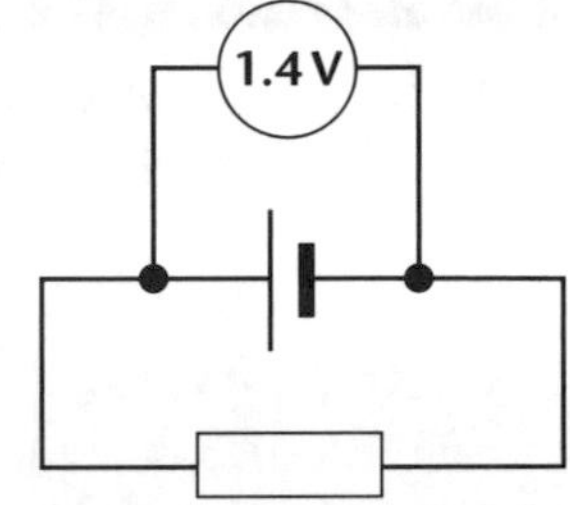

0.1 J of energy is needed to move each coulomb of charge through the cell, leaving 1.4 J to move the charge round the rest of the circuit.

> **KEY POINT**
> The relationship between e.m.f. and terminal p.d. is:
> e.m.f. = terminal p.d. + 'lost volts'
> $E = V + Ir$

Compare this equation with that on pages 59–60.

Kirchhoff's laws

In a circuit, the flow of **charge** transfers **energy** from the source to the components. Neither of these physical quantities becomes used up in the process. Both are **conserved. Kirchhoff's laws** are re-statements of these fundamental laws in terms that apply to electric circuits.

Mass, charge and energy are fundamental conserved quantities. They cannot be created or destroyed.

> **KEY POINT**
> **Kirchhoff's first law** states that **the total current that enters a junction is equal to the total current that leaves the junction.**

This is illustrated in the diagram below and is simply a statement that charge is conserved at the junction.

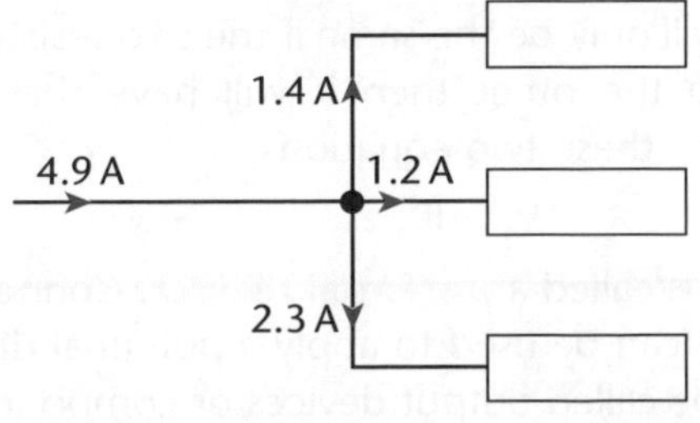

> **Kirchhoff's second law is about conservation of energy. It states that around any closed loop (i.e. complete series path), the total e.m.f. is equal to the sum of the potential differences, $E = \Sigma(IR)$.**
>
> Note that R here is the total circuit resistance including the internal resistance of the source.
>
> KEY POINT

Remember that e.m.f. relates to energy transfer to the charge and p.d. relates to energy transfer **from** the charge, so this law is stating that when a quantity of charge makes a complete circuit, the energy transfers to the charge and from the charge are the same. The second law can be applied to a series circuit or any series path within a parallel circuit.

The variable resistor

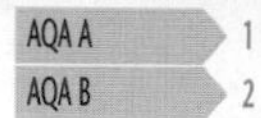

A variable resistor allows the length of wire through which current must flow to be varied.

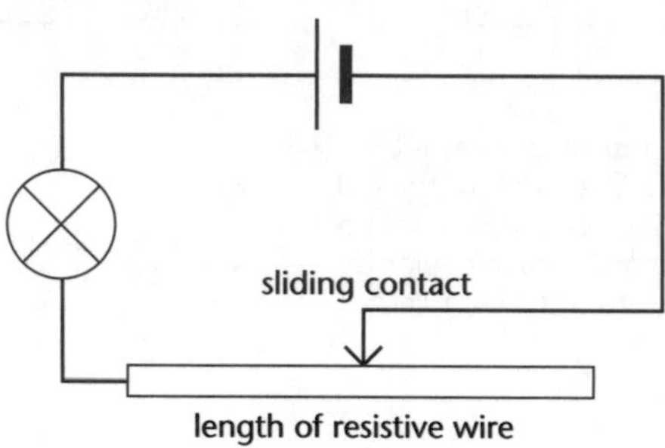

The potential divider

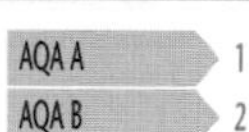

A potential difference can be applied to a pair of resistors. When a current flows, provided that there are no junctions between the resistors, the current is the same in them both.

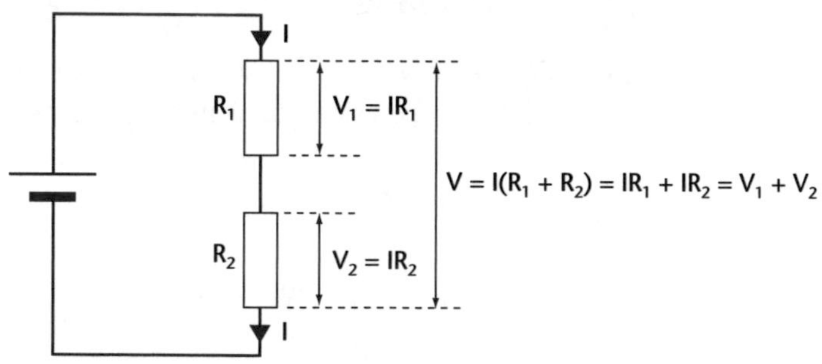

> Remember: Potential difference is a DIFFERENCE between two points in a circuit. Here, the two differences, V_1 and V_2, one after the other, add up to the total difference, V.

V is the total potential difference across both resistors. V_1 and V_2 are the p.d.s between the ends of resistor R_1 and resistor R_2.

$$V = V_1 + V_2$$

V_1 and V_2 will only be the same if the two resistances are equal. If one resistance is bigger than the other then it will have the bigger potential difference, as is suggested by these two equations:

$$V_1 = IR_1 \qquad V_2 = IR_2$$

The system is called a **potential divider**. Connections to the two ends of one of the resistors can be used to apply a potential difference to further components. These can be called output devices or components.

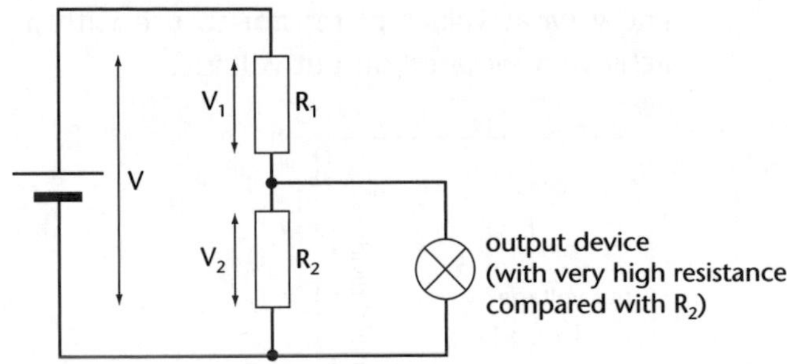

The potential difference between the ends of the lamp is the same as that between the ends of resistor R_2.

Input and output voltage

AQA A 1
AQA B 2

The total potential difference applied to the two resistors of a potential divider system can be called the input voltage. The potential difference that is then applied to an output device is called the output voltage.

If one of the resistors in a potential divider is an LDR, for example, then changes in light intensity cause its resistance to change. This causes the output voltage to change. This means that the behaviour of the output components changes when light intensity changes.

Thermistors and other 'environmental sensing' devices can be used in place of, or even as well as, LDRs. The behaviour of the output component then depends on temperature. So potential dividers can form the basis of 'sensing circuits'.

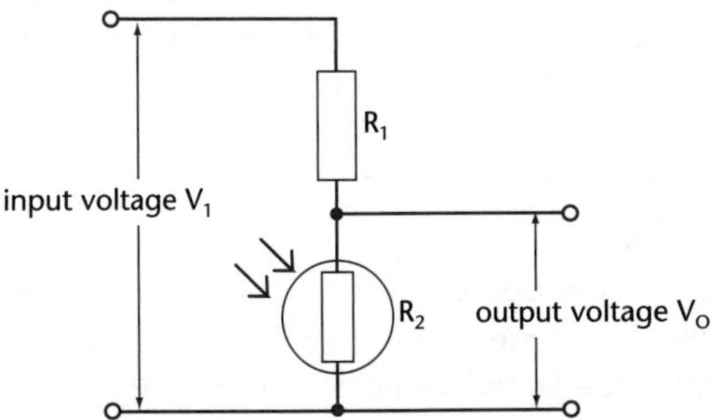

In an investigation in which potential difference is controlled (as the independent variable), a continuous wire with a sliding connection can act as a potential divider. When the slider is at one end, the 'output' voltage is equal to the total potential difference applied to the wire. When the slider is at the other end then the output voltage is zero.

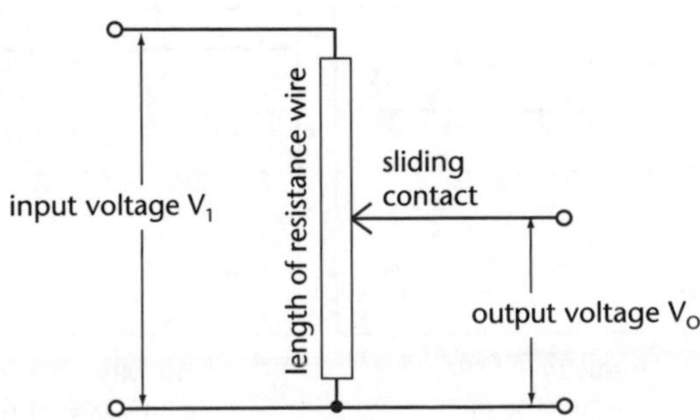

There is an equation that can be used as a predictive tool, so that it is possible to know what values of resistor to use with a particular input voltage, in order to achieve a required output voltage.

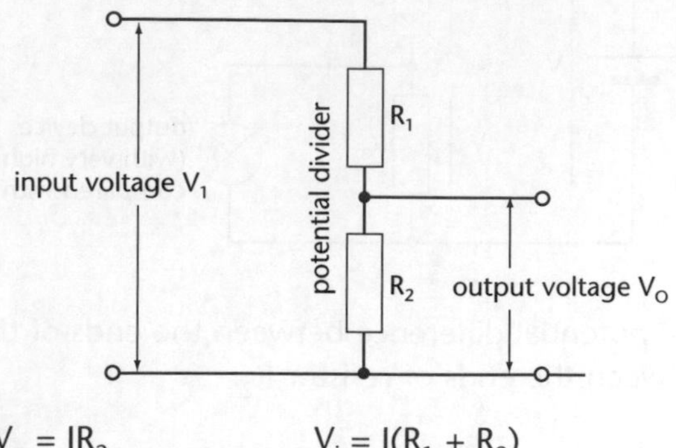

$V_o = IR_2$ $\qquad V_i = I(R_1 + R_2)$

These equations together tell us that $\frac{V_o}{V_i}$ and $\frac{IR_2}{I(R_1 + R_2)}$ are identical.

$$\frac{V_o}{V_i} = \frac{IR_2}{I(R_1 + R_2)} = \frac{R_2}{(R_1 + R_2)}$$

So, $V_o = \frac{R_2}{R_1 + R_2} V_i$

Using an oscilloscope to measure voltage

AQA A 1

Potential difference results in a force acting on charged bodies. If a potential difference is applied to two metal plates then electrons between the plates experience force. If they are moving in a beam then the beam is deflected. The amount of deflection depends on the size of the potential difference. This is what happens in an oscilloscope, which has two parallel horizontal metal plates connected to input terminals. A beam of electrons hits the screen to create a glowing dot.

The oscilloscope can be set to produce a single dot at the centre of the screen. When a voltage is then applied to the input terminals and so to the plates, the beam is deflected vertically and the dot moves.

The amount of deflection provides a measure of voltage, and the screen has a grid to make this easier. The **sensitivity** can be varied.

The oscilloscope also has a **time base** control. Switching the time base on applies a varying potential difference to two vertical plates and this makes the dot scan horizontally across the screen. Most scanning speeds are fast, so that the dot appears as a single horizontal straight line.

These diagrams compare the traces obtained for a 3.0 V direct voltage input with different sensitivity and time base settings.

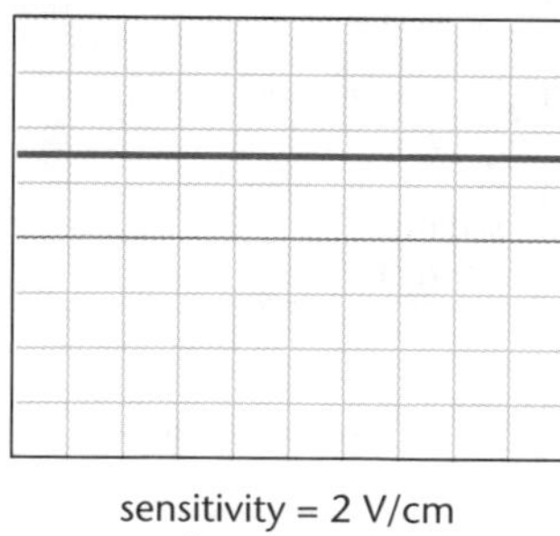

sensitivity = 2 V/cm
time base on

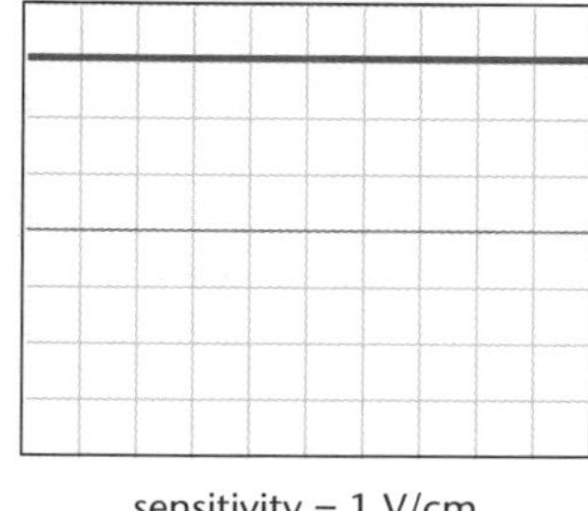

sensitivity = 1 V/cm
time base on

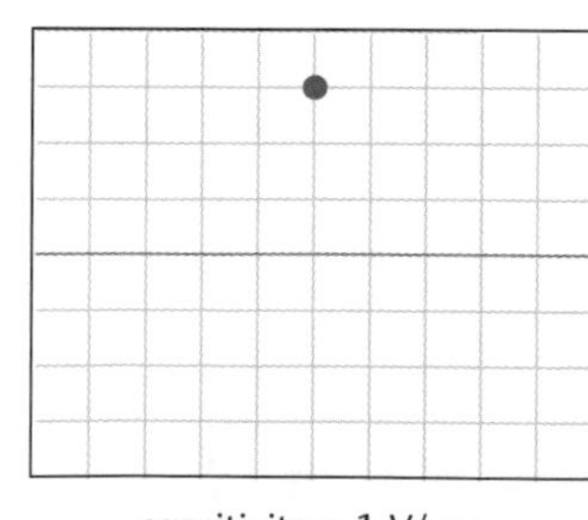

sensitivity = 1 V/cm
time base off

Using an oscilloscope to measure time and frequency

AQA A 1

If the voltage varies, the dot moves up and down. If the time base is switched on and set to a particular sensitivity the horizontal distance across the screen acts rather like a time axis.

If the voltage that is applied to the oscilloscope input terminals varies periodically, in a regular pattern, then the time for one cycle can be read from the screen, provided that the setting of the time base control is known. From the time for one cycle, or period, the frequency can be calculated.

- A time interval is measured by multiplying the appropriate horizontal distance on the screen by the time base setting.
- A frequency is measured by calculating (1/time) for one complete cycle.

Remember to convert times in ms or μs into s before calculating a frequency.

In the example shown in the diagram, the time base is set to 1 ms cm^{-1}, or 1.0×10^{-3} s cm^{-1}.
One complete cycle of the wave occupies a distance of 4.0 cm on the screen, so the time taken for one cycle is 4.0×10^{-3} s.
Since *frequency* = 1/*time period* or $f = 1/t$, the frequency of the wave is $1 \div (4.0 \times 10^{-3})$ s = 250 Hz.

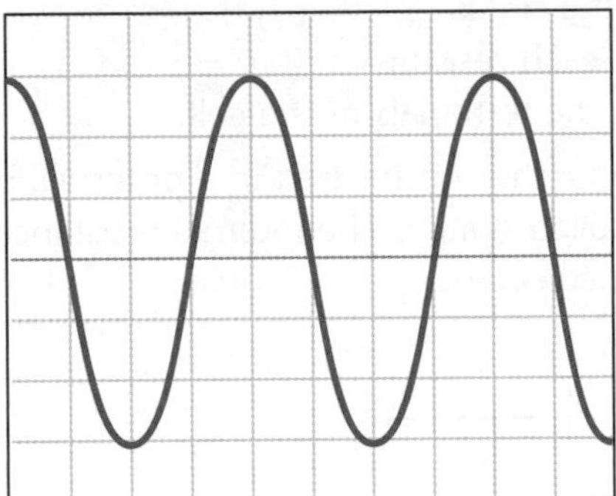

Progress check

> In **b**, work out the resistance of the parallel combination first.

1 Calculate the effective resistance of each of the following combinations of resistors.

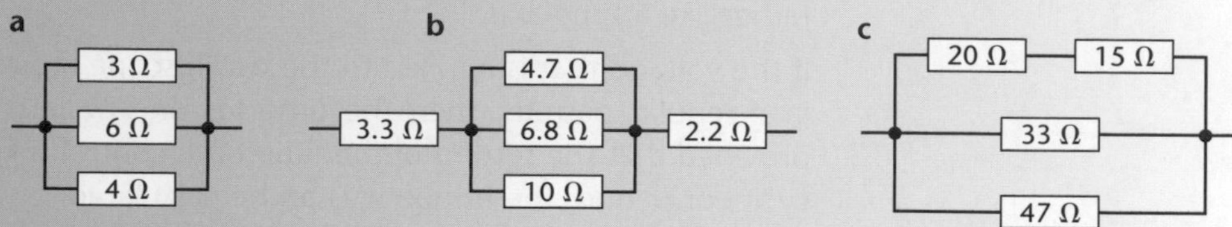

> Remember that the p.d. across the terminals is equal to that across the external resistor.

2 A cell has an e.m.f. of 1.6 V and internal resistance 0.5 Ω.
Calculate the current in the cell and the p.d. across the terminals when it is connected to:

a a 0.5 Ω resistor

b a 3.5 Ω resistor.

3 A battery of e.m.f 4.5 V and internal resistance 0.6 Ω is connected to a 3.3 Ω resistor in series with a 5.1 Ω resistor.
Calculate:

a the current in the circuit

b the p.d. across each resistor

c the p.d. across the terminals of the cell.

4 The diagram shows a thermistor used in a potential divider as part of a temperature-controlled switch. The internal resistance of the power supply is small and can be neglected.

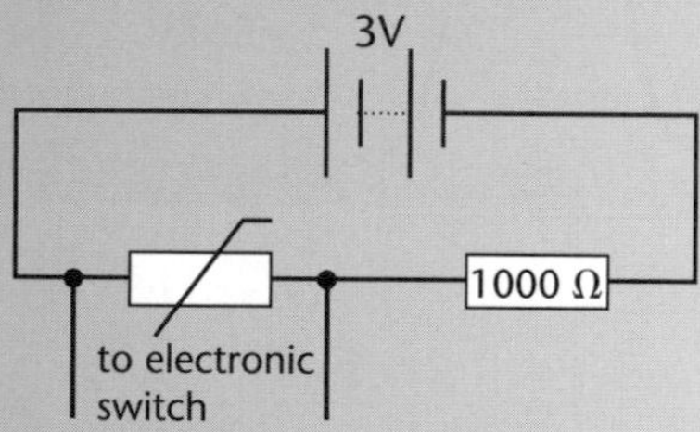

The electronic switch turns on a heater when the p.d. across the thermistor rises to 0.6 V.

> The quickest way to answer **a** is to use ratios.

a Calculate the p.d. across the thermistor when its resistance is 200 Ω.

b Explain how the heater becomes switched off.

c Calculate the resistance of the thermistor when the heater switches off.

5 The diagram shows an oscilloscope display of an alternating voltage.

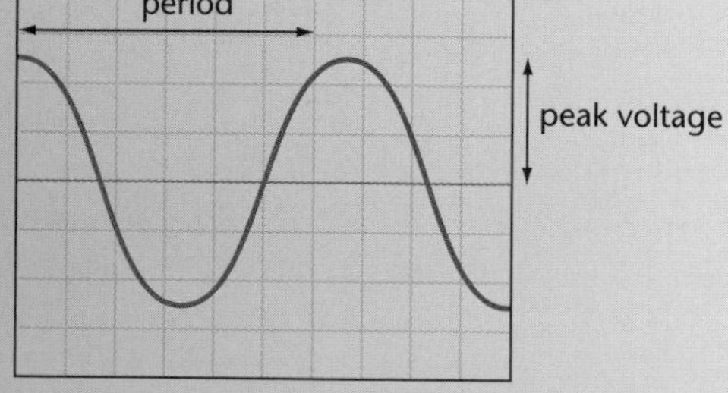

The sensitivity is set to 10 V/cm and the time base to 5 ms/cm. Calculate:

a the peak value of the voltage

b the frequency of the voltage.

1 **a** 1.33 Ω **b** 7.67 Ω **c** 12.5 Ω
2 **a** 1.6 A; 0.8 V **b** 0.4 A; 1.4 V
3 **a** 0.50 A **b** 1.65 V and 2.55 V **c** 4.2 V
4 **a** 0.5 V **b** The temperature of the thermistor rises, causing its resistance and the p.d. across it to fall.
c 250 Ω.
5 **a** 25 V **b** 33.3 Hz

Sample question and model answer

(a) What is the unit of resistivity? [1]

Since $\rho = AR/l$, the units of ρ must equal those of AR/l.
Units of $AR/l = m^2 \times \Omega \div m = \Omega m$. 1 mark

> This technique relies on the fact that a relationship between physical quantities must be homogeneous, i.e. the units on each side are the same.

(b) A cable consists of seven straight strands of copper wire each of diameter 1.35 mm as shown in the diagram. [6]

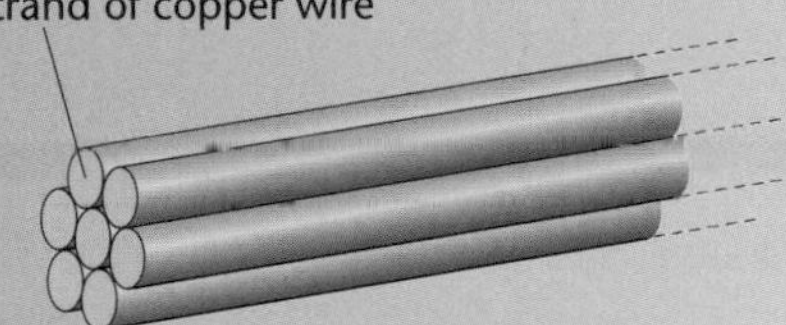

Calculate:

(i) the cross-sectional area of **one strand** of copper wire.

Cross-sectional area $= \frac{1}{4}\pi d^2$
$= \frac{1}{4} \times \pi \times (1.35 \times 10^{-3})^2 = 1.43 \times 10^{-6}\ m^2$ 1 mark

> It is important here to convert the diameter from mm to m in order to maintain a consistent set of units. Many AS candidates, faced with this type of calculation, work out the answer in mm² and then apply a wrong conversion factor to change it to m².

(ii) the resistance of a 100 m length of the cable, given that the resistivity of copper is $1.6 \times 10^{-8}\ \Omega$ m.

The resistance of one strand:
$R = \rho l/A$ 1 mark
$= 1.6 \times 10^{-8}\Omega m \times 100 m \div (1.43 \times 10^{-6}) m^2 = 1.12\ \Omega$ 1 mark
The resistance of 100 m of the cable is $\frac{1}{7}$ of this, i.e. 0.16Ω. 1 mark

> The emphasis here is on reading the question thoroughly. A common error is to calculate the resistance of one strand only, ignoring the fact that the resistance of the cable is emphasised in the question.

(iii) The cable carries a current of 20 A. What is the potential difference between the ends of the cable? [2]

$V = IR = 20A \times 0.16\Omega = 3.2V$. 1 mark

(iv) If a single strand of the copper wire in part (b) carried a current of 20 A, what would be the potential difference between its ends?

$20A \times 1.12\Omega = 22.4V$. 1 mark

> The calculations in (**c**) are straightforward applications of the resistance equation. Do not worry if your answers to (**b**) were wrong. You still gain full marks in (**c**) for correct working with the wrong resistance values.

Practice examination questions

1 (a) Write down the formula that relates the resistivity of a material to the resistance of a particular sample. [1]

(b) Explain the difference between *resistance* and *resistivity*. [2]

(c) The diagram shows a cuboid made from carbon.

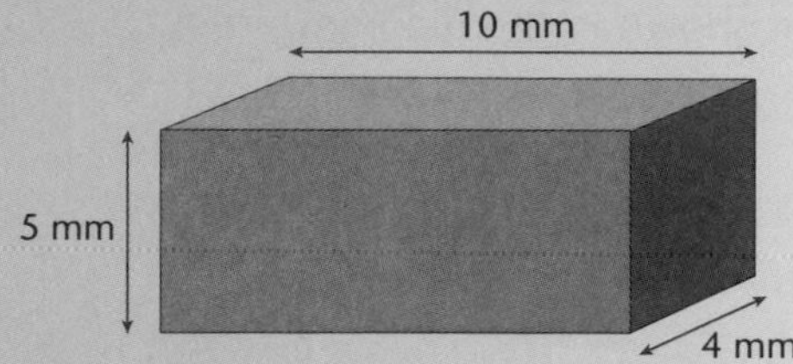

The resistivity of carbon is $3.00 \times 10^{-5}\ \Omega\,m$.
Calculate the resistance of the cuboid, measured between the faces that are 10 mm apart. [3]

2 In the circuit shown in the diagram, the internal resistance of the cell can be neglected.

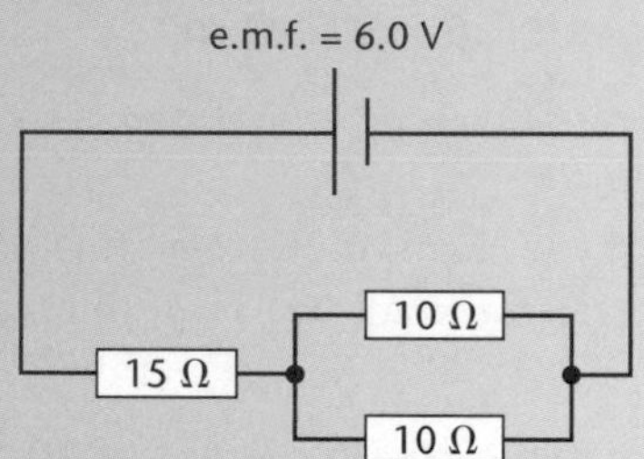

Calculate:

(a) the effective resistance of the two resistors in parallel [2]

(b) the current in the 15 Ω resistor [3]

(c) the current in each 10 Ω resistor [2]

(d) the potential difference across each 10 Ω resistor. [2]

3 (a) (i) Draw a sketch graph to show how the resistance of a thermistor varies with increasing temperature. [2]

(ii) State TWO ways in which this graph differs from a graph that shows how the resistance of a metallic conductor varies with increasing temperature. [2]

(iii) Explain why Ohm's law does not apply to either graph. [2]

(b) The diagram shows a thermistor in a series circuit with a 100 Ω resistor.

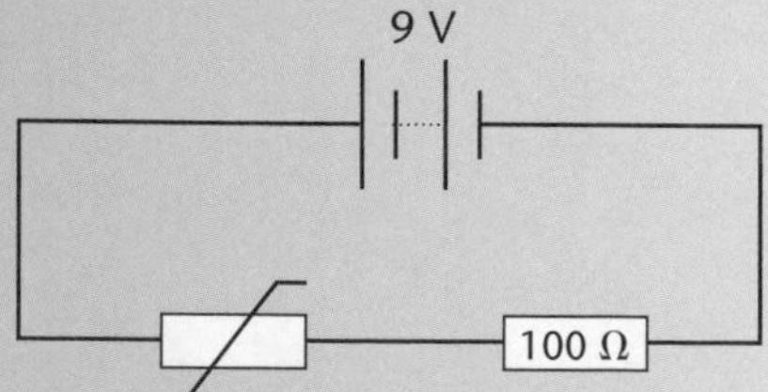

Practice examination questions (continued)

The table gives some data about the resistance of the thermistor.

Temperature/°C	*Thermistor resistance/Ω*
5	500
20	300
40	75

(i) Calculate the potential difference across the thermistor at a temperature of 5°C. [3]

(ii) Explain how the potential difference across the thermistor changes as the temperature rises. [2]

(iii) The potential difference across the resistor is used to switch off a heater when the temperature reaches 40°C. Calculate the potential difference at which the switch operates. [3]

(iv) The circuit is adapted to switch off the heater when the temperature reaches 20°C. Calculate the value of the fixed resistor required. [3]

4 A cell has an e.m.f. of 1.60 V. When connected to a digital voltmeter of resistance 10 MΩ ($1.0 \times 10^7\,\Omega$), the reading on the voltmeter is 1.60 V. When the cell is connected to a moving coil voltmeter of resistance 1 kΩ ($1.0 \times 10^3\,\Omega$) the reading on the voltmeter is 1.55 V.

(a) Explain why there is a difference in the voltmeter readings. [3]

(b) Calculate the internal resistance of the cell. [3]

5 A 12 V car battery is used to light four 6 W parking lamps connected in parallel.

(a) Calculate the current in the battery. [1]

(b) How much charge flows through the battery in one minute? [3]

(c) Calculate the effective resistance of the four lamps. [3]

6 (a) A 4.7 Ω resistor has a power rating of 10 W.

(i) Calculate the maximum current that should pass in the resistor. [3]

(ii) Calculate the maximum potential difference across the resistor. [2]

(iii) The resistor is manufactured from constantan wire of cross-section $2.0 \times 10^{-7}\,m^2$ and resistivity $5.0 \times 10^{-7}\,\Omega\,m$.
Calculate the length of wire used to make the resistor. [3]

(b) Low-power resistors are often made from carbon. The resistivity of carbon is $3.0 \times 10^{-5}\,\Omega\,m$. What is the advantage of making resistors from carbon? [2]

Practice examination questions (continued)

7 The diagrams show two circuits that can be used to investigate the current–voltage characteristics of component X.

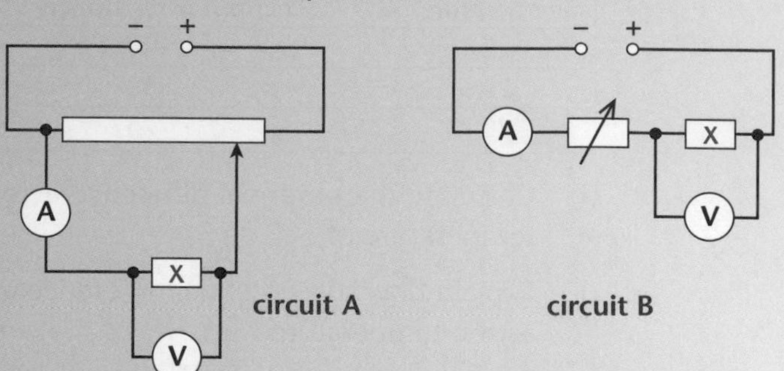

(a) What name is given to circuit A? [1]

(b) Describe what circuit A is able to do. [1]

(c) What is the advantage of using circuit A rather than circuit B to investigate the characteristics of a diode? [2]

(d) The table shows data obtained by varying the voltage across a diode.

voltage/V	0.65	0.71	0.76	0.78	0.80
current/A	0.06	0.20	0.44	0.68	1.00
resistance/Ω					

(i) Complete the table. [2]

(ii) Draw a graph of resistance against voltage. [4]

(iii) Use the graph to estimate the voltage at which the resistance of the diode is 1 Ω. [1]

(iv) Suggest why the data only span a narrow voltage range. [1]

8 In a simple lighting circuit, a 60 W lamp is connected to a 12 V battery using copper cable of cross-section 1.25 mm^2 (1.25×10^{-6} m^2). The lamp filament is made of tungsten of cross-section 3.0×10^{-4} mm^2 (3.0×10^{-10} m^2). The number of free electrons per unit volume for copper is 8.0×10^{28} m^{-3} and that for tungsten is 3.4×10^{28} m^{-3}. The electronic charge e = – 1.60×10^{-19} C.

(a) Calculate the drift velocity of the electrons in the copper cable. [3]

(b) Give TWO reasons why the drift velocity of the electrons in the tungsten is much greater than that of electrons in the copper. [2]

(c) Suggest why the tungsten becomes heated while the copper remains cool. [2]

9 A cell of e.m.f. 3.0 V and internal resistance 0.5 Ω is connected to a 3.0 Ω resistor placed in series with a parallel combination of a 6.0 Ω and a 12.0 Ω resistor.

Calculate:

(a) the current in the circuit [3]

(b) the p.d. across the terminals of the cell. [2]

Chapter 3

Waves, imaging and information

The following topics are covered in this chapter:

- *Wave behaviours*
- *Sound and superposition*
- *Light and imaging*
- *Information transfer*

3.1 Wave behaviours

After studying this section you should be able to:

LEARNING SUMMARY

- *explain that waves transmit energy and information from point to point without transmitting energy between the two points*
- *describe process of transmission and absorption, refraction, diffraction and polarisation of waves*
- *distinguish clearly between longitudinal and transverse waves*
- *use the equation $v = f\lambda$ in calculations and describe spectra of electromagnetic radiation and sound*
- *demonstrate how an oscilloscope can be used to measure period and frequency*
- *explain that 'inverse square law' behaviour is a result of spreading of power over increasing area*
- *describe the law of reflection*
- *interpret refraction using rays and wavefronts*
- *describe how the diffraction of a wave at a gap depends on the wavelength and the size of the gap*
- *explain the meaning of phase and phase difference*
- *understand the health hazards presented by different bands of UV radiation*

Source, journeys and interactions

Mechanical waves, like sound waves, originate from sources that physically vibrate and the travel of the sound needs a medium in which there are particles to vibrate. For light waves, no physical material is involved, and the vibrations involve electric and magnetic fields.

Some kinds of absorption have a particularly significant impact. They change the material which they reach, as happens in photographic film, in the CCDs of digital cameras, and in human eyes. The absorption gives rise to detection.

Emission involves transfer of energy away from a source. Waves can transfer energy to material that stops, or absorbs, them. **Absorption** is the end of a journey for the energy, and so for the waves themselves. A defining feature of waves is that, although they carry energy, and can carry information in the process, no material travels from source to absorber.

The nature of the information carried by a wave depends on the nature of the source, and also on what happens during the journey. It is in this way, by light waves and sound waves, that we know of the world around us.

Different processes can happen on journeys of waves before they are absorbed. They may simply be **transmitted**, continuing on their journey with no change. Boundaries between materials may also **reflect**, **refract** or **diffract** waves. Within material, light can experience **polarisation** and **scattering**. Absorption itself also happens in materials, and it can happen either relatively abruptly soon after the waves cross boundaries into a material, or it can happen gradually.

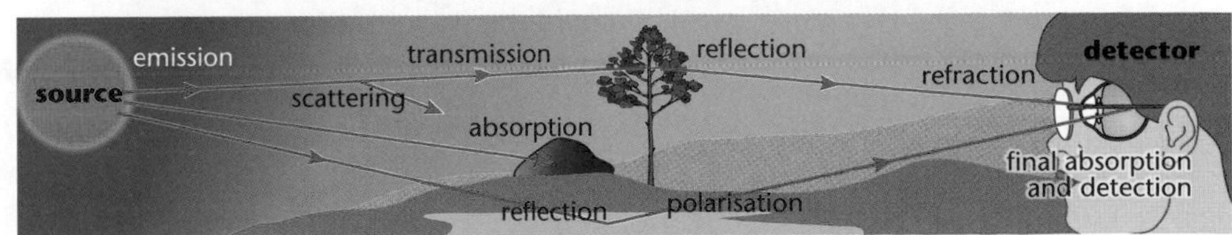

Wave motion

AQA A 2
AQA B 1

Waves are used to transfer a signal or energy. They do this without an accompanying flow of material, although some waves, such as sound, can only be transmitted by the particles of a substance.

The waves on a water surface are almost transverse, the particles move in an elliptical path.

Sound and other compression waves are classified as **longitudinal** because the vibrations of the particles carrying the wave are along, or parallel to, the direction of wave travel. All electromagnetic waves are **transverse**; the vibrations of the electric and magnetic fields in these waves are at right angles to the direction of wave travel.

The diagram shows a longitudinal wave being transmitted by a spring and a transverse wave on a rope.

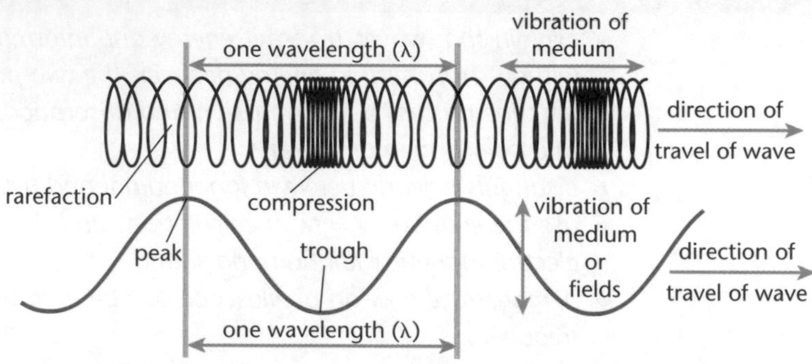

Wave measurements

AQA A 2
AQA B 1

These measurements apply to all **progressive** waves. A progressive wave is one that has a profile that moves through space.

- **Wavelength** (symbol λ) is the length of one complete cycle – see diagram above.
- **Amplitude** (symbol a) is the maximum displacement from the mean position, see diagram on page 92.
- **Frequency** (symbol f) is the number of vibrations per second, measured in hertz (Hz).
- **Speed** (symbol v) is the speed at which the profile moves through space.
- **Period** (symbol T) is the time taken for one vibration to occur. It is related to frequency by the equation $T = 1/f$.

KEY POINT

For all waves, the wavelength, speed and frequency are related by the equation:

speed = frequency × wavelength

$$v = f \times \lambda$$

Representing waves

AQA A 2
AQA B 1

The diagram above shows representations of **longitudinal** and **transverse** waves. There are other ways of representing waves and their motion. The diagram here shows waves, which could be longitudinal or transverse, spreading out from a source.

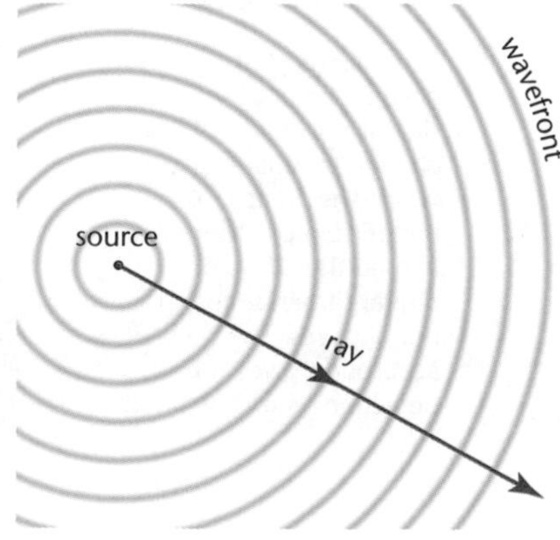

The **wavefront** can be thought of as the line of a crest or trough of a transverse wave or as a compression or rarefaction of a longitudinal wave. A **ray** shows the direction of travel. Note that as the waves spread from the source shown in the diagram on page 88, the rays 'radiate'. They behave rather like 'radii'.

Diagram **a**, below, shows a bird floating on water. The horizontal direction on the page corresponds to horizontal distance across the water. The vertical direction corresponds to vertical displacement. Diagram **b** shows a displacement–time graph for the motion of the bird. The basic shape of the curve, called a **sinusoidal curve**, is the same in both parts, but the way that information is represented is very different.

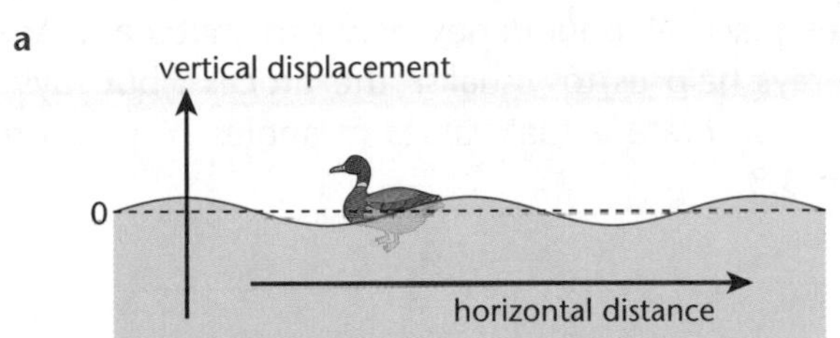

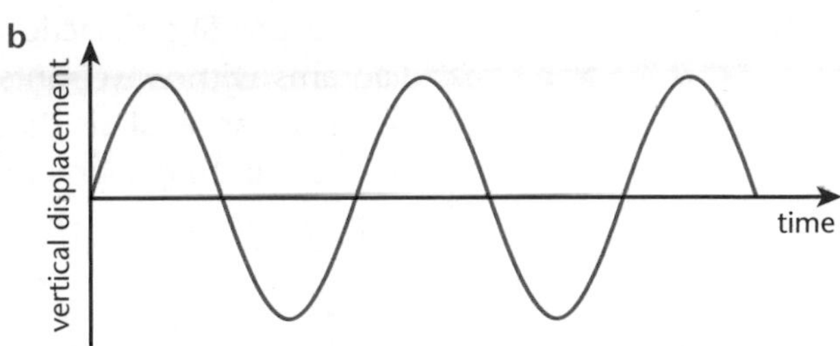

Wave representations on a cathode ray oscilloscope

AQA A 2
AQA B 1

A **cathode ray oscilloscope** (CRO) can be used to display a representation of wave motion. The CRO plots the displacement of one point on the wave, such as the location of the bird in the diagram above, against time.

> A common error is to misinterpret a CRO display as a graph of displacement against distance. This leads to a horizontal measurement, which represents time, being interpreted as a distance.

To measure the **period** and **frequency** of a wave from a CRO display, the time–base control, which controls the rate at which the dot sweeps the screen horizontally, needs to be in the 'cal' (calibration) position. The diagram below shows a CRO display of a sound wave, with the time–base set to $2\,\text{ms}\,\text{cm}^{-1}$.

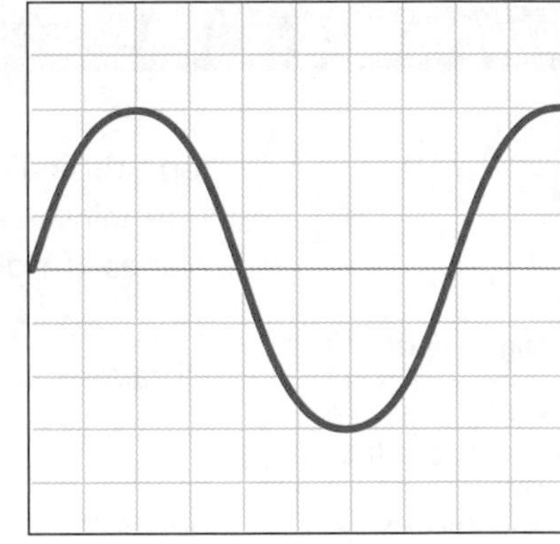

One cycle of the wave occupies a distance of 8 cm on the screen, so the period, $T = 8 \times 2\,\text{ms} = 16\ \text{ms} = 1.6 \times 10^{-2}$ s.

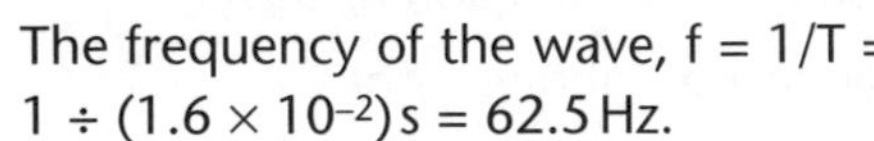

The frequency of the wave, $f = 1/T = 1 \div (1.6 \times 10^{-2})\,\text{s} = 62.5\,\text{Hz}$.

Intensity and the inverse square law

AQA B 1

The signal strength of a radio broadcast decreases the further away the receiver is from the transmitting aerial. In a similar way, a sound appears fainter and a light seems dimmer further away from their origins. This is due to energy becoming spread over a wider area as the wave travels away from its source.

> A point source such as a star radiates energy in all directions. In this case the variation of intensity with distance follows an inverse square pattern, with:
>
> $I \propto 1/r^2$
>
> $I = P_o/4\pi r^2$
>
> where P_o is the power of emission of a localised source.

The **intensity** with which a wave is received depends on the power incident on the area of the detector.

KEY POINT

> Intensity = power per unit area measured at right angles to the direction of travel.
>
> $$I = P/A$$
>
> Intensity is measured in $\text{W}\,\text{m}^{-2}$.

The intensity of a wave is related to its amplitude:
intensity, $I \propto$ amplitude^2

Radiation flux is an alternative name for intensity.

The diagram opposite shows the sound spreading out from a loudspeaker. The ear that is further away detects a quieter sound because the power is spread over a wider area.

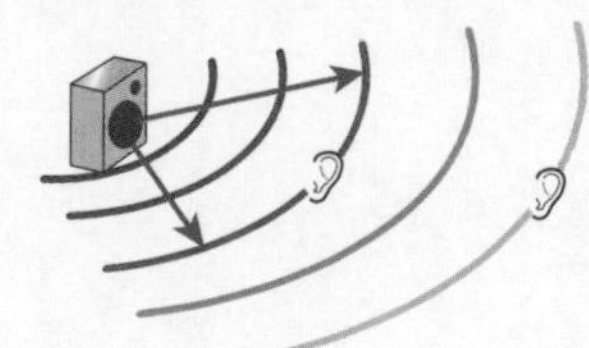

Note that **radiation flux** is a measure of the power at which radiation travels through unit area. It is also measured in $W\,m^{-2}$, and obeys the inverse square law.

Refraction

AQA A 2

Refraction, like reflection, takes place at boundaries between materials. Again, diagrams with wavefronts and rays help us to visualise the process, but rays are better for an analysis that can incorporate actual values of angles. You can read more about refraction in section 3.3.

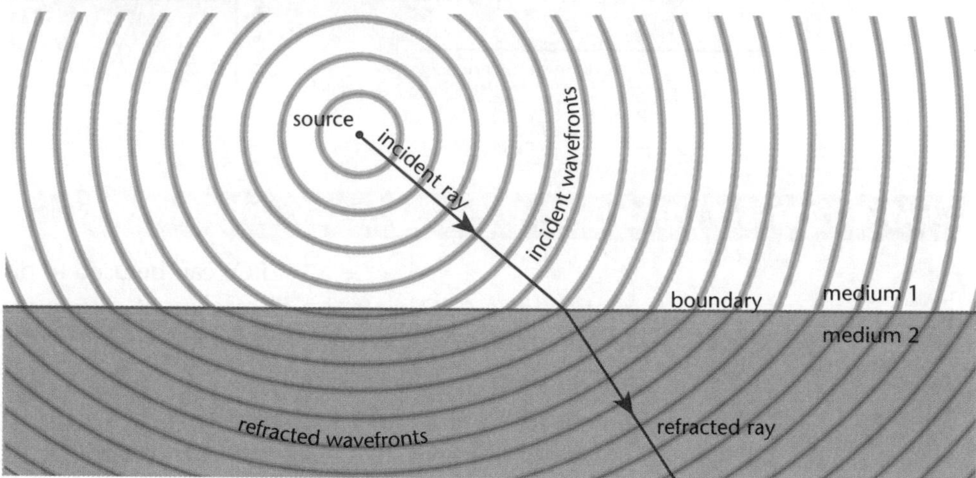

Polarisation

AQA A 2
AQA B 1

Polarisation is a phenomenon that affects transverse waves only, and not longitudinal waves. Light waves are transverse, of course, and they are not normally polarised when emitted. In an unpolarised wave the vibrations are in all planes at right angles to the direction of travel.

Polarising material is used in some types of sunglasses and in filters for camera lenses. It cuts out reflected light from a flat surface such as an expanse of water.

The diagram opposite shows the vibrations in unpolarised and polarised waves. Light waves can become polarised as they pass through some materials and they are partially polarised when reflected.

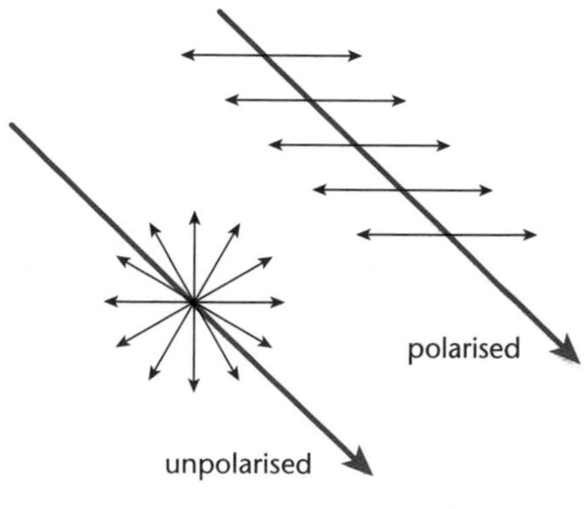

In order to receive a television or radio broadcast, the receiving aerial has to be lined up with that of the transmitter. This is because radio waves broadcast from aerials are **polarised**: the vibrations are only in one plane. For radio and television broadcasts the **plane of polarisation** is usually either vertical or horizontal.

When polarised light passes through a polarising material, its plane of polarisation is rotated, and some absorption takes place. **Malus' law** of polarisation provides a rule for predicting the intensity of transmitted light:

$$I = I_0 \cos^2\theta$$

where I_0 is the initial intensity and θ is the angle between the plane of polarisation of the initial light and the **polarising axis** of the material.

Diffraction

The spreading out of waves as they pass through openings or the edges of obstacles is called **diffraction**. All waves can be diffracted.

The diagrams overleaf show what happens when waves pass through gaps of different sizes. The amount of spreading when the wave has passed through the gap depends on the relative sizes of the gap and the wavelength.

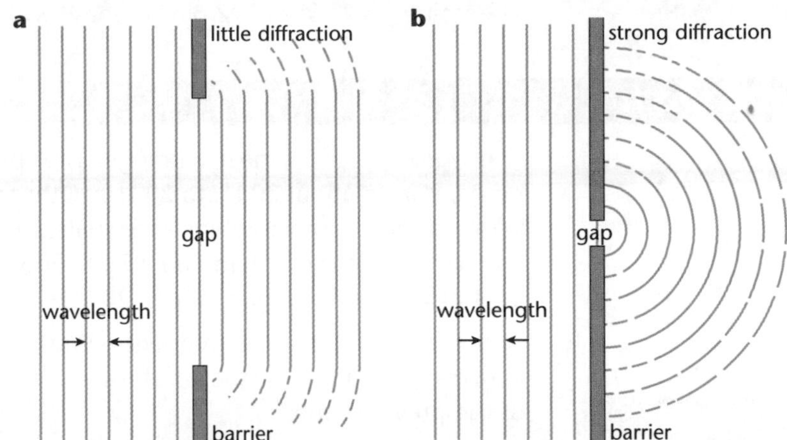

When answering questions about diffraction, always emphasise the size of the gap compared to the wavelength.

- For a gap that is many wavelengths wide, see diagram **a**, little spreading takes place.
- The maximum amount of spreading corresponds to a gap that is the same size as the wavelength as in diagram **b**.

Diffraction explains why you can 'hear' round corners but you cannot 'see' round corners.

Diagram **a** models what happens when light (wavelength approximately 5×10^{-7} m) passes through a doorway, which is much larger than the wavelength. Diagram **b** models sound (wavelength approximately 1 m) passing through the same doorway. There is much more diffraction when wavelength and gap are similar in size.

This also explains why long wavelength radio broadcasts can be detected in the shadows of hills and buildings, but short wavelength broadcasts cannot.

Phase and phase difference

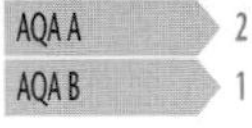

Both transverse and longitudinal waves can be represented by graphs of displacement against position. The position along a wave pattern can be measured in two ways:

- as a distance from a point in space
- as an angle.

In angular measure, one complete cycle of the wave is represented by an angle of 360°. This is shown in the diagram below.

A displacement–position or displacement–time graph does not distinguish between a transverse wave and a longitudinal one.

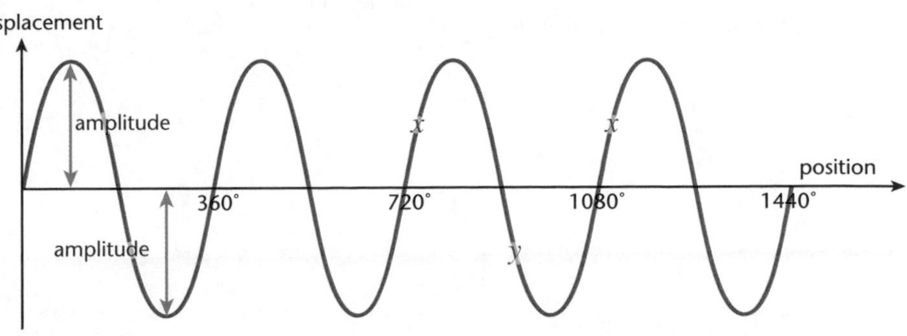

A displacement–position graph is like a slice through the wave at one moment, like a snapshot. At first sight it looks like a displacement–time graph, but it says very little about time. It is easy to confuse the two kinds of graph, but they provide very different information.

Using angular measure is a convenient way of comparing the **phase** of different parts of a wave. Two points on a wave are **in phase** if they have the same displacement and velocity. Points marked x on the diagram are in phase.

The point marked y is exactly **out of phase** with the points x as it has the opposite displacement and its velocity has the same value but in the opposite direction.

The **phase difference** between two points on a wave describes their relative displacement and velocity and is normally expressed in degrees or wavelengths. The point y is 180° or $\lambda/2$ out of phase with the points x.

Sound spectra

AQA A	2
AQA B	1

The word **spectrum** means a range. The spectrum of human hearing, the audible range, is from about 20 Hz to 20 kHz. Other animals have different and often much wider ranges. Some dolphins and whales, for example, have audible ranges from a few hertz up to more than 200 kHz. Sound with frequency that is above the upper limit of the human range, above 20 kHz, is called **ultrasound**.

The hertz, or Hz for short, is the SI unit of frequency. One hertz is one complete 'event' or one cycle per second. $1\ \text{Hz} = 1\ \text{s}^{-1}$.

A pure note has a single frequency. Western musical notation divides sounds into octaves, with the frequency of the start of one octave being twice that of the start of the previous one.

Take care – the spectrum of a particular sound is not the same as the complete spectrum of all possible sound frequencies.

Many sounds are not pure notes, but can be analysed as combinations of several frequencies. The frequency combination of a particular sound is sometimes called the 'spectrum' of that particular sound.

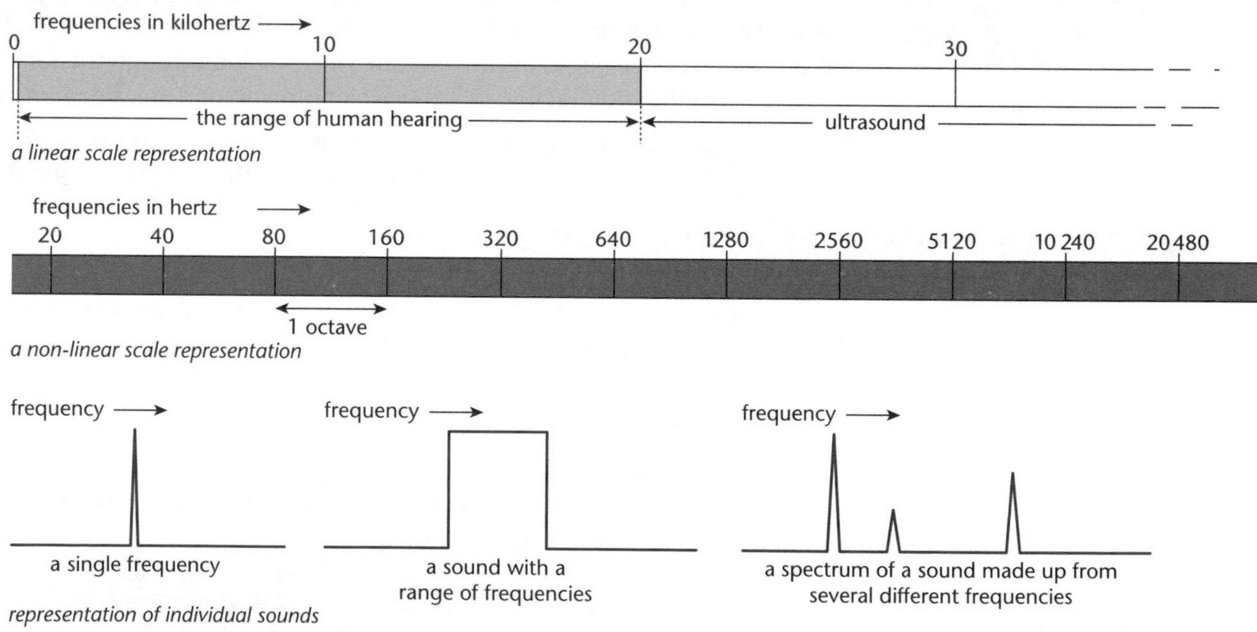

The light spectrum

AQA A 2
AQA B 1

Visible light is a small part of a very wide spectrum. The electromagnetic spectrum consists of a whole family of waves, with some similarities and some differences in their behaviour.

Similar properties include:

- they consist of electric and magnetic fields oscillating at right angles to each other
- they travel at the same speed in a vacuum
- they are transverse waves, and they can be polarised
- they show the same pattern of behaviour in reflection, refraction, interference and diffraction.

The differences include:

- the shorter wavelength waves undergo a greater change of speed when being refracted
- diffraction is easier to observe in the behaviour of the longer wavelength waves
- interactions of light with matter, transmission (which is really an absence of interaction), absorption and reflection, vary considerably.

The diagram below shows the range of wavelengths and frequencies of the waves that make up the spectrum.

The speed of all electromagnetic waves in a vacuum is 2.999×10^8 m s^{-1}. The value 3.00×10^8 m s^{-1} is normally used for both a vacuum and air.

The different changes of speed when light consisting of a range of wavelengths is refracted is responsible for dispersion, the splitting of light into a spectrum.

Microwaves are given here as a separate part of the spectrum, although they can be considered as short-wavelength radio waves.

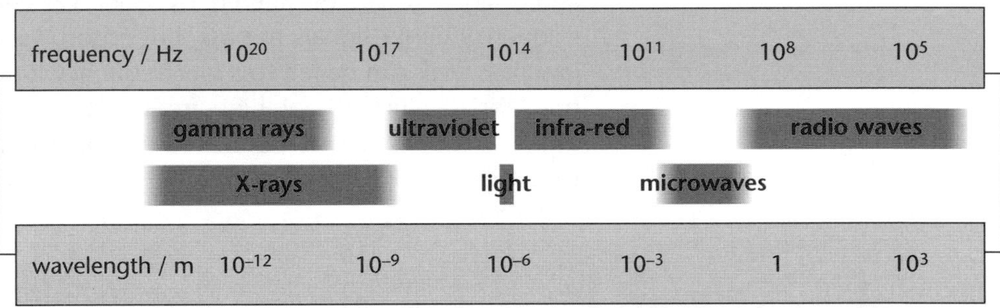

The frequency and wavelength ranges of X-rays and gamma rays overlap. The difference between them is in the origin of the waves: an X-ray is emitted when a high-speed electron is suddenly brought to rest and a gamma ray is emitted from a nucleus, usually along with alpha or beta emission.

The table below shows typical wavelengths, the origins of the waves and the uses of the main parts of the electromagnetic spectrum.

Excited here means having energy to lose.

All objects give out infra-red radiation, no matter what the temperature. The hotter the object, the more power is emitted and the greater the range of wavelengths.

Name of radiation	*Typical wavelength/m*	*How produced*	*Used for*
gamma	1×10^{-12}	an excited nucleus releasing energy in radioactive decay	tracing the flow of fluids and treating cancer
X-rays	1×10^{-10}	high-speed electrons being stopped by a target	seeing inside the body
ultraviolet	1×10^{-8}	very hot objects and passing electricity through gases	security marking
light	1×10^{-6}	hot objects and electrically excited gases	vision and photography
infra-red	1×10^{-5}	warm and hot objects	heating and cooking
microwave	1×10^{-1}	microwave diode or oscillating electrons in an aerial	communications and cooking
radio	1×10^{2}	oscillating electrons in an aerial	communications

Transmission and absorption of ultraviolet, or UV, radiation by atmosphere, skin and sunscreen

UVc radiation from the Sun (wavelength approximately 100 to 280 nanometres) is mostly absorbed by the atmosphere. UVb (wavelength 280 to 320 nanometres) is partly but not completely absorbed by the atmosphere. It is then absorbed by the outer layers of skin, and this localised absorption means that it transfers energy rapidly to the skin. This can cause chemical damage, with short-term burning of the skin and increased risk of skin cancer in the longer term. UVa (wavelength 320 to 400 nanometres) is transmitted by the atmosphere, and is absorbed more gradually than UVb by the skin. So although it travels further through the skin and is still a cause of problems, it is less dangerous than UVb.

Commercial sunscreens are designed to absorb UVb, and the 'factor' provided by the supplier relates to UVb absorption and not to UVa. Some 'broad spectrum' sunscreens are available, however, and these are better absorbers of UVa.

Filters

AQA B 1

Filters transmit some frequencies but allow others to pass through. Sunscreen filters UVb out of a stream of radiation. A red filter transmits low frequency visible light but absorbs higher frequencies. It makes the world seem rose-tinted.

It is more difficult to filter sounds, but electrical signals that represent sounds, or audio signals, can be filtered by electrical circuitry. Low pass filters allow low frequency audio signals to pass, but absorb higher frequencies. The signal from a low pass filter can be fed to a low frequency loudspeaker, or woofer, so that it does not vibrate efficiently at high frequencies. Likewise, a high pass filter transmits higher frequencies that can be fed to a high frequency loudspeaker, or tweeter, so that it can work most efficiently.

Progress check

1 Describe the difference between a longitudinal and a transverse wave and give one example of each.

2 Calculate the frequency of a VHF radio broadcast that has a wavelength of 2.75 m. The speed of radio waves, $c = 3.00 \times 10^{8}\,\mathrm{m\,s^{-1}}$.

3 Two points on a wavefront have a phase difference of 180°. Describe their relative displacement and velocity.

1 In a longitudinal wave the vibrations are parallel to the direction of travel, e.g. sound or any other compression wave.
In a transverse wave the vibrations are at right angles to the direction of travel, e.g. light or any other electromagnetic wave.

2 1.09×10^{8} Hz.

3 The points have the same amount of displacement but in opposite directions, i.e. the displacement of each one is the negative of the displacement of the other one.
Similarly, the velocities are equal in size but opposite in direction.

3.2 Sound and superposition

After studying this section you should be able to:

- *explain superposition*
- *describe how stationary waves are created on strings*
- *explain the harmonics of a string, and how they are related to the string's length and the sound's wavelength*
- *explain how beats are created*
- *explain how wave superposition causes interference patterns and describe the conditions needed for patterns to be observable*
- *compare the decibel scale with the range of audible intensities of sound*

LEARNING SUMMARY

Superposition and stationary waves

Sounds are made by vibrating sources, and most musical sounds are made by vibrating strings and pipes.

Simple mechanical waves can travel along ropes and along strings. A flick of a rope is enough to start a wave. If the wave reaches a fixed end of the rope or string, then its energy must be conserved. Some energy might pass out of the string, and be absorbed by a body that it is fixed to. Much of the energy can be reflected back along the spring. Such reflections can combine with the original wave to create **stationary or standing waves**. This process of combination, in which the total displacement of a point on the string is the sum of displacements caused by more than one wave, is called **superposition**. Superposition can take place with any type of wave, including light waves.

A stationary wave has a static profile. The waves on the vibrating strings and in the vibrating air columns of musical instruments are stationary waves.

Stationary waves are caused by the superposition of two waves of the same wavelength travelling in opposite directions. They often arise when a wave is reflected at a boundary, but the waves can come from two separate sources. The diagram shows the vibrations of a stationary wave on a string.

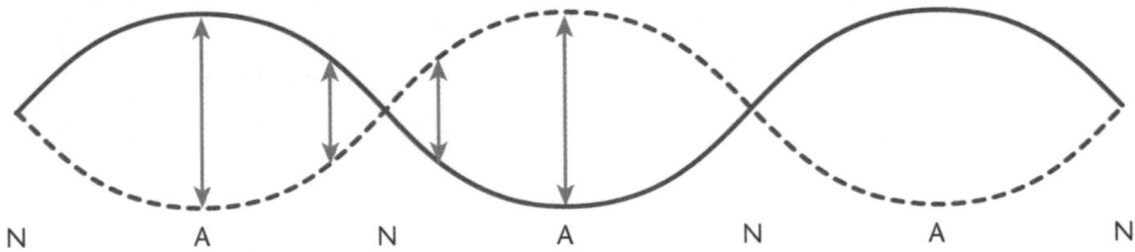

Stationary waves differ from progressive waves in a number of respects:

- there is no flow of energy along a stationary wave, although stationary waves often radiate energy
- within each loop of a stationary wave, all particles vibrate in phase with each other, and exactly out of phase (180° phase difference) with the particles in adjacent loops
- the amplitude of vibration varies with position in the loop
- there are **nodes** (points marked N in the diagram), where the displacement is always zero and **antinodes** (marked A in the diagram) which vibrate with the maximum amplitude.

A convenient way of measuring the wavelength of a progressive wave is to use it to set up a stationary wave, for example by superimposing the wave on its reflection from a barrier, and measuring the distance between a number of successive nodes or antinodes.

The wavelength of a stationary wave is twice the distance between two adjacent nodes or antinodes. The diagram opposite shows how a stationary wave (shown with a solid line) is formed from two progressive waves (shown with broken lines) travelling in opposite directions.

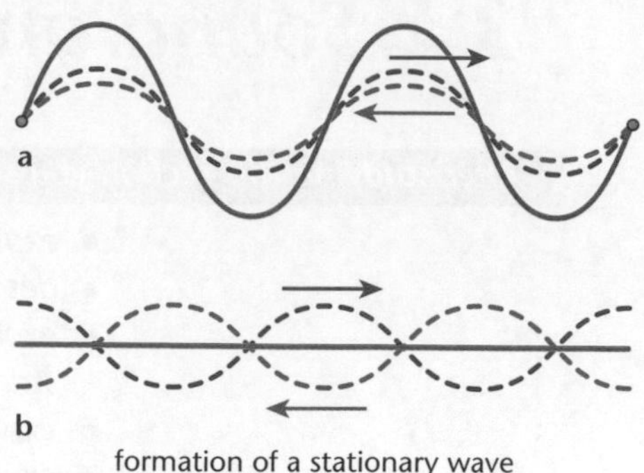

formation of a stationary wave

In **a** the waves' superposition takes place constructively, so each point on the wave has its maximum displacement.

In **b** each progressive wave has moved one quarter of a wavelength; the superposition is now destructive, resulting (just for a moment) in no displacement at all points on the wave.

Harmonics

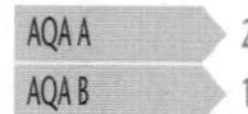

A string that is fixed at its ends will have a node at each end. The simplest kind of stationary wave on a string has a node at each end and an antinode on the middle. It is called the string's **fundamental** note.

Along the length of the string, however, there could be one more node, or two more, or any number more. This means that the stationary wave can have different wavelengths, and different frequencies of vibration. These different ways of vibrating are called the **harmonics** of the string.

In practice, different harmonics can exist at the same time, to produce sounds that are more complex than simple notes. However, when a string is plucked, the higher frequency harmonics tend to fade most rapidly, so that the note changes, becoming more 'pure' as the single fundamental or first harmonic note becomes more dominant.

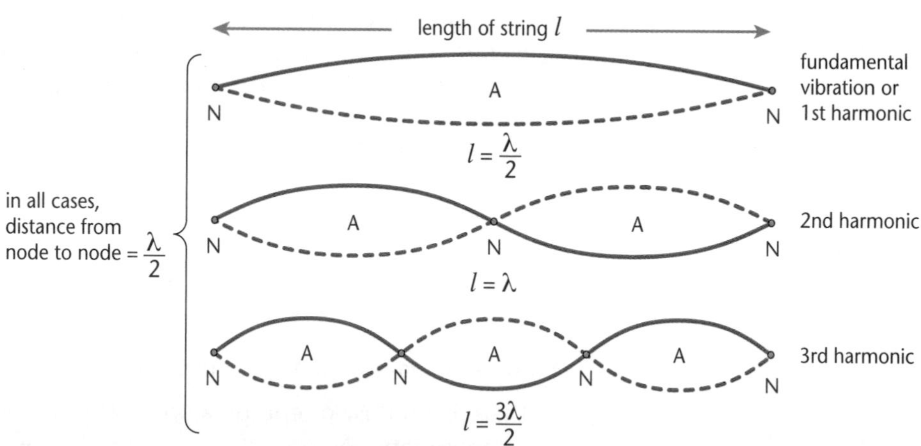

Factors affecting fundamental frequency of a string – Melde's apparatus

AQA B 1

Melde's apparatus consists of a string or wire fixed at one end to a vibration generator, which provides a source of vibrations for which frequency can be continuously varied. The string passes over a pulley, and its other end carries a load, normally a set of slotted masses which can be used to vary the tension in the whole string. A small movable 'bridge' creates a fixed point for vibration of the string, so that the vibrating length of string can be measured.

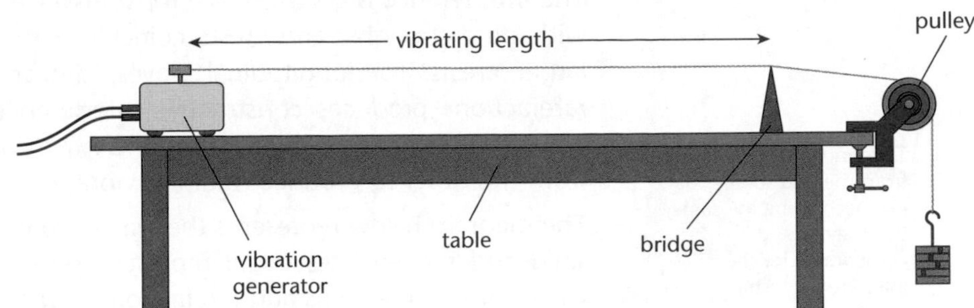

With the vibrating length fixed, the relationship between tension and fundamental frequency can be investigated. This provides a precise relationship in support of what every string musician knows: fundamental frequency, f, increases as tension, T, increases. In more detail:

$$f \propto \sqrt{T}$$

Similarly with fixed tension, the effect of length can be found:

fundamental frequency, f, decreases as length, l, increases, so that the proportionality is now inverse:

$$f \propto 1/l$$

By using strings of different mass per unit length, μ, it is possible to find out that:

$$f \propto 1/\sqrt{\mu}$$

The frequency of a vibrating string is related to its various physical circumstances by the equation:

$$f = \frac{1}{2l}\sqrt{\frac{T}{\mu}}$$

Superposition and beats

AQA B 1

Superposition of two notes of similar but not identical frequencies, f_1 and f_2, produces a fluctuation of sound intensity and loudness. The fluctuations are called **beats**.

The frequency of beat fluctuations is simply the difference between the frequencies of the two sources. If these frequencies become the same then the beats cease to exist. This point is used to tune instruments, using the instrument as one source of sound, f_1, and a tuning fork or other source of known and fixed frequency, f_2, as the other.

beat frequency = $f_1 - f_2$

When the beat frequency becomes zero the two frequencies, f_1 and f_2, are the same.

Superposition and interference

AQA A 2
AQA B 1

Superposition of waves from two loudspeakers can produce a pattern of varying amplitudes of vibration called an **interference** pattern. Two identical sources produce remarkably neat patterns of alternating maxima and minima.

Interference can occur with any type of wave, including water ripples and light waves.

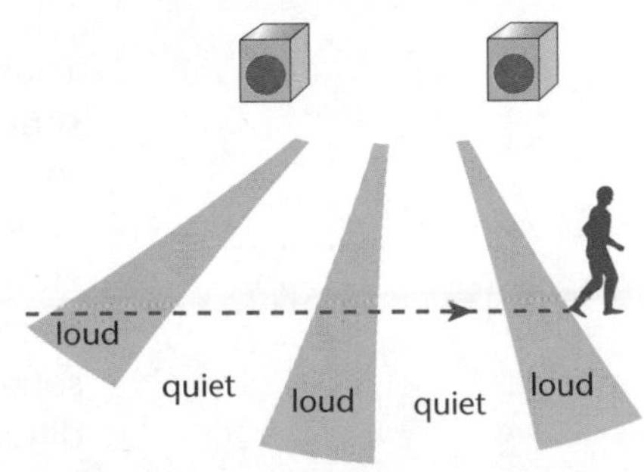

Superposition of sound waves

The interference is **constructive** for transverse waves when troughs add together with other troughs, and crests coincide with crests, making larger troughs and larger crests. For longitudinal waves, similar addition of compressions and of rarefactions produces constructive interference. The interference is **destructive** when crests or compressions from one wave combine with troughs or rarefactions from the other to produce resulting vibration with minimised amplitude.

The amplitudes of waves from the two sources do not need to be the same, but they should be comparable for the interference to be observed. If one wave has a much greater amplitude than the other then the effects of cancellation and reinforcement will not be noticeable.

Try drawing similar patterns and investigating the effect of changing the wavelength and the separation of the sources.

The diagram below represents the waves from two sources vibrating in phase. The solid and broken lines could represent peaks and troughs in the case of water waves or compressions and rarefactions in the case of sound waves.

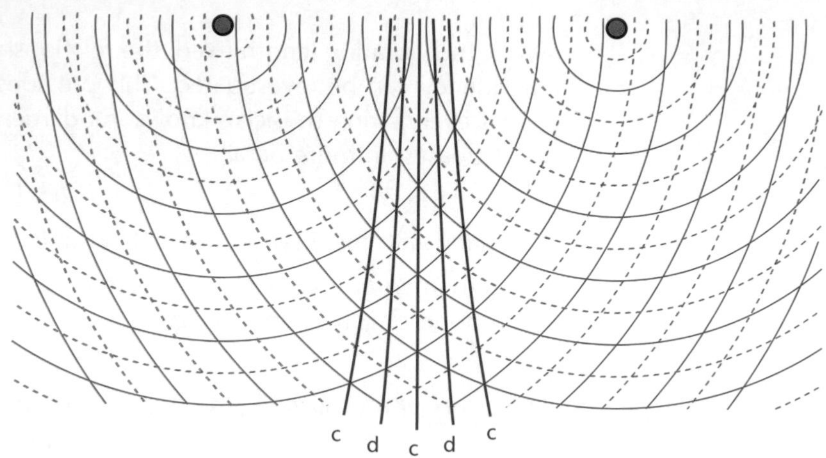

Constructive and destructive interference from two wave sources

Some lines of constructive interference **c** and destructive interference **d** have been marked in.

If the sources in the diagram are in phase, then at any point equidistant from both sources the waves arrive in phase, giving constructive interference. This is the case for points along the central **c** line in the diagram. At all other points in the interference pattern, there is a **path difference**, i.e. the wave from one source has travelled further than that from the other source.

- **Constructive interference** takes place when the path difference is a whole number of wavelengths.
- **Destructive interference** takes place when the path difference is one and a half wavelengths, two and a half wavelengths, etc.

The conditions can be written as $n\lambda$ path difference for constructive interference and $(2n+1)\lambda/2$ path difference for destructive interference, where n is an integer 0, 1, 2, 3 ... and so on.

This is shown in the diagram.

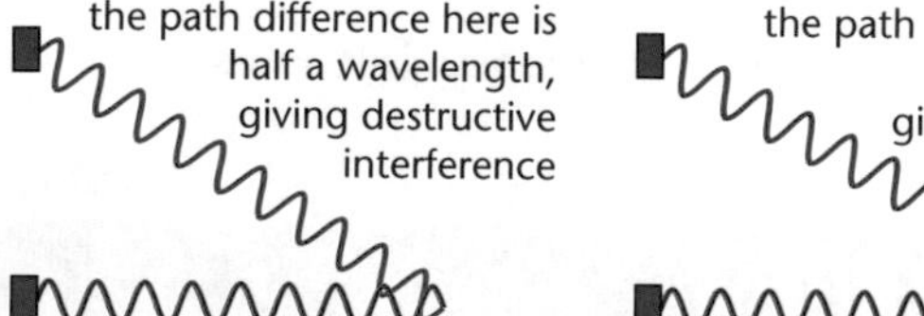

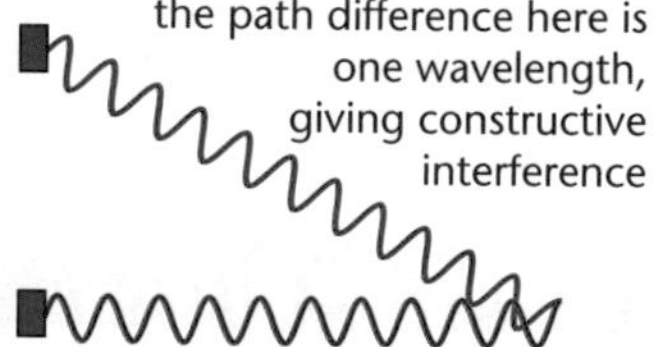

These conditions for constructive and destructive interference only apply to two sources that are in phase. If the phase of one of the oscillators in the diagram at the top of this page is reversed so that the sources are in antiphase, then the conditions for constructive and destructive interference are also reversed. The effect is to shift the interference pattern so that lines of constructive interference **c** become destructive and vice versa. Any value of phase difference between the two sources gives an interference pattern that is stationary, provided that the phase difference is not changing. Two sources with a **fixed phase difference** are said to be **coherent**. Coherent sources must have the same wavelength and frequency.

Intensity, loudness and the decibel scale

AQA B 1

Doubling the intensity (measured in Wm^{-2}) of a sound does not double its loudness. The loudness we hear is not proportional to the intensity. Apparent loudness is better (but still not perfectly) expressed in terms of **intensity level**, for which the unit is the **decibel**, dB.

On the decibel scale, the quietest sound that a person with normal hearing can perceive is 0 dB. It is sometimes called the threshold of hearing. On the decibel scale, an increase of a factor of 10 in the intensity adds 10 decibels, as in the table:

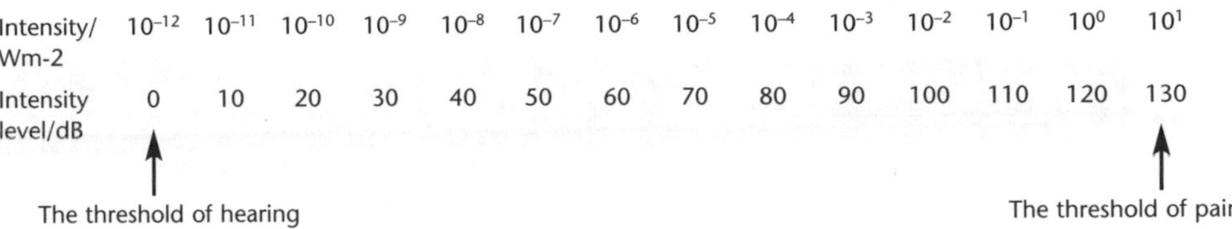

Intensity/ Wm-2	10^{-12}	10^{-11}	10^{-10}	10^{-9}	10^{-8}	10^{-7}	10^{-6}	10^{-5}	10^{-4}	10^{-3}	10^{-2}	10^{-1}	10^{0}	10^{1}
Intensity level/dB	0	10	20	30	40	50	60	70	80	90	100	110	120	130

Note that the range of intensities that the human ear can detect is very large, from the threshold of hearing at about $10^{-12}Wm^{-2}$ up to the threshold of pain at $10^{1}Wm^{-2}$. Instant perforation of the eardrum takes place at about $10^{4}Wm^{-2}$.

The relationship suggested in the table can be expressed more precisely as a formula:

$$\text{intensity level} = 10\log_{10} I/I_0$$

where I is intensity and I_0 is the intensity of the threshold of hearing

The situation is made more complicated by the fact that the threshold of hearing, and sensitivity of the ear in general, is different at different frequencies. Subjective judgement of loudness does not correlate perfectly with the objective decibel scale of sound intensity level. This is shown in **curves of constant loudness**.

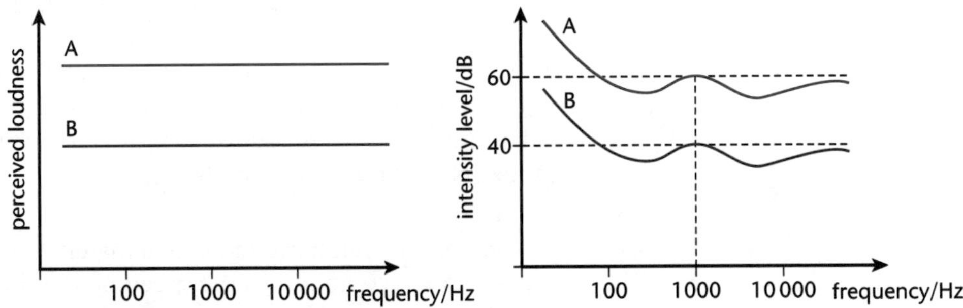

Sounds of different frequency with the same apparent loudness, shown by the two lines on the left, do not necessarily have the same intensity level, as shown in the curves on the right.

Progress check

1 When light passes through a slit which is 2 mm wide, it produces a narrow beam. Explain why there is no observable diffraction.

2 For an observable interference pattern between two sources, the sources must be coherent.

a What is meant by two coherent sources?

b What TWO other conditions should be met for an observable interference pattern?

1 The width of the slit is very large compared to the wavelength of light.

2 a Coherent means they have a fixed phase relationship.

b They should be the same type of wave and similar in amplitude.

3.3 Light and imaging

After studying this section you should be able to:

LEARNING SUMMARY

- *use the relationships between refractive index, speeds of light, and angles of incidence and refraction*
- *describe how optical fibres are used in communication and medicine*
- *perform calculations based on 'double-slit' interference of light*

Refraction and refractive index

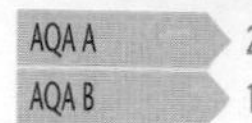

When waves cross a boundary between two materials there is a change of speed, called **refraction**. If the direction of wave travel is at any angle other than along the normal line, this change in speed causes a change in direction. The diagram opposite shows what happens to light passing through glass. Note that there is always some reflection at a boundary where refraction takes place.

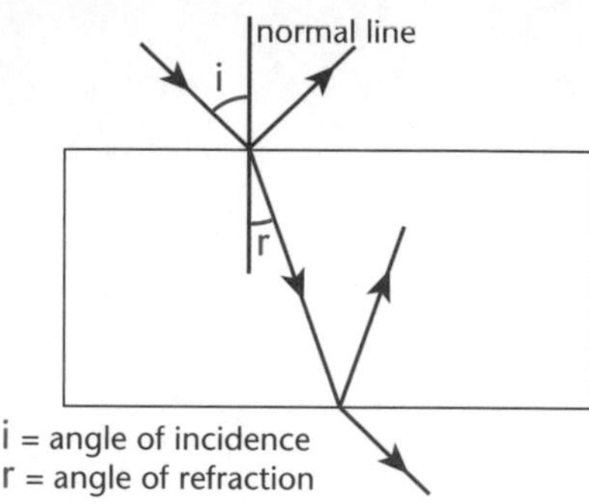

The change in speed is described by the **refractive index**.

KEY POINT

The refractive index of a boundary, ${}_1n_2$, is the ratio of the speed of light in the first medium, c_1, to the speed of light in the second medium, c_2.

$${}_1n_2 = c_1 / c_2$$

Refractive index is a ratio, it does not have a unit.

The **absolute refractive index** of a material is the factor by which the speed of light is reduced when it passes from a vacuum into the material. This is sometimes called the **absolute** refractive index (n) to distinguish it from the **relative** refractive index (n^{u2}) between two materials.

KEY POINT

The absolute refractive index of a material, n, is the ratio of the speed of light in a vacuum, c_v, to the speed of light in the material, c_m.

$$n = c_v / c_m$$

The refractive index of air is 1.0003, so there is no difference in the refractive index of a material relative to air and relative to a vacuum when working to three significant figures.

The greater the change in speed when light is refracted, the greater the change in direction. At a given angle of incidence, there is a greater change in direction when light passes into glass, with absolute refractive index $n_g = 1.50$, than when light passes into water, with absolute refractive index $n_w = 1.33$. The values quoted here are absolute values, but in practice there is little difference between the refractive index of a material relative to a vacuum and that relative to air.

When light is refracted:

- the incident light, refracted light and the normal all lie in the same plane
- **Snell's law** relates the change in direction to the change in speed that takes place.

KEY POINT

Snell's law states that:

$$\frac{\sin i}{\sin r} = \frac{c_1}{c_2} = {}_1n_2$$

So the sines of the angles are in the same ratio as the speeds in the media when refraction takes place.

Total internal reflection and critical angle

AQA A 2
AQA B 1

When light speeds up as it crosses a boundary, the change in direction is away from the normal line. For a particular angle of incidence, called the **critical angle**, the angle of refraction is 90°.

The diagrams below shows what happens to light meeting a glass–air boundary at angles of incidence that are **a** less than, **b** equal to, and **c** greater than, the critical angle.

The intensity of the internal reflection increases as the angle of incidence increases.

a

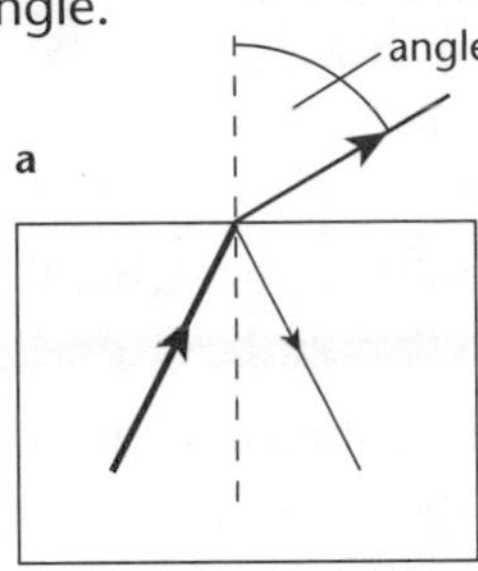

there is a weak reflection

b

critical angle

there is a stronger reflection

c

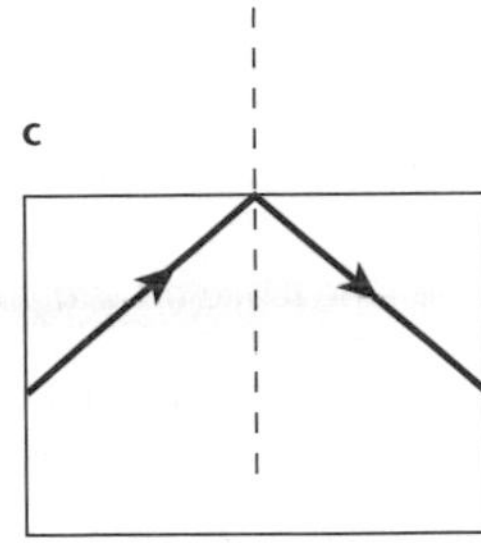

all light is reflected

These diagrams show that:

- at angles of incidence less than the critical angle, both reflection and refraction take place
- at the critical angle, the angle of refraction is 90°
- at angles of incidence greater than the critical angle, the light is **totally internally reflected**.

The abbreviation TIR is often used for total internal reflection.

A diamond has a very high refractive index and a low critical angle. Multiple internal reflections make diamonds sparkle.

Snell's law can be used to show that the relationship between the critical angle and refractive index is:

KEY POINT

$$n = 1/\sin c$$

where c is the critical angle and n is the absolute refractive index.

Fibre optics

Total internal reflection is used in prismatic binoculars, car and cycle reflectors and cats' eyes, but its greatest impact has been in the fields of medicine and communications.

Optical fibres enable light to travel round curves by repeated total internal reflection at the boundaries of the fibre, which is made from glass or plastic. Provided that light hits the boundary at an angle greater than the critical angle, none passes out of the fibre. The diagram opposite shows how light can be made to travel 'round the bend' of a fibre.

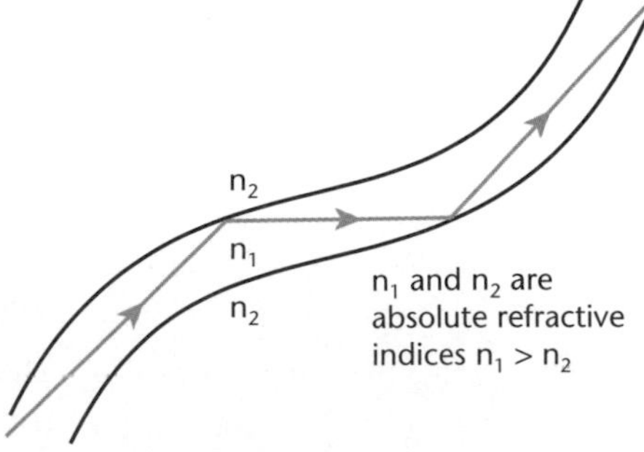

In medicine, optical fibres are used in **endoscopy** to look inside a patient. This can be done using a natural body opening or by 'keyhole surgery', which requires a small incision just wide enough to take the fibre. Two bundles of fibres are used in endoscopy: one to transmit light into the patient and the other to carry the reflected light to a small television camera.

Although they are commonly referred to as light, the pulses used are usually in the infra-red region of the electromagnetic spectrum.

In communications, optical fibres are used to transmit telephone conversations, television and radio programmes in the form of **digital signals**. The signals used can only have the values 1 or 0, where 1 is represented by a pulse of light from a laser diode and 0 is the absence of light. The diagram opposite shows how optical fibres are used in telephone communications.

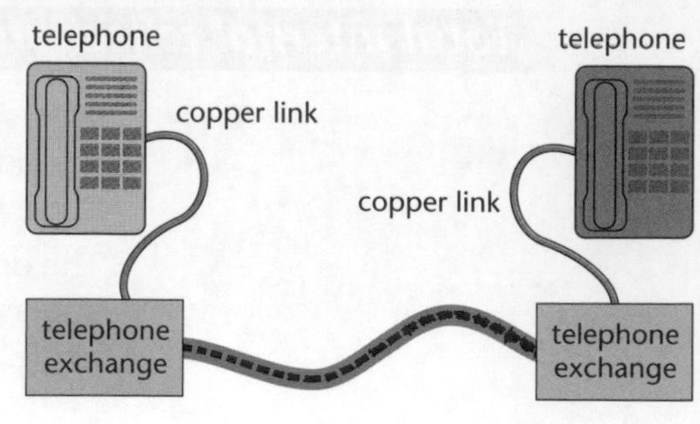

The range of a signal in a copper cable is only a few km, compared to up to 100 km in a glass fibre.

The advantages of using optical fibres rather than copper cables in communications are:

- the range of a signal (the distance it can travel) in an optical fibre is much greater than that in a copper cable, so amplification is needed less frequently
- optical fibres do not pick up **noise** from changing magnetic fields.

However, the signal in an optical fibre is still subject to distortion as it travels along the fibre. This is due to **multimode dispersion**, a process which causes the pulse to become elongated as some parts of the pulse travel further than others, for example when the cable goes round corners. In modern fibres the effect of this is reduced by concentrating the light pulses along a very narrow core of fibre, so that there is only one effective route for the light. These are **monomode** fibres.

A similar effect occurs if 'light' containing more than one wavelength is used, as the different wavelengths travel at different speeds in the fibre.

Multimode dispersion and its effect on a digital signal

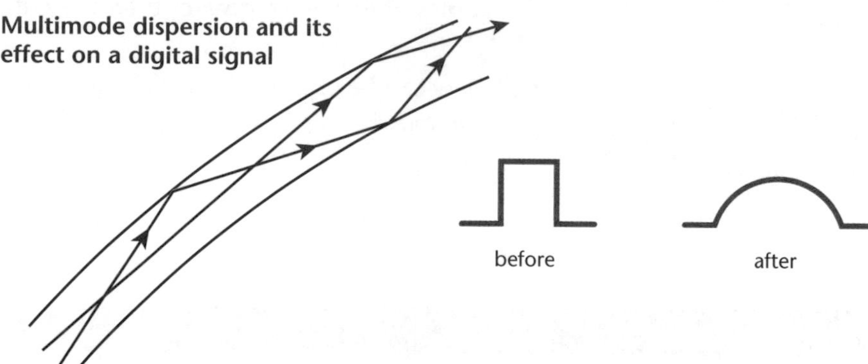

Noise and analogue and digital signals

AQA B 1

Noise or distortion can be easily removed from digital signals in a process called regeneration.

Noise affects the amplitude of a signal. In AM (amplitude modulated) transmissions, the information is carried in the form of variations in amplitude of a wave. Any noise cannot be distinguished from the signal and so cannot be removed. When the signal is amplified, the noise is also amplified.

This is not the case with a digital signal, which only has certain allowed values. If the amplitude varies around these values, it can be restored. This is shown in the diagram on the next page.

The 'allowed values' used in communications are normally 1 and 0, so regeneration is a straightforward task.

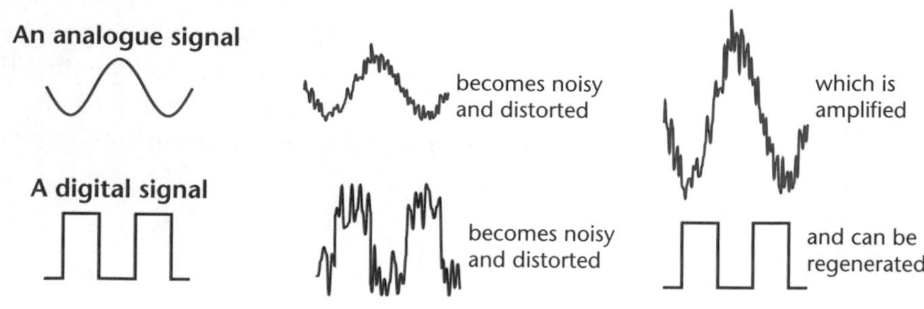

Double-slit interference

AQA A 2
AQA B 1

Interference patterns are easy to set up with water waves and sound because two sources can be driven from the same oscillator, giving coherence. Because of the way in which light is emitted in random bursts of energy from a source, it is not possible to have two separate sources that are coherent. Instead, diffraction is used to obtain two identical copies of the light from a single source. This is done by illuminating two narrow slits with a lamp that is parallel to the slits so that the same wavefronts arrive at each slit. A suitable arrangement is shown in the diagram below.

Diffraction of light is, as for any other kind of wave, fundamentally a quite simple process. Waves spread out when passing through a gap or passing an obstacle. However, different parts of a beam of light act as different sources, so that interference takes place. This happens both at gaps and at small obstacles.

A colour filter can be used between the lamp and the slits to reduce the range of wavelengths interfering, but this also reduces the intensity of the interference pattern.

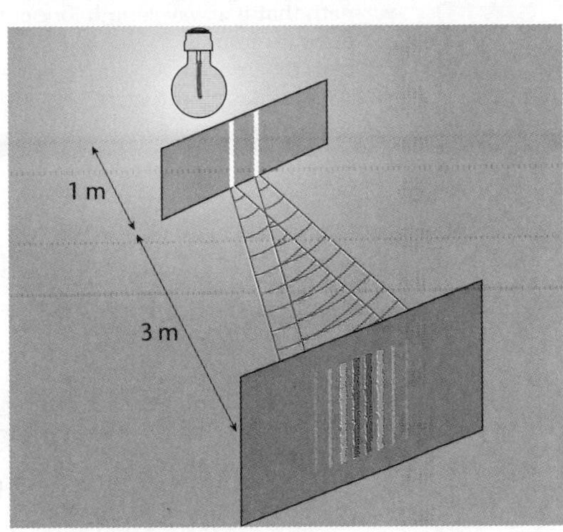

Bright and dark fringes on the screen are due to constructive and destructive interference between the two overlapping beams of light. The separation of the fringes depends on:

- the separation of the slits
- the wavelength of the light
- the distance between the slits and the screen.

This formula enables the wavelength of light to be measured from a simple experiment. The fringe spacing, x, should be obtained by measuring the separation of as many fringes as are visible and dividing by the number of fringes.

KEY POINT

The separation of the fringes is related to the other variables by the formula:

$$\text{fringe spacing} = \frac{\text{wavelength} \times \text{distance from slits to screen}}{\text{slit separation}}$$

$$x = \lambda D/a$$

where x is the distance between adjacent bright (or dark) fringes
λ is the wavelength of the light
D is the distance between the slits and the screen
a is the distance between the slits.

As the wavelength of light is very small, the slits need to be close together and separated from the screen by a large distance for the interference fringes to be seen.

Diffraction gratings

AQA A 2
AQA B 1

A diffraction grating is a piece of material, such as a glass slide, with parallel lines engraved very close together onto its surface. Light passing through such a grid or grating of lines is experiencing many gaps and obstacles. Diffraction takes place, so that the grating acts as an array of sources of light. Interference then takes place between light from the gaps, producing patterns of maximum and minimum intensity. The close spacing of the gaps results in large spacing of the maxima and minima.

The relationship between the spacing of the gaps and the angles at which maxima appear is:

$n\lambda = d \sin\theta$

λ is the wavelength of the light

$n = 0$ for the central maximum, 1 for the next maximum, and so on

d is the grating separation

θ represents the angles at which the maxima appear

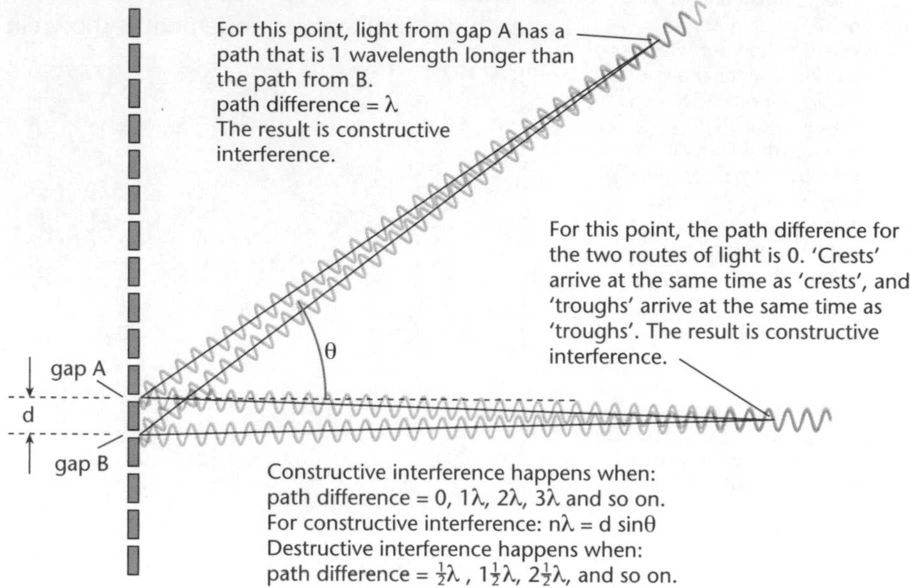

Note that the appearance of wavelength in the equation tells us that different colours of light produce maxima and minima at different angles. Thus diffraction can disperse white light into a spectrum. This is the source of the colours seen when looking at a CD surface – the surface produces diffraction by reflection rather than transmission in this case, but the principle is the same.

Light of a single wavelength or frequency has a single colour, and is called **monochromatic**.

Progress check

The table gives the speed of light in different media.

Medium	*Speed of light/m s⁻¹*
vacuum	3.00×10^8
ice	2.29×10^8
ethanol	2.21×10^8
quartz	1.94×10^8

1 **a** Calculate the values of the absolute refractive indexes of ice, ethanol and quartz.
b Calculate the refractive index for light travelling from ice into ethanol.

2 A piece of ice is placed in ethanol.
a In which direction is light travelling for total internal reflection to take place?
b Calculate the value of the critical angle.

3 The critical angle for a certain glass is 39°. Calculate the refractive index of the glass.

4 In a two-slit interference experiment using light, the slits are separated by a distance of 1.2 mm. The distance from the slits to the screen is 3.0 m and the separation of the bright fringes is 1.4 mm.
Calculate the wavelength of the light.

1 **a** ice 1.31 ethanol 1.36 quartz 1.55 **b** 1.04
2 **a** from ethanol to ice **b** 75°
3 1.59
4 5.6×10^{-7} m

3.4 Information transfer

After studying this section you should be able to:

- *explain the terms 'bandwidth', 'AM', 'FM', 'bit', 'sampling' and 'multiplexing'*
- *compare analogue and digital signals and how they deal with the problem of noise*
- *distinguish between different ways of encoding and transmitting information*

LEARNING SUMMARY

Carrier waves, AM and FM

AQA B 1

Waves can transfer information as well as energy. A simple wave carries information about its frequency, wavelength and amplitude. It can carry much more information in the form of patterns of amplitude or patterns of frequency. The first type of pattern is **amplitude modulation** and the second is **frequency modulation**.

For amplitude modulation, or AM, the frequency (and wavelength) of the wave can be fixed. A radio set must be tuned in to that frequency when AM is used for radio transmission.

Frequency modulation radio transmission, or FM, can use a constant amplitude of wave, but needs to use a band or range of frequencies so that varying patterns of frequency can carry the encoded signal. The greater the band of frequencies, the more encoding possibilities there are and the faster information can be transmitted. This introduces the concept of **bandwidth**.

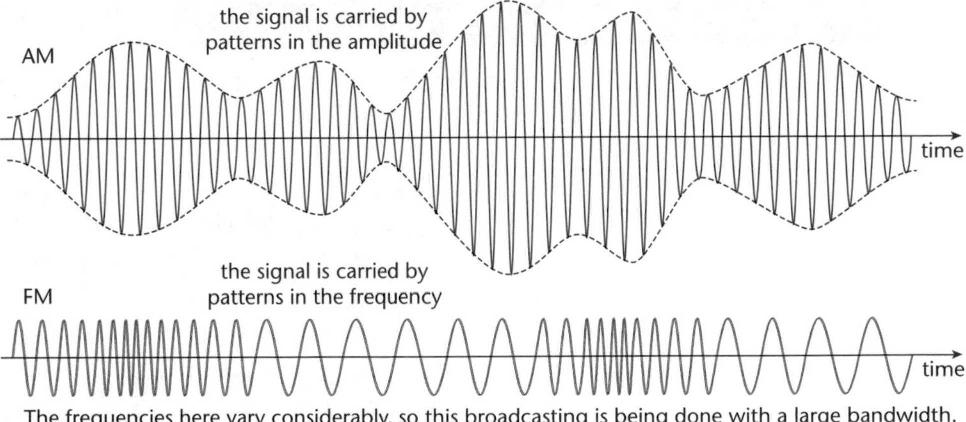

The frequencies here vary considerably, so this broadcasting is being done with a large bandwidth.

Analogue and digital signals

AQA B 1

Analogue measurement uses variation in one quantity to represent another. An ordinary clock uses the angle turned through by a hand of the clock to represent time. In amplitude modulation communication, the amplitude of the wave represents something else, such as voltage, which may in turn represent the changing pressure of air at a microphone. In general, analogue values can vary continuously, to represent the continuous variation in the original quantity such as voltage or air pressure.

The binary system of counting is simpler than the one we are familiar with – it is based on just two symbols, 0 and 1, unlike the familiar system that needs 10 symbols: 0, 1, 2, 3, 4, 5, 6, 7, 8, 9. Electrical systems can deal with just two possibilities much better than they can deal with 10 possibilities.

Digital signals do not vary continuously, but use just two values of a variable, such as light intensity or voltage. The two values are usually given the names 0 and 1. These are called binary digits, or **bits**.

A bit can have two values, so the information it can carry is very limited, but a series of bits can carry much more. A series of 8 bits is called a **byte**.

In one byte, there are 256 possible patterns of 0 and 1. 256 is 2^8.

In general:

number of possible different patterns = $2^{\text{number of bits}}$

A kilobyte is 1024 bytes. That is 1024 × 8 bits. $2^{1024\times8}$ is a very large number, and it can carry a lot of information.

Digital images, such as those produced by digital cameras, are broken into small squares or pixels. The information about each pixel can be stored as sets of numbers. One set of numbers can refer to the brightness of the pixel. The set of numbers could be 0 to 255, allowing 256 levels of brightness. Then one byte of information is enough to carry that information. Likewise, colours can be encoded as numbers.

Information storage

AQA B 1

Information can be stored in analogue format or in digital format. Vinyl records, for example, use the shape of the groove in the plastic as a direct analogue representation of the sound.

CDs also use shape, having pits and spaces that reflect or do not reflect a laser beam. The detector either receives a reflected beam or it does not – it receives a binary digit 1 or a binary digit 0. A CD is a digital storage medium. DVDs are similar, but have more closely packed pits so that they can store more information.

Analogue to digital – sampling

AQA B 1

An analogue signal can be turned into a digital signal by measuring at regular intervals, or **sampling**.

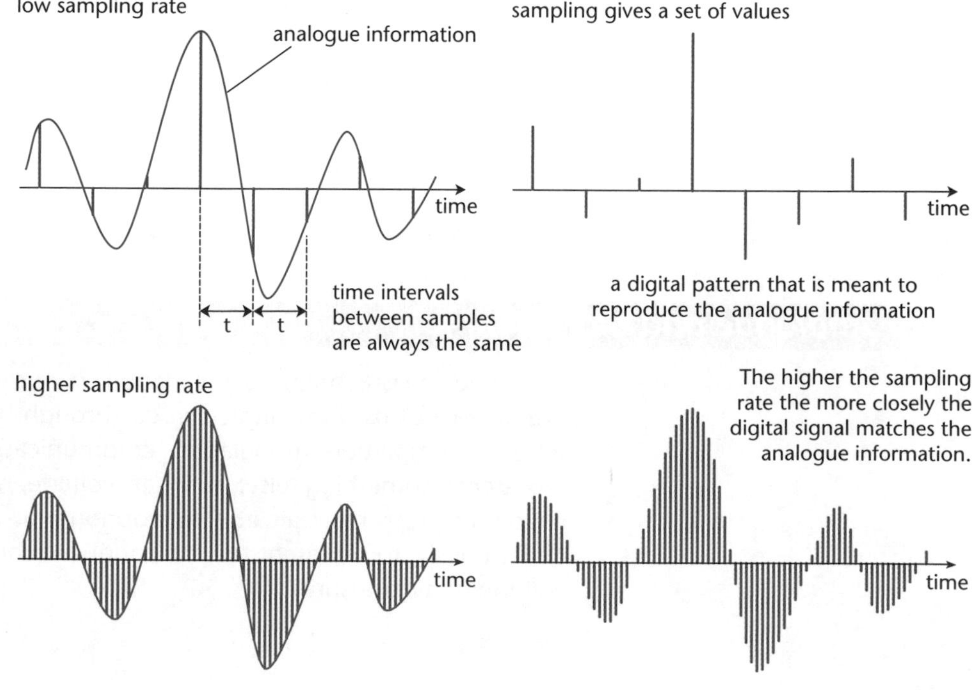

Transmission rate

AQA B 1

Transmission rate or **bit rate**, measured in bits per second, provides a measure of speed of transfer of information. A film has sound and images, so the information needed for a movie display, by DVD or the internet, must be available more quickly than for listening to sound only. The rate of transmission must be higher for the DVD. Transmission rate can be measured in bits per second.

A CD player can take information from a disc at a rate of roughly 1 megabit s^{-1}. A Blu-ray player can transfer information at about 50 megabit s^{-1}, using a blue laser beam because its shorter wavelength allows it to read from very tightly spaced pits on the disc surface.

Time division multiplexing

AQA B 1

Multiplexing is the sending of many streams of information from different sources through a single system.

In **frequency division multiplexing**, the different streams use different small ranges of frequency of a wider band. The different streams travel at the same time.

In time division multiplexing, streams are broken up into very short sections, and each of these takes turns with sections from other streams.

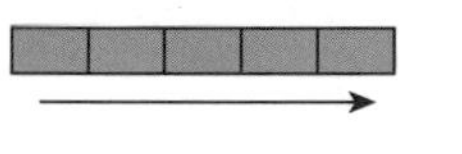

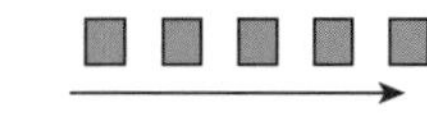

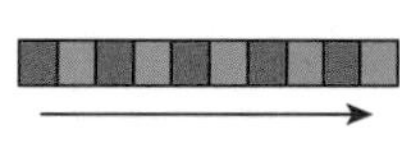

Compression

AQA B 1

Computer memory space is limited, and so is the rate at which information can travel. Smaller amounts of information use less space, and travel more quickly. File compression is a process of reducing the amount of information from an original source. A principal way to do this is by looking for redundancy, which means needless repetition of information. In the text in this section, several words, such as 'information', appear more than once. Multiple appearances become redundant if the first one is given a code and later appearances use the same code.

Progress check

1 In what ways, when you speak, do you use,
   a amplitude modulation,
   b frequency modulation?

2 Many mobile phones take photographs that can then be transmitted to another phone. Explain the effect on this of:
   a Having high resolution images, with a large number of pixels per square centimetre.
   b Having images with a very large range of possible brightness values.

3 a Explain why a larger bandwidth allows a higher transmission rate.
   b Explain why a higher transmission rate allows a more faithful reproduction of information.

1 a To vary loudness b To vary pitch

2 a The more pixels per square centimetre, for a given light collecting area and for a given amount of information per pixel, the more information there is. This means that transmission rate must be high or time for sending and receiving will be long.
   b The larger the range of brightnesses the more bits must be used to send information about each pixel. Again, this requires high transmission rate or long time.

3 a Larger bandwidth allows more frequencies to be used for sending signals, so information can be sent more quickly.
   b Higher transmission rate allows a higher sampling rate, so that the digital signal more closely matches the original information.

Sample question and model answer

More students lose marks through poor communication than through poor maths. Use large sketches with clear labels as much as possible. Labels can be easier to write and sequence than long sentences. Examiners appreciate concise answers. Note that part (c) can be answered fully with a well-labelled sketch.

(a) What do ultrasound and ultraviolet light have in common when compared with audible sound and visible light, other than that we cannot detect them directly? [2]

(Ultrasound and ultraviolet light both have) higher, — 1 mark
frequency (than the corresponding audible and visible radiations). — 1 mark

(b) What is the role of diffraction in the formation of an interference pattern by a 'diffraction grating'? [2]

A diffraction grating has many/alternating lines/gaps that transmit light. Each line/gap is small/narrow enough for strong diffraction effect. Each line/gap acts as a source of coherent light. — any 2 points, 1 mark each

(c) Explain how a diffraction grating produces bands of colour, with large angular separations that are dependent on wavelength. [5]

$n\lambda = d\ \sin\theta$ — 1 mark

Significance of n as 0th, 1st, 2nd and so on order maxima, stated or shown on sketch. — 1 mark
Significance of d as line separation, stated or shown on sketch. — 1 mark
Significance of θ as angular separation (from 0th order diffraction or system axis), stated or shown on sketch. — 1 mark
Small value of d (relative to wavelength) results in large value of θ for any given value of n, stated or shown on sketch. — 1 mark

Practice examination questions

1 (a) Distinguish between a **transverse** and a **longitudinal** wave and give one example of each. [4]

(b) Light can be **polarised** when it is reflected at a surface.

(i) Describe the difference between polarised light and unpolarised light. [2]

(ii) Explain why sound cannot be polarised. [1]

2 (a) State the laws that describe the refraction of light. [2]

(b) Light of frequency 6.00×10^{14} Hz is incident on an air–water boundary at an angle of 63°. The speed of light in air is $3.00 \times 10^8 \text{ m s}^{-1}$ and the speed of light in water is $2.25 \times 10^8 \text{ m s}^{-1}$.

(i) Calculate the wavelength of the light in air. [3]

(ii) Calculate the refractive index for light passing from air into water. [2]

(iii) Calculate the angle of refraction in the water. [2]

3 The diagram shows how the profile of a wave on a rope changes. In a period of 2.50×10^{-4} s it moves from position 1 to position 2.

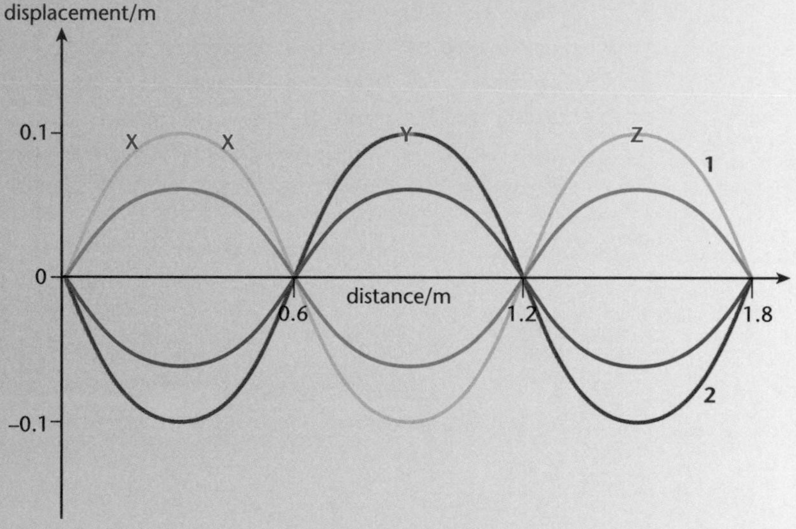

(a) (i) Write down the wavelength of the wave. [1]

(ii) Calculate the frequency of the wave. [2]

(iii) Calculate the speed of the wave along the rope. [3]

(b) (i) How can you tell from the diagram that the wave is a stationary wave? [1]

(ii) Suggest how the stationary wave is formed in this example. [2]

(c) What is the phase difference between:

(i) the two points marked X on the diagram? [1]

(ii) the points marked Y and Z on the diagram? [1]

Give your answers in degrees.

Practice examination questions (continued)

4 The diagram shows two identical loudspeakers, A and B, placed 0.75 m apart. Each loudspeaker emits sound of frequency 2000 Hz.
Point C is on a line midway between the speakers and 5.0 m away from the line joining the speakers. A listener at C hears a maximum intensity of sound. If the listener then moves from C to E or D, the sound intensity heard decreases to a minimum. Further movement in the same direction results in the repeated increase and decrease in the sound intensity.

speed of sound in air = 330 m s^{-1}

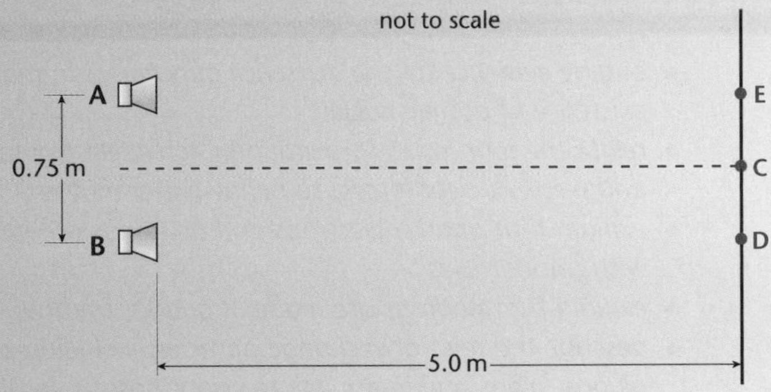

(a) Explain why the sound intensity is:
  (i) a maximum at C
  (ii) a minimum at D or E. [2]

(b) Calculate:
  (i) the wavelength of the sound
  (ii) the distance CE. [2]

5 The diagram shows an optical fibre made of a plastic-coated glass.
The refractive index of the glass is 1.58 and that of the plastic is 1.24.

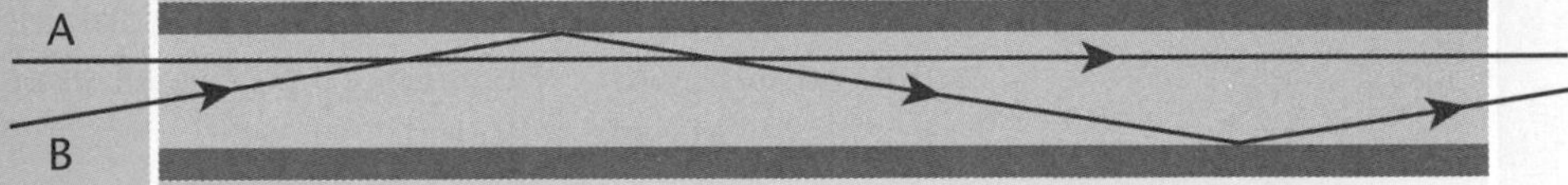

(a) Calculate the refractive index for light striking the glass–plastic interface. [2]

(b) Calculate the value of the critical angle at this interface. [2]

(c) The diagram shows two paths of light through the fibre.
Explain how this can affect a pulse of light as it passes along a fibre. [2]

Chapter 4

Waves, particles and the Universe

The following topics are covered in this chapter:

- The particle structure of matter
- Nuclei and radioactivity
- Light and matter
- The wider Universe

4.1 The particle structure of matter

LEARNING SUMMARY

After studying this section you should be able to:

- outline evidence for the existence of particles in matter, including the existence of atomic nuclei
- relate the four forces (gravitational force, electromagnetic force, strong and weak nuclear forces) to behaviour of matter
- explain that quarks, electrons and neutrinos are fundamental particles with antiparticles
- explain that hadrons are made of quarks, and name examples of hadrons
- describe the roles of exchange particles, including photons, gravitons, gluons, pions and intermediate vector bosons, in all forces between bodies
- use Feynman diagrams to represent particle exchange

The existence of particles

AQA A 1, 2
AQA B 1, 2

The kinetic model pictures a gas as being made up of large numbers of individual particles in constant motion. At normal pressures and temperatures the particles are widely spaced compared to their size. The motion of an individual particle is:

- **rapid** – a typical average speed at room temperature is 500 m s^{-1}
- **random** in both speed and direction – these are constantly changing due to the effects of collisions.

Gases exert **pressure** on the walls of their container and any other objects that they are in contact with. This pressure acts in all directions and is due to the forces as particles collide and rebound.

Brownian motion was first observed by Robert Brown while studying pollen grains suspended in water. The explanation of Brownian motion is due to Einstein.

Evidence for this movement of gas particles comes from **Brownian motion**; the random, lurching movement of comparatively massive particles such as smoke specks when suspended in air. This movement can only be attributed to the bombardment by much smaller air particles which are too small to be seen by an optical microscope and must therefore be moving very rapidly.

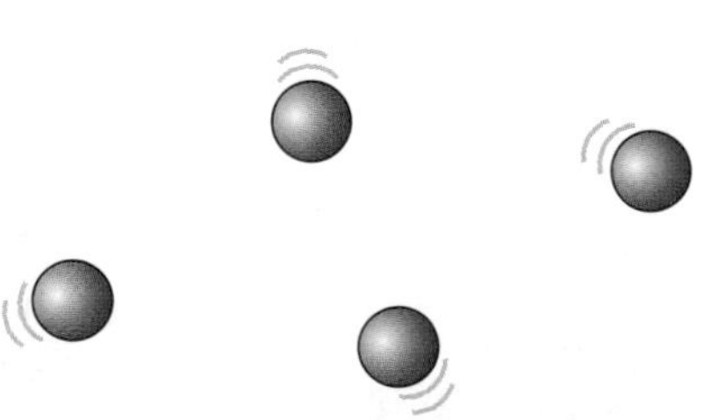

Gas particles are widely spaced. Their movement is random in both speed and direction.

Further evidence comes from diffusion – materials can mix because they do not have continuous structure, and all that is needed is mobility of their particles. The fixed and relatively simple ratios of masses of materials taking part in chemical equations is also explained by supposing that the materials have particle structures.

Particles can change their arrangements, resulting in different states or phases of matter.

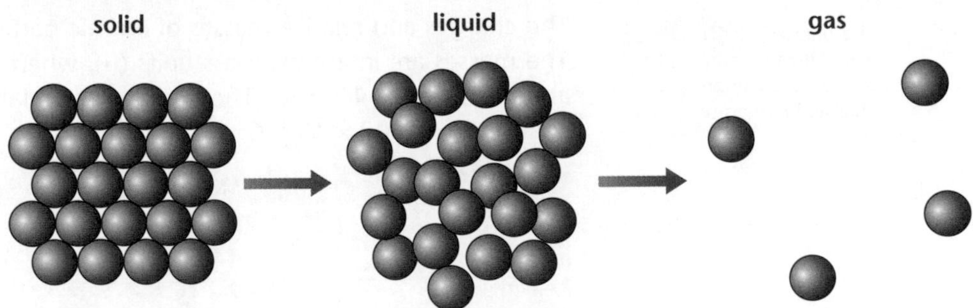

The diagrams above show particles that could be complex molecules or simple atoms. The particles are shown as spheres, for the sake of simplicity. But it is possible to explore inside atoms, to find out about their structure.

Nuclei and electrons

AQA A 1
AQA B 1

Atoms are the building blocks of everyday material. In gold, for example, all of the atoms are essentially the same. About a century ago, people were trying to find out about the structures of atoms. They were trying to 'model' atoms.

The discovery of electrons, as part of matter, led to the idea of these being embedded in a spongy ball of positively charged matter. This idea was called the **plum pudding model** of the atom.

A later atomic model pictures the atom as a tiny, positively charged nucleus surrounded by negatively charged particles (electrons) in orbit. Evidence for this model comes from the scattering of alpha particles as they pass through a thin material such as gold foil. The results of these experiments, first carried out under the guidance of Ernest Rutherford in 1911, can be summarised as:

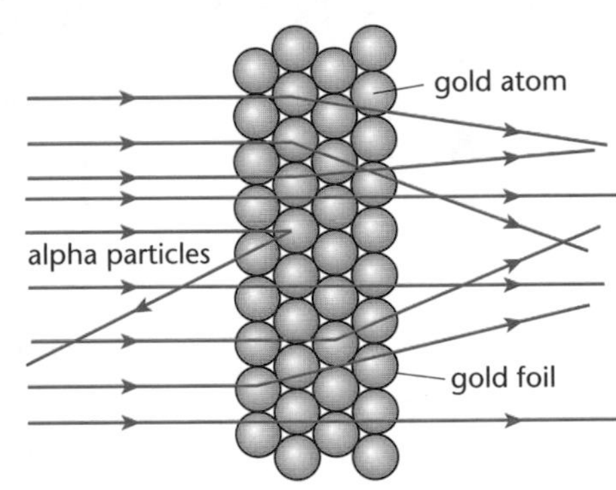

An alpha particle is a positively charged particle consisting of two protons and two neutrons.
The experiments were carried out in a vacuum so that the alpha particles were not scattered by air particles.

- most of the alpha particles travel straight through the foil with little or no deflection
- a small number are deflected by a large amount
- a tiny number are scattered backwards.

Rutherford concluded that this **alpha scattering** showed that the atoms of gold are mainly empty space, with tiny regions of concentrated charge. This charge must be the same sign as that of alpha particles (positive) for the back-scattering to be due to the repulsion between similar-charged objects.

There are two types of particle in the nucleus, **protons** and **neutrons.**

> **KEY POINT**
> The nucleus of an element is represented as $^{A}_{Z}\mathrm{El}$.
> Z is the atomic number, the number of protons.
> A is the mass number, the number of nucleons (protons and neutrons).

You will not find El in the periodic table – it is fictitious.

The element is fixed by Z, the number of protons. A neutral atom has equal numbers of protons in the nucleus and electrons in orbit. Different atoms of the same element can have different values of A due to having more or fewer neutrons. They are called **isotopes** of the element. As this does not affect the number of electrons in a neutral atom, the chemical properties of these atoms are the same. The most common form of carbon, for example, is carbon-12, $^{12}_{6}\mathrm{C}$, which has six protons and six neutrons in the nucleus. Carbon-14, $^{14}_{6}\mathrm{C}$, has the same number of protons but two extra neutrons. These two forms of carbon are isotopes of the same element.

The phrase 'relative to the value' means compared to the actual amount of charge, ignoring the sign.

The charges and relative masses of atomic particles are shown in the table below. The masses are in atomic mass units (u), where $1u = \frac{1}{12}$ the mass of a carbon-12 atom = 1.661×10^{-27} kg. The charges are relative to the value of the electronic charge, $e = 1.602 \times 10^{-19}$ C.

Atomic particle	*Mass*	*Charge*
proton	1.01	+1
neutron	1.01	0
electron	5.49×10^{-4}	–1

The ratio of charge to mass of a body, q/m, is called its specific charge. The specific charge of electrons is much larger than that of protons, because electrons have much less mass.

Representing atoms

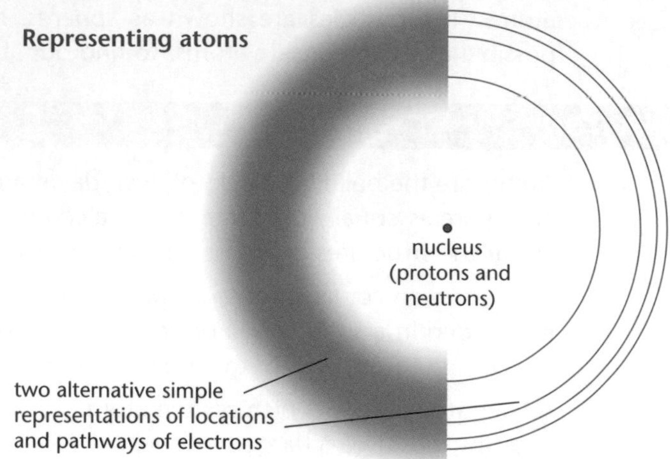

Evidence from alpha-scattering experiments shows that the diameter of a nucleus is approximately one ten-thousandth (1×10^{-4}) that of the atom and the nuclear volume is 1×10^{-12} that of the atomic volume. As almost all the atomic mass is in the nucleus, it must be extremely dense.

Electron diffraction experiments allow more precise estimates of the size of nuclei to be made. The results of these experiments show that:

- all nuclei have approximately the same density, about 1×10^{17} kg m^{-3}
- the greater the number of nucleons, the larger the radius of the nucleus.

The four forces

AQA A 1
AQA B 1

The forces that we experience every day can be classified as one of two types:

- **gravitational forces** affect all objects that have mass – they are very weak and have an infinite range
- **electromagnetic forces** also have an infinite range and are much stronger than gravitational forces – they include all forces due to static or moving charges.

On the nuclear scale, gravitational forces are so weak that they can be ignored.

The nucleus is very concentrated in terms of both mass and charge. The electrostatic force between the protons is immense and the gravitational attraction is tiny, so if these were the only forces acting, the nucleus would be unstable. In addition to the forces above, there are two other fundamental forces.

- The **strong nuclear force** affects protons and neutrons but not electrons. Its limited range, 1×10^{-15} m, means that it only acts between nucleons that are very close, i.e. next to each other in the nucleus.
- The **weak nuclear force** affects all particles. It has an even shorter range than the strong force, 1×10^{-18} m, and is responsible for beta decay.

It is the strong nuclear force that keeps the nucleus together. The strong attraction between the nucleons balances the electrostatic repulsion.

The strong force decreases very rapidly with increasing separation of the particles.

The word 'particle' is now used for comparatively huge bodies such as molecules as well as 'fundamental' bodies that can't be split into anything smaller, such as electrons.

Fundamental and non-fundamental particles

AQA A 1
AQA B 1

In an inelastic collision, kinetic energy is not conserved.

Electrons are **fundamental** particles, they cannot be split up. Protons and neutrons are themselves made up of other particles so they are not fundamental. Evidence for the structure of protons and neutrons comes from **deep inelastic scattering** experiments. In these experiments, very-high-energy electrons are fired at nucleons. The electrons are not affected by the strong force and they penetrate the nucleons in an inelastic collision, resulting in the electrons being scattered through a range of angles, with some being back-scattered in the same way that alpha particles can be back-scattered by gold foil.

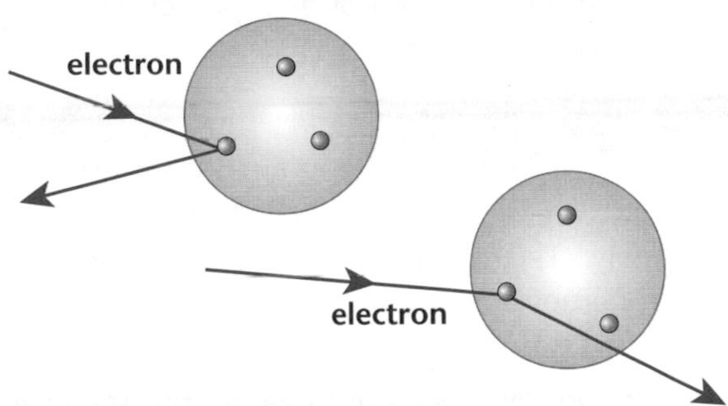

Deep inelastic scattering experiments show that nucleons contain small regions of intense charge.

This gives evidence that a nucleon contains small, dense regions of charge that are themselves fundamental particles. These particles are called **quarks**. The diagram above illustrates some of the findings of deep inelastic scattering. In the diagram below each nucleon is shown as containing three quarks.

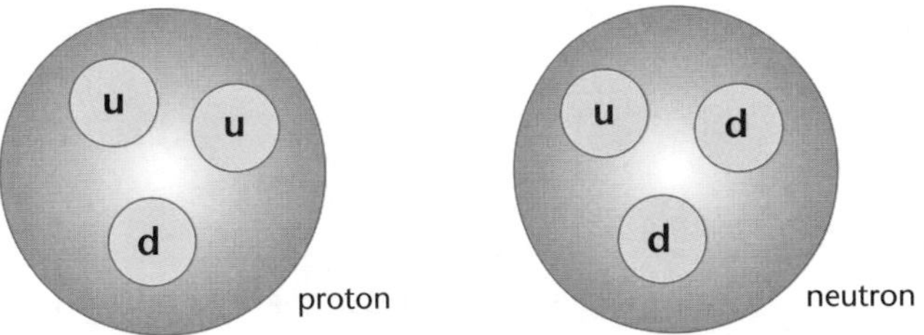

The quarks contained in nucleons

The symbol e is used throughout this section to represent the amount of charge on an electron, ignoring the sign. Positive (+) and negative (–) show the sign of a charge.

Of the six types of quark, two occur in protons and neutrons. These are the **up** and **down** quarks. The up quark, symbol u, has a charge of $+\frac{2}{3}$e and that on the down quark, symbol d, is $-\frac{1}{3}$e. The diagram above shows the quarks in a proton and a neutron.

Particles and their antis

AQA A 1
AQA B 1

Where a particle has zero charge, for example the neutron, its antiparticle also has zero charge.

For every type of particle there is an antiparticle. An antiparticle:

- has the same mass as the particle, but has the opposite charge if it is the antiparticle of a charged particle
- annihilates its particle when they collide, the collision resulting in energy in the form of gamma radiation or the production of other particles. This process is called **pair annihilation**
- can be created, along with its particle, when a gamma ray passes close to a nucleus. This process is called **pair production**.

Anti-electrons, or **positrons**, are emitted in some types of radioactive decay and can be created when cosmic rays interact with the nuclei of atoms. Like all antiparticles, they cannot exist for very long on Earth before being annihilated by the corresponding particles.

The symbols β^- and β^+ are normally used to refer to electrons and positrons given off as a result of radioactive decay.

KEY POINT

Using symbols for particles

An electron is represented by the symbol e^- or β^-.
The positron can be represented by either of the symbols e^+ and β^+.
The proton and antiproton are written as p^+ and p^-.
More generally, a line drawn above the symbol for a particle refers to an antiparticle, so u and $\bar{u}$ (read as u-bar) refer to the up quark and its antiquark.

The greek letter ν is pronounced as 'new'.

The **antineutrino,** $\bar{\nu}$, is a particle of antimatter emitted along with an electron when a nucleus undergoes β^- decay. It is a particle that has almost no mass and no charge. Its corresponding particle the **neutrino,** ν, is emitted with a positron in β^+ decay.

The quark family

AQA A 1
AQA B 1

There are six quarks which, together with their antiquarks, make up the quark family. The name of each quark describes its type, or **flavour**. Some of the properties of quarks are:

- charge – this is $+\frac{2}{3}e$ or $-\frac{1}{3}e$ for quarks and is always conserved
- baryon number – like charge, this is conserved in all interactions
- strangeness – this property describes the strange behaviour of some particles that contain quarks. It is conserved in strong and electromagnetic interactions but can be changed in weak interactions.

The values of these properties are shown in the table.

The six quarks can be thought of as three pairs, u–d, c–s and t–b.

Quark	*Symbol*	*Charge/e*	*Baryon number,* B	*Strangeness,* S
up	u	$+\frac{2}{3}$	$\frac{1}{3}$	0
down	d	$-\frac{1}{3}$	$\frac{1}{3}$	0
charm	c	$+\frac{2}{3}$	$\frac{1}{3}$	0
strange	s	$-\frac{1}{3}$	$\frac{1}{3}$	–1
top	t	$+\frac{2}{3}$	$\frac{1}{3}$	0
bottom	b	$-\frac{1}{3}$	$\frac{1}{3}$	0

The properties of antiquarks are similar to those of the corresponding quark but with the opposite sign, so the antiquark $\bar{s}$ has charge $+\frac{1}{3}$, baryon number $-\frac{1}{3}$.

The hadrons

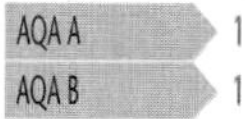
AQA A 1
AQA B 1

The **hadrons** and **leptons** are two groups of particles. Hadrons are affected by both the strong and the weak nuclear force, but leptons are affected by the weak force only.

Hadrons are not fundamental particles; they are made up of quarks:

- **baryons** are made up of three quarks
- **mesons** consist of a quark and an antiquark.

The familiar baryons are the proton, whose quark structure is uud, and the neutron, udd. Protons are particularly stable – they rarely change into anything else. There are other baryons which have different combinations of quarks but they are all unstable and have short lifetimes. All baryons have a baryon number +1, as each quark contributes a baryon number of $+\frac{1}{3}$.

Two families of mesons are the **pi-mesons**, or **pions**, and the **kaons**. Because a meson consists of a quark and an antiquark, it has baryon number 0.

The table shows the structure and some properties of the kaons and pions.

Particle	Structure	Charge/e	Baryon number, B	Strangeness, S
π^0	$u\bar{u}$ or $d\bar{d}$	0	0	0
π^+	$u\bar{d}$	+1	0	0
π^-	$\bar{u}d$	−1	0	0
K^0	$d\bar{s}$	0	0	+1
K^+	$u\bar{s}$	+1	0	+1
K^-	$\bar{u}s$	−1	0	−1

The leptons

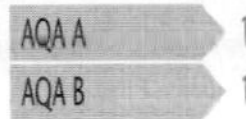

Leptons are fundamental particles. There are three families of leptons:

- the electron and its antiparticle the positron, e^- and e^+, together with the neutrinos given off in β^- and β^+ decay. These neutrinos are called the electron-antineutrino, $\bar{\nu}_e$, and the electron-neutrino, ν_e, to distinguish them from the neutrinos associated with the other leptons
- the muons, μ^- and μ^+, together with their neutrinos ν_μ and $\bar{\nu}_\mu$. Muons have 200 times the mass of electrons, but are unstable, decaying to an electron and a neutrino
- the tauons, τ^- and τ^+ and their neutrinos ν_τ and $\bar{\nu}_\tau$. The tauons are the most massive leptons, having twice the mass of a proton. Like the muons, they are unstable.

Muons are produced by cosmic rays. Their properties are similar to those of electrons and positrons but they are much more massive. Tauons are the elephants of the lepton tribe; they are produced by the interaction of high-energy electrons and positrons.

Exchanging particles

The concept of a **field** is used to describe the effects of gravitational and electromagnetic forces and how they vary with distance. This is a useful concept that enables the size and direction of forces to be calculated, but it does not explain what causes them.

The interaction between particles that results in attractive and repulsive forces is due to a continual exchange of other particles. These **exchange particles** have a very short lifetime and owe their existence to borrowed energy, so they are often referred to as **virtual particles**.

There are different exchange particles associated with each of the four fundamental forces. For electromagnetic forces the exchange particles are photons, short bursts of electromagnetic radiation.

A photon is a 'packet' of electromagnetic radiation. If you are unfamiliar with the concept of a photon, refer to section 4.3.

When two electrons approach each other they exchange virtual photons. The closer the electrons are to each other, the shorter the wavelength of the virtual photons exchanged.

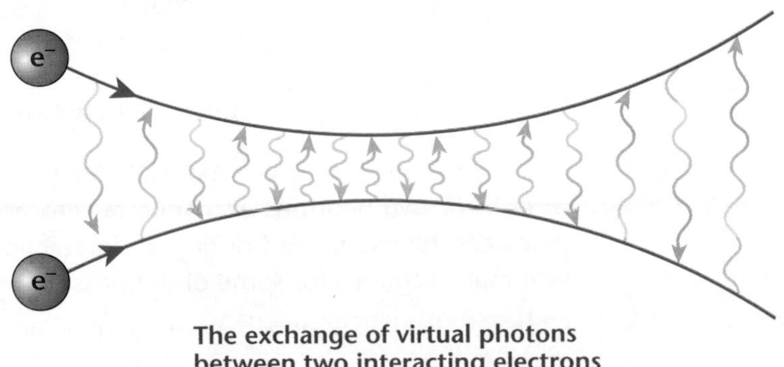

The exchange of virtual photons between two interacting electrons

This exchange of virtual photons results in the electrons moving away from each other in the same way as two ice skaters on a collision course would move away from each other if they were to keep throwing things at each other!

Richard Feynman was an American physicist. He is famous for his books *Lectures on Physics* and for his way of explaining Physics concepts by using diagrams.

The diagram on page 107 represents the exchange of virtual photons between two interacting electrons. Interactions between particles can be represented by **Feynman diagrams**. In these diagrams:

- straight lines are used for the interacting particles
- wavy lines represent the exchange particles
- the straight lines extend beyond the diagram, showing the existence of these particles before and after the interaction
- the wavy lines are contained within the diagram, showing that the exchange particles are short-lived.

This is a Feynman diagram for the interaction between two electrons shown in the diagram on page 107. The arrows do not represent directions of travel: arrows pointing in show the particles before the interaction and those pointing out show the particles after the interaction. The symbol γ is used to represent the exchange of virtual photons.

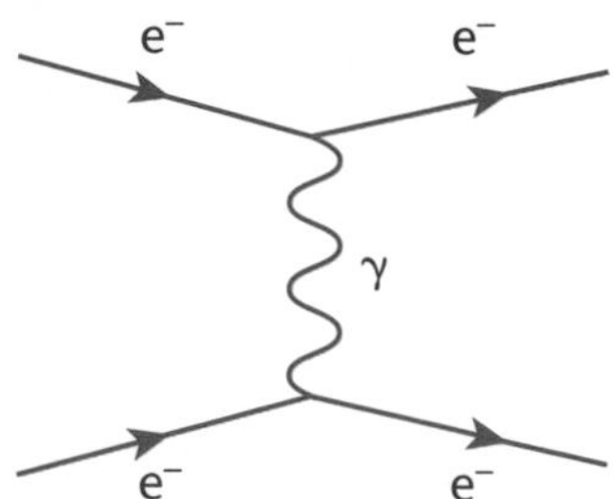

The exchange particles for gravitational forces are called **gravitons**, symbol g, though these have not yet been observed.

The strong force

AQA A 1
AQA B 1

It is the strong force that holds nucleons together in a nucleus and holds quarks together in a nucleon.

The exchange particles between quarks are called **gluons**, symbol g. The quarks that make up a proton or neutron are constantly exchanging gluons as they interact with each other.

The same symbol, g, is used for both gluons and gravitons.

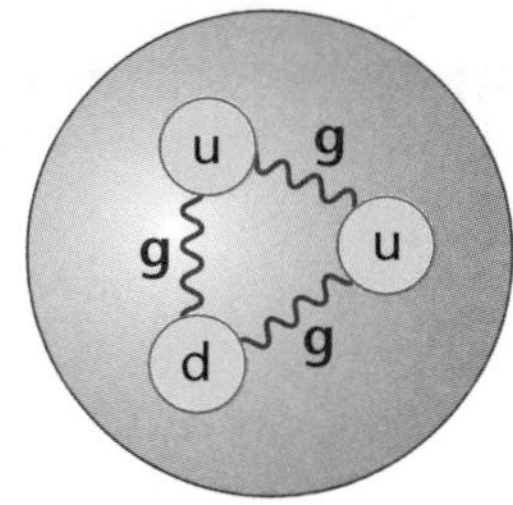

The exchange of gluons between the quarks in a proton

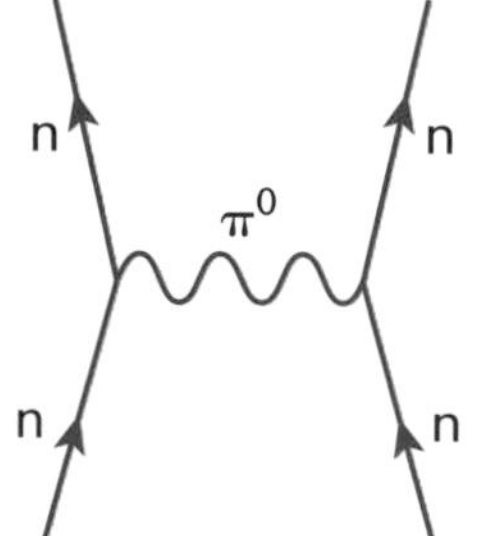

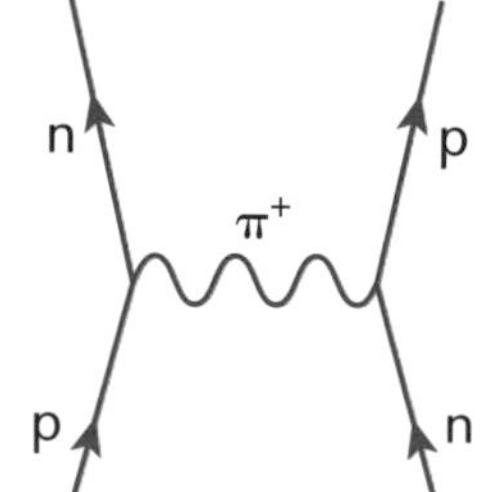

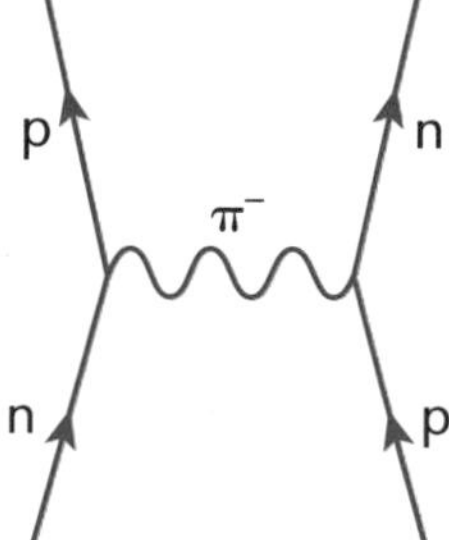

Interactions between nucleons

Pions, or pi-mesons, are responsible for the strong force between nucleons. Two protons or two neutrons exchange pi-zero, π^0, particles, but any one of the three pions can be exchanged during an interaction between a proton and a neutron. Feynman diagrams for some of the possible interactions are shown above.

Notice that charge and baryon number are conserved at each junction of the diagrams.

The weak force

AQA A 1
AQA B 1

Weak interactions involve the exchange of one of three particles called **intermediate vector bosons**. Like the pi-mesons, their symbols, W^+, W^- and Z^0 indicate the charge on each boson. The W^+ boson carries charge +e and the W^- boson carries charge −e.

The intermediate vector bosons form another group of particles, separate from the hadrons and leptons.

- The W^+ boson transfers charge +e; it is exchanged in an interaction between a neutrino and a neutron, resulting in an electron and a proton.
- The W^- boson is responsible for β^- decay of a neutron.
- Weak interactions where there is no transfer of charge involve the Z^0 boson.

Feynman diagrams for these interactions are shown below.

Can you interpret these Feynman diagrams? Try to describe the interaction that each one represents.

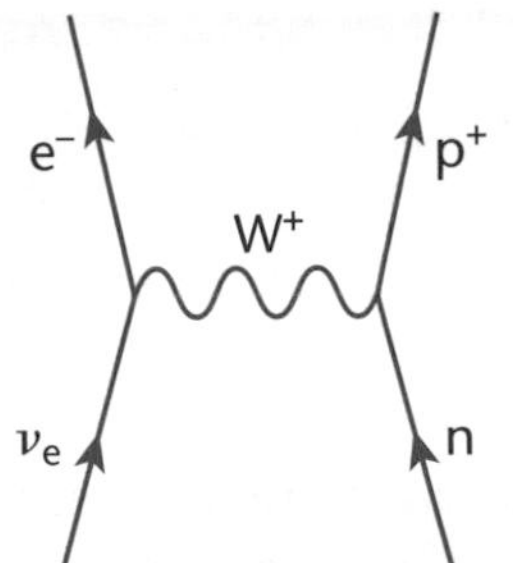

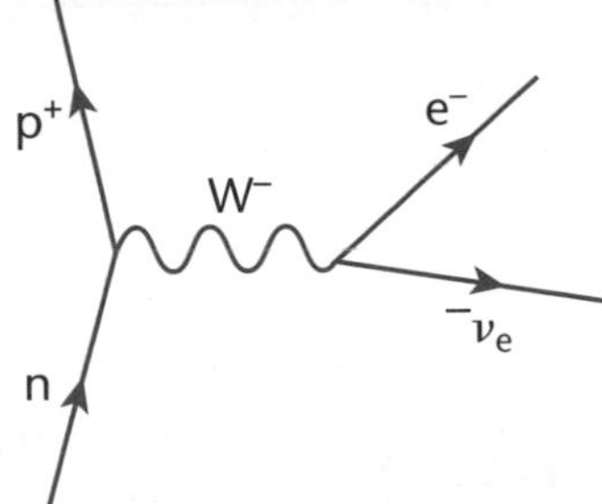

In two of these examples the type or flavour of a quark in the neutron has been changed from d to u by the weak interaction involving a charge-carrying boson. In β^- decay a down quark emits a W^- particle which decays to an electron and its antineutrino. This is shown in the diagram below.

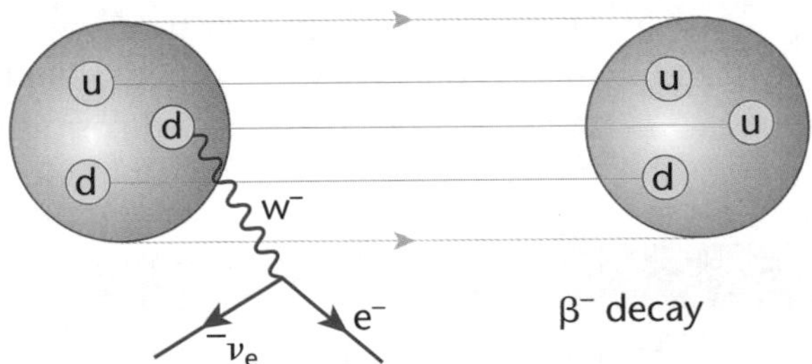

Progress check

1 State one piece of evidence for:
   a the particle nature of matter
   b the existence of nuclei.

2 Describe the differences between:
   a Hadrons and leptons
   b Baryons and mesons.

3 What change takes place in a proton in β^+ decay?

4 a What feature of the results of alpha-particle scattering experiments shows that the nucleus is positively charged?
   b Explain how the results of these experiments suggest that the nucleus occupies a very small proportion of the volume of the atom.

1 Possible answers include: **a** Brownian motion **b** alpha scattering
2 a Hadrons are made up of quarks.
   Leptons are fundamental particles.
   b Baryons contain 3 quarks. Mesons contain a quark and an antiquark.
3 A u quark changes to a d quark, changing the quark structure of the nucleon from uud to udd.
4 a The fact that some alpha particles are back-scattered shows that they are repelled from the nucleus, so they must have the same charge.
   b Most alpha particles pass through thin gold foil without being deviated.

4.2 Nuclei and radioactivity

After studying this section you should be able to:

LEARNING SUMMARY

- *explain that radioactivity is a natural phenomenon, involving emission of different types of ionising radiation*
- *describe the properties of alpha, beta-minus, beta-plus and gamma radiations*
- *apply rules of conservation of charge and nucleon number in writing balanced 'equations' to represent nuclear decay*
- *describe patterns of nuclear size and instability*

Radiation all around us

AQA A 1
AQA B 1

Radioactive decay occurs when an atomic nucleus changes to a more stable form. This is a random event that cannot be predicted. The emissions from these nuclei are collectively called **radioactivity** or **radiation**.

The term 'background radiation' is also used to describe the microwave radiation left over from the 'Big Bang'. This is a different type of radiation to nuclear radiation.

We are subjected to a constant stream of radiation called **background radiation**. Most of this is **natural** in the sense that it is not caused by the activities of people living on Earth. Sources of background radiation include:

- the air that we breathe – radioactive radon gas from rocks can concentrate in buildings
- the ground and buildings – all rocks contain radioactive isotopes
- the food that we eat – the food chain starts with photosynthesis. Radioactivity enters the food chain in the form of carbon-14, an unstable isotope of carbon that is continually being formed in the atmosphere
- radiation from space, called cosmic radiation
- medical and industrial uses of radioactive materials.

The emissions

AQA A 1
AQA B 1

Radioactive emissions are detected by their ability to cause **ionisation**, creating charged particles from neutral atoms and molecules by removing outer electrons. This results in a transfer of energy from the emitted particle, which is effectively absorbed when all its energy has been lost in this way. The four main emissions are **alpha** (α), **beta-minus** (β^-), **beta-plus** (β^+) and **gamma** (γ). Of these, alpha radiation is the most intensely ionising and has the shortest range, with the exception of beta-plus, while gamma radiation is the least intensely ionising and has the longest range.

The range of a beta-plus particle is effectively zero, since it is annihilated when it collides with an electron.

- In alpha emission the nucleus emits a particle consisting of two protons and two neutrons. This has the same make-up as a helium nucleus.
- Beta-minus emission occurs when a neutron decays into a proton, emitting an electron in the process.
- In beta-plus emission a proton changes to a neutron by emitting a positron.
- A gamma emission is short-wavelength electromagnetic radiation.

Alpha radiation is the most intensely ionising and can cause a lot of damage to body tissue. Although they cannot penetrate the skin, alpha emitters can enter the lungs during breathing.

Some properties of these radiations are shown in the table overleaf.

Radioactive emission	*Nature*	*Charge/e*	*Symbol*	*Penetration*	*Causes ionisation*	*Affected by electric and magnetic fields*
alpha	two neutrons and two protons	+2	$^{4}_{2}\text{He}$ or $^{4}_{2}\alpha$	absorbed by paper or a few cm of air	intensely	yes
beta-minus	high-energy electron	–1	$^{0}_{-1}\text{e}$ or $^{0}_{-1}\beta$	absorbed by 3 mm of aluminium	weakly	yes
beta-plus	positron (antielectron)	+1	$^{0}_{+1}\text{e}$ or $^{0}_{+1}\beta$	annihilated by an electron	yes	yes
gamma	short-wavelength electromagnetic radiation	none	γ	reduced by several cm of lead	very weakly	no

Notice that the electron emitted in beta-minus decay and the positron emitted in beta-plus decay have been allocated the atomic numbers –1 and +1. This is because of the effect on the nucleus when these particles are emitted.

Balanced equations

When a nucleus decays by alpha or beta emission, the numbers of protons and neutrons are changed. Gamma emission does not change the make-up of the nucleus, but corresponds to the nucleus losing excess energy. Gamma emission often occurs alongside alpha and beta emissions, though some artificial radioactive isotopes emit gamma radiation only.

Nucleon is a name for protons and neutrons (just as 'parent' is a name for mothers and fathers). An alpha particle has four nucleons – two protons and two neutrons – so its nucleon number, A, is 4. An electron has no nucleons, and its nucleon number is 0.

The changes that take place due to alpha and beta emissions are:

- **alpha** – the number of protons decreases by two and the number of neutrons also decreases by two
- **beta-minus** – the number of neutrons decreases by one and the number of protons increases by one
- **beta-plus** – the number of neutrons increases by one and the number of protons decreases by one.

In writing equations that describe nuclear decay, both charge (represented by Z) and the number of nucleons (represented by A) are conserved. The table below summarises these changes and gives examples of each type of decay.

Check that the equations given as examples are balanced in terms of charge and number of nucleons.

Particle emitted	*Effect on A*	*Effect on Z*	*Example*
alpha	–4	–2	$^{226}_{88}\text{Ra} \rightarrow ^{222}_{86}\text{Rn} + ^{4}_{2}\text{He}$
beta-minus	unchanged	+1	$^{14}_{6}\text{C} \rightarrow ^{14}_{7}\text{N} + ^{0}_{-1}\text{e}$
beta-plus	unchanged	–1	$^{11}_{6}\text{C} \rightarrow ^{11}_{5}\text{B} + ^{0}_{+1}\text{e}$

Stable and unstable nuclei

Carbon-11, carbon-12 and carbon-14 are three isotopes of carbon. Of these, only carbon-12 is stable. It has equal numbers of protons and neutrons. The graph shows the relationship between the number of neutrons (N) and the number of protons (Z) for stable nuclei.

Remember, isotopes of an element all have the same number of protons but different numbers of neutrons in the nucleus.

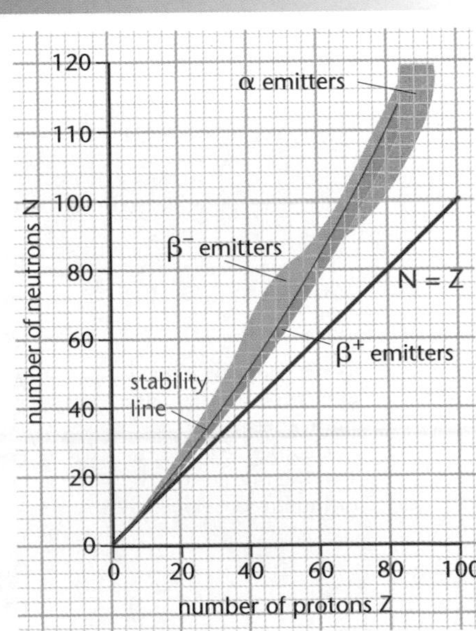

It can be seen from this graph that the condition for a nucleus to be stable depends on the number of protons (Z):

- for values of Z up to 20, a stable nucleus has equal numbers of protons and neutrons
- for values of Z greater than 20, a stable nucleus has more neutrons than protons.

For stable nuclei with more than 20 protons, the neutron:proton ratio increases steadily to a value of around 1.5 for the most massive nuclei.

Unstable nuclei above the stability line in the graph are **neutron-rich**; they can become more stable by decreasing the number of neutrons. They decay by β^- emission; this leads to one less neutron and one extra proton and brings the neutron–proton ratio closer to, or equal to, one. An example is:

$$^{24}_{11}\text{Na} \rightarrow {}^{24}_{12}\text{Mg} + {}^{0}_{-1}\text{e}$$

Unstable nuclei below the stability line decay by β^+ emission; this increases the neutron number by one at the expense of the proton number. An example is:

$$^{11}_{6}\text{C} \rightarrow {}^{11}_{5}\text{B} + {}^{0}_{+1}\text{e}$$

The N = Z line corresponds to a neutron–proton ratio of 1. As an alpha particle consists of two neutrons and two protons, its emission would hardly affect a neutron–proton ratio that is nearly 1.

The emission of an alpha particle has little effect on the neutron–proton ratio for isotopes that are close to the N = Z line and is confined to the more massive nuclei. For these nuclei, emission of an alpha particle changes the balance of the proton–neutron ratio in the favour of the neutrons. In the decay of thorium-228 shown below, the neutron–proton ratio increases from 1.53 to 1.55.

$$^{228}_{90}\text{Th} \rightarrow {}^{224}_{88}\text{Ra} + {}^{4}_{2}\text{He}$$

A carbon-12 nucleus consists of six protons and six neutrons. The mass of the atom is precisely 12 u, by definition, so after taking into account the mass of the electrons, that of the nucleus is 11.9967 u. The mass of the constituent neutrons and protons is:

$$6\ m_p + 6\ m_n = 6(1.0073\ \text{u} + 1.0087\ \text{u}) = 12.0960\ \text{u}$$

Einstein's equation $E = mc^2$ gives the method for working out how much mass is associated with energy. Try using it to work out the increase in your mass when you walk upstairs.

The nucleus has less mass than the particles that make it up. This appears to contravene the principle of conservation of mass. Einstein established that *energy has mass*. The mass that you gain due to your increased energy when you walk upstairs is infinitesimally small, but at a nuclear level this mass cannot be ignored. Any change in energy is accompanied by a change in mass, and vice versa.

When dealing with the nucleus and nuclear particles, energy and mass are so closely linked that their equivalence, and that of their units, has been established:

An electronvolt is a tiny amount of energy compared to the joule, which is too large a unit to use on an atomic scale. 1 eV = 1.60×10^{-19} J.

KEY POINT

1 u = 930 MeV

where 1 eV (one electronvolt) is the energy transfer when an electron moves through a potential difference of 1 volt.

Using this relationship, the separate conservation rules regarding mass and energy can be combined into one so that (mass + energy) is always conserved in nuclear interactions.

To split a nucleus up into its constituent nucleons requires energy. It follows that a nucleus has less energy than the sum of the energies of the corresponding number of free neutrons and protons. So the fact that a nucleus has less energy than its nucleons would have in isolation, means that it also has less mass.

KEY POINT

The difference between the sum of the masses of the individual nucleons and the mass of the nucleus is called the **mass defect** or **nuclear binding energy**. It represents the energy required to separate a nucleus into its individual nucleons.

In the case of carbon-12 the mass defect, or nuclear binding energy, is equal to 0.093 u = 89.3 MeV.

As would be expected, the greater the number of nucleons, the greater the binding energy. The figure shows how the **binding energy per nucleon** varies with nucleon number. The most stable nuclei have the greatest binding energy per nucleon.

The greater the binding energy per nucleon, the more energy (per nucleon) is required to split the nucleus up, giving it more stability.

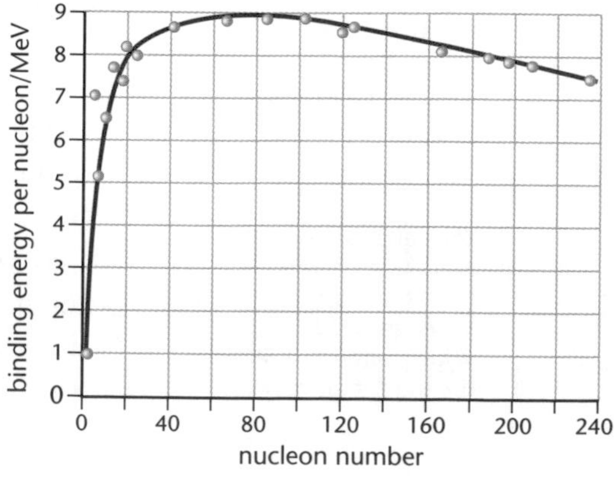

Energy spectra

AQA B 1

When a nucleus decays, the daughter nucleus has less mass and so more binding energy than the parent. The energy difference is the **decay energy**, the energy released by the nucleus as a result of its decay. This energy is transferred to:

- kinetic energy of the decay products
- energy of a gamma ray, if one is emitted
- energy of a neutrino or antineutrino, in the case of β emission.

A nucleus can only decay naturally to a state where it has less energy.

The kinetic energy of the decay products includes that of the recoiling nucleus. This is most significant in the case of alpha decay.

Energy being removed from a nucleus in alpha emission

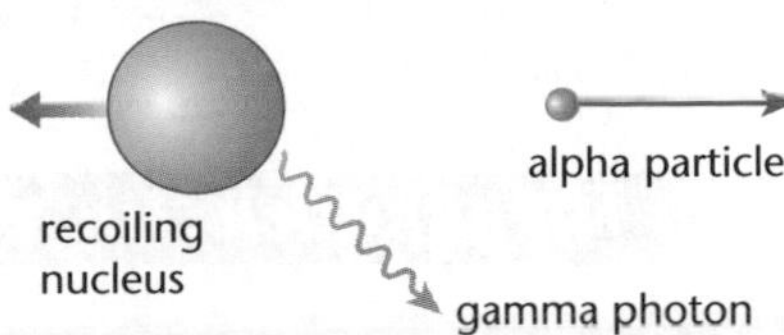

There are different patterns in the energies of alpha, beta and gamma emissions.

- The energy of alpha particles depends on the source; some sources emit alpha particles that all have the same energy, other sources emit alpha particles with two or more possible energies.
- The energy of beta particles varies continuously over a range from zero up to a maximum value that depends on the source.
- In gamma emission, each source emits a line spectrum; the photons have only one or a small number of energies.

The energies of the emitted particles are different for different sources. The shorter the half-life, the greater the energy of the emissions.

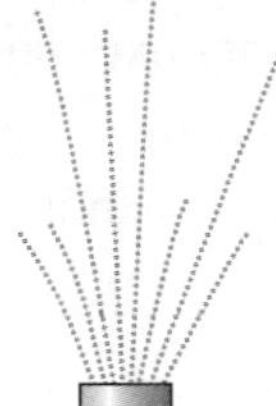

These cloud chamber tracks are from a source that emits alpha particles of two separate energies.

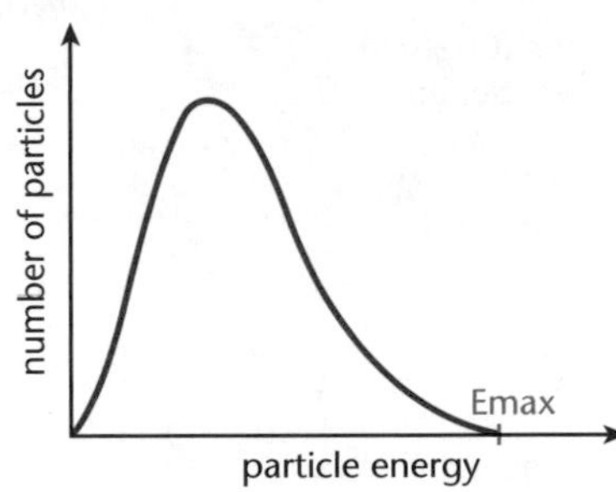

The range of energies in beta emission

While alpha and gamma emissions correspond to certain energy values, in beta emission most of the particles are emitted with less than the maximum energy available. This observation led to the discovery of the neutrino, ν, and antineutrino, $\bar{\nu}$. The antineutrino is a particle emitted along with the electron in β^- decay and a neutrino is emitted along with a positron in β^+ decay. The energies of these particles account for the difference between the energy released by the nucleus and the kinetic energies of the beta particle and the recoiling nucleus together with the energy of any gamma ray photon that is emitted.

Remember that gamma emission is usually due to a nucleus being left in an excited state following the emission of an alpha or beta particle.

Progress check

1 Tritium, ^3_1H, is a radioactive isotope of hydrogen. Use the N–Z graph to explain how it decays and to write an equation for the decay.

2 Strontium-90, $^{90}_{38}Sr$, decays to yttrium, symbol Y, by beta-minus emission. Write a balanced nuclear equation for this decay.

3 a In what way is the energy spectrum of beta emissions different to that of alpha and gamma emissions?
  b Explain how this is accounted for.

1 As tritium lies above the N–Z line, it decays by beta-minus emission.
$^3_1H \rightarrow {}^3_2He + {}^0_{-1}e$

2 $^{90}_{38}Sr \rightarrow {}^{90}_{39}Y + {}^0_{-1}e$

3 a The beta emission spectrum is continuous, whereas those of alpha and gamma only contain certain values of energy.
  b This is explained by the energy of the neutrino or antineutrino that accompanies a beta particle.

4.3 Light and matter

After studying this section you should be able to:

- *describe evidence for quantised transfer of energy by light*
- *explain why gases have spectra of discrete wavelengths*
- *explain what is meant by wave–particle duality*

LEARNING SUMMARY

The photoelectric effect

AQA A 1
AQA B 1

Light and matter interact. Matter emits and absorbs electromagnetic radiation, losing energy to radiation or taking energy away from it. These processes tell us a lot about the nature of the world. One type of absorption (or transfer of energy from light to matter) that helped to change our view of the nature of light is the photoelectric effect.

The **photoelectric effect** provides evidence that electromagnetic waves have a particle-like behaviour which is more pronounced at the short-wavelength end of the spectrum. In the photoelectric effect electrons are emitted from a metal surface when it absorbs electromagnetic radiation.

The **threshold wavelength**, λ_0, is the wavelength of the waves that have the threshold frequency:
$\lambda_0 = c \div f_0$

The results of photoelectricity experiments show that:

- there is no emission of electrons below a certain frequency, called the **threshold frequency**, f_0, which is different for different metals
- above this frequency, electrons are emitted with a range of kinetic energies up to a maximum, $(\frac{1}{2}mv^2)_{max}$
- increasing the frequency of the radiation causes an increase in the maximum kinetic energy of the emitted electrons, but has no effect on the photoelectric current, i.e. the rate of emission of electrons
- increasing the intensity of the radiation has no effect if the frequency is below the threshold frequency; for frequencies above the threshold it causes an increase in the photoelectric current, so more electrons are emitted per unit time.

The photoelectric effect can be demonstrated using a zinc plate connected to a gold leaf electroscope. An ultraviolet lamp discharges a negatively charged plate but has no effect on a positively charged plate.

The wave model cannot explain this behaviour; if electromagnetic radiation is a continuous stream of energy then radiation of all frequencies should cause photoelectric emission, it should only be a matter of time for an electron to absorb enough energy to be able to escape from the attractive forces of the positive ions in the metal.

The word quantum refers to the smallest amount of a quantity that can exist. A quantum of electromagnetic radiation is the smallest amount of energy of that frequency.

The explanation for the photoelectric effect relies on the concept of a **photon**, a quantum or packet of energy. We picture electromagnetic radiation as short bursts of energy, the energy of a photon depending on its frequency.

A lamp emits random bursts of energy. Each burst is a photon, a quantum of radiation.

KEY POINT

The relationship between the energy, E of a photon, or quantum of electromagnetic radiation, and its frequency, f, is:

$$E = hf$$

where h is Planck's constant and has the value 6.63×10^{-34} J s.

The energy of a photon can be measured in either joules or **electronvolts**. The electronvolt is a much smaller unit than the joule.

The conversion factor for changing energies in eV to energies in joules is: 1.60×10^{-19} J eV^{-1}

> **KEY POINT**
> One electronvolt (1 eV) is the energy transfer when an electron moves through a potential difference of 1 volt.
> $$1 \text{ eV} = 1.60 \times 10^{-19} \text{ J}$$

The work function is the **minimum** energy needed to liberate an electron from a metal. Some electrons need more than this amount of energy.

Radiation below the threshold frequency, f_0, no matter how intense, does not cause any emission of electrons.

Einstein's explanation of photoelectric emission is:

- an electron needs to absorb a minimum amount of energy to escape from a metal. This minimum amount of energy is a property of the metal and is called the **work function**, ϕ
- if the photons of the incident radiation have energy hf less than ϕ, then there is no emission of electrons
- emission becomes just possible when $hf = \phi$
- for photons with energy greater than ϕ, the electrons emitted have a range of energies, those with the maximum energy being the ones that needed the minimum energy to escape
- increasing the intensity of the radiation increases the number of photons incident each second. This causes a greater emission of electrons, but does not affect their maximum kinetic energy.

> **KEY POINT**
> Einstein's photoelectric equation relates the maximum kinetic energy of the emitted electrons to the work function and the energy of each photon:
> $$hf = \phi + (\tfrac{1}{2}mv^2)_{max}$$

At the threshold frequency, the minimum frequency that can cause emission from a given metal, $(\frac{1}{2}mv^2)_{max}$ is zero and so the equation becomes $hf_0 = \phi$.

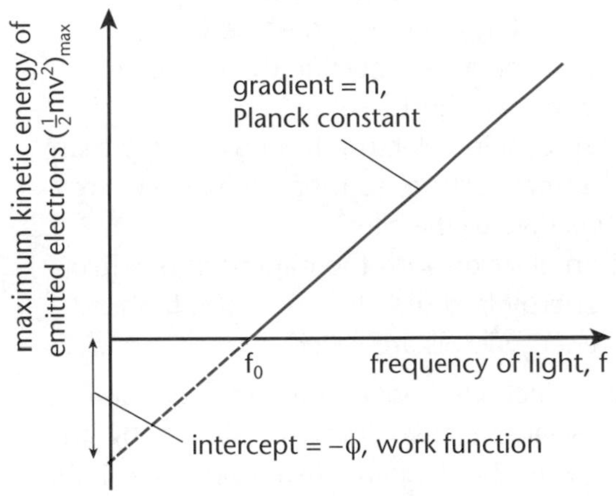

The equation:
$hf = \phi + (\frac{1}{2}mv^2)_{max}$
can be rewritten as:
$(\frac{1}{2}mv^2)_{max} = hf - \phi$
This is a graph of the form:
$y = mx + c$
where y is $(\frac{1}{2}mv^2)_{max}$
m is h
x is f
c is $-\phi$
A graph of the form $y = mx + c$ is always a straight line with gradient m and intercept c.

The word often used to describe separate distinct values, rather than continuous variation, is 'discrete'.

> **KEY POINT**
> The most important point about the photoelectric effect is that it tells us that light does not transfer energy smoothly and continuously, but in separate 'packets' or quanta. A photon is a name for a quantum of energy carried by light.

Line spectra

AQA A	1
AQA B	1

With the exception of gamma rays, the emission of electromagnetic radiation by matter is associated with electrons losing energy. A hot solid can radiate a wide range of wavelengths through the infra-red and part of the visible spectrum, the extent of the range depending on its temperature.

When an electric current is passed through an ionised gas, only radiation with a small number of specific wavelengths is emitted. The wavelengths are characteristic of the gas used and are called a **line spectrum**.

Spectra can be seen by isolating a gas at a low pressure in an enclosed glass container, and applying a potential difference to it to supply energy. The light that the gas emits is allowed to escape from the container through a narrow slit, to act as a small and coherent source. A diffraction grating separates the different wavelengths of light that the gas emits, and because the light comes from a narrow slit, the spectrum appears as a set of lines parallel to the slit. Different gases have different spectra. The diagram below shows part of a hydrogen spectrum. The same pattern of wavelengths, or spectrum, is seen with all samples of hydrogen.

The diagram shows some of the lines present in the hydrogen spectrum and their wavelengths.

wavelength/μm

0.4 0.5 0.6 0.7 X

Not all of these lines lie in the visible spectrum, which extends from 0.4 μm to 0.65 μm.

In some circumstances, radiation from solids can include line spectra. In an **X-ray tube**, high energy electrons hit a solid target. The rapid deceleration of electrons produces radiation with a continuous spectrum, but superimposed on this are peaks with specific wavelengths that are characteristic of the material that the target is made of.

The existence of line spectra provides evidence that the electrons in orbit around a nucleus can only have certain values of energy, the values being characteristic of an atom. Energy can only be emitted or absorbed in amounts that correspond to the differences between these allowed values.

An **energy level diagram** shows the amounts of energy that an electron can have. The diagram below is an energy level diagram for hydrogen. Note that on an energy level diagram:

- the energies are measured relative to a zero that represents the energy of an electron at rest outside the atom, i.e. one that is just free
- an orbiting electron has less energy than a free electron, so it has negative energy relative to the zero
- an electron with the minimum possible energy is in the **ground state**; higher energy levels are called **excited states**.

Line spectra are observed for all elements, though of course many have to be heated to become gases.

Movement of an electron to the ground state results in the emission of a photon with an energy in excess of 10 eV. Photons with this amount of energy give spectral lines in the ultraviolet region of the spectrum.

An electron movement (or transition) between the two levels as shown by the arrow in the diagram always produces light of a particular wavelength, of just over 0.65 μm. This corresponds to line X shown in the spectrum above.

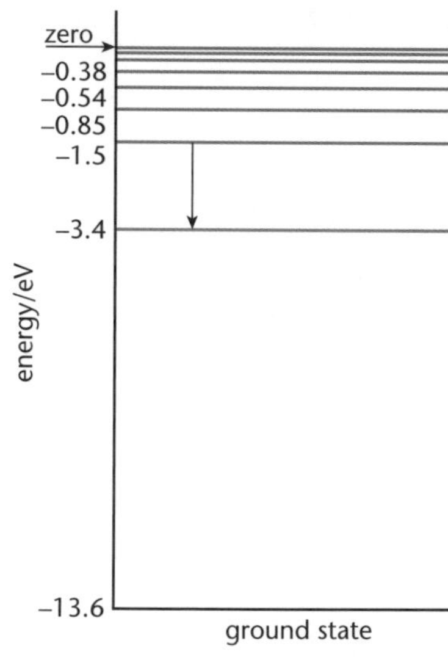

Energy levels in a hydrogen atom

The lines of the emission spectrum often appear black in the absorption spectrum. The photons of these energies are absorbed as the electrons move to more excited states, and then released when the electrons lose energy. The emitted photons are radiated in all directions, so very little energy is detected in any one direction.

Energy is emitted in the form of a photon when an electron moves from an excited state to a lower energy level. The energy of the photon is equal to the difference in the values of the energy levels. For example, an electron moving from an energy level with –0.38 eV of energy to one with –0.85 eV loses energy equal to (–0.38 – –0.85) eV = 0.47 eV. This corresponds to emitting a photon of frequency 1.13×10^{14} Hz, which lies in the infra-red part of the electromagnetic spectrum.

KEY POINT

When an electron moves from an energy level E_1 to a lower energy level E_2, the energy (hf) of the photon emitted is given by:

$$hf = E_1 - E_2$$

Electrons can gain energy by absorbing photons. As with emission, the only photons that can be absorbed are those that correspond to allowed movements, or transitions, of electrons. An **absorption spectrum** is produced by shining white light through a sample of a gaseous element. The spectrum that emerges is the full spectrum with the element's emission spectrum missing or of low intensity. This is due to the electrons absorbing photons of just the right energy to allow them to move to a more excited state.

Ionisation

AQA A 1
AQA B 1

Note that because a hydrogen atom is so simple, a hydrogen ion is just a proton.

Hydrogen atoms are the smallest and simplest, with one proton in the nucleus and with just one electron. A hydrogen atom in its ground state has its electron in the lowest possible energy level. The electron would need 13.6 eV of energy in order to escape from the atom to create a free electron and a hydrogen ion. That is, 13.6 eV is needed to ionise the atom, so it is called **ionisation energy**.

In a fluorescent lighting tube, a high potential difference causes some ionisation, and the ions and the freed electrons then accelerate in opposite directions. This causes violent collisions, and more ionisation as well as excitation of atoms. When electrons return towards the ground state they lose energy, which is emitted as electromagnetic radiation. Domestic lamps use mercury as the active gas inside the tube, and this emits UV radiation. The white coating inside the tube absorbs the UV and emits visible light.

Particles or waves

AQA A 1
AQA B 1

If waves can show particle-like behaviour in photoelectric emission, can particles also behave as waves? Snooker balls bounce off cushions in the same way that light bounces off a mirror, so reflection is not a test for wave-like or particle-like behaviour. Diffraction and interference are properties unique to waves, so particles can be said to have a wave-like behaviour if they show these properties.

Try calculating the de Broglie (pronounced de Broy) wavelength of a moving snooker ball. Is it possible for the ball to show wave-like behaviour?

All particles have an associated wavelength called the **de Broglie** wavelength.

The wavelength, λ, of a particle is related to its momentum, mv, by the de Broglie equation:

$$\lambda = h/mv$$

where h is the Planck constant.

The de Broglie wavelength of such an electron is of the order of 1×10^{-10} m.

An electron that has been accelerated through a potential difference of a few hundred volts has a wavelength similar to that of X-rays and gamma rays.

This wavelength is also similar to the spacing of the atoms in crystalline materials, so these materials provide suitable-sized gaps to cause diffraction.

You may have seen this pattern formed on a fluorescent screen in a vacuum tube.

Diffraction patterns formed by a beam of electrons after passing through thin foil or graphite show a set of bright and dark rings on photographic film, similar to those formed by X-ray diffraction.

A diffraction pattern formed by passing a beam of electrons through graphite

Electrons can also be made to interfere when two coherent beams overlap. They produce an interference pattern similar to that of light, but on a much smaller scale.

Particles and waves are the models that we use to describe and explain physical phenomena. It is not surprising that the real world does not fit neatly into our models.

Other particles such as protons and neutrons also show wave-like behaviour.

There are two separate models of how matter behaves. The particle model explains such phenomena as ionisation and photoelectricity, while the wave model explains interference and diffraction. It is not appropriate to classify matter as waves or particles because photons and electrons can fit either model, depending on the circumstances.

> The use of two models to think about light and matter at small scales is called **wave–particle duality**.
>
> KEY POINT

Progress check

1 A photon of green light has a wavelength of 5.0×10^{-7} m.
- **a** Calculate the frequency of the light.
- **b** Calculate the energy of the photon:
  - **i** in J
  - **ii** in eV.

2 Calculate the de Broglie wavelength of an electron, $m_e = 9.1 \times 10^{-31}$ kg, travelling at a speed of $3.0 \times 10^{7}\,\text{m s}^{-1}$.

3 An electron in a hydrogen atom undergoes a transition from an energy level of –0.54 eV to one of –3.40 eV.
- **a** Calculate the frequency of the photon emitted.
- **b** In what part of the electromagnetic spectrum is the emitted radiation?

1 a 6.0×10^{14} Hz b i 3.98×10^{-19} J ii 2.49 eV
2 2.43×10^{-11} m
3 a 6.90×10^{14} Hz b visible (light)

4.4 The wider Universe

After studying this section you should be able to:

- *explain how we know about stars' chemical compositions, temperature and luminosity*
- *describe how information on temperature and luminosity is used to classify stars*
- *explain why stars emit photons and neutrinos*
- *explain what red shift is and why it is important*
- *explain how we can estimate the age of the Universe*
- *consider the challenge of dark matter*

LEARNING SUMMARY

Stars' absorption spectra

AQA B 1

Stars are a long way away, and all we have is their light by which to study them. The first thing we can do is to disperse the light into spectra. That shows a continuous spectrum with a peak of intensity, and with some discrete colours being absent.

Take the absent colours first. These appear as dark lines across the spectrum. They are absorption lines, and their patterns provide matches with **absorption lines** seen by shining light through different materials here on Earth. The absorption takes place as light passes through material in the outer layers of stars at the start of its long journey into space. The patterns, matched with those seen on Earth, show us what chemical elements those outer layers are made of. So light from a star tells us, in detail, about the chemical composition of its outer layers.

Stars' continuous spectra

AQA B 1

A perfect black body is one for which the emission of light depends only on temperature, and not on other factors such as shininess or reflecting properties. Stars behave much like black bodies.

Experiments on black bodies here on Earth show that their spectra have particular patterns, with a peak of intensity that is related to temperature. The relationship is called **Wien's law**, and it can be written as:

temperature, T, in Kelvin (K) = $2.9 \times 10^{-3} / \lambda_{peak}$

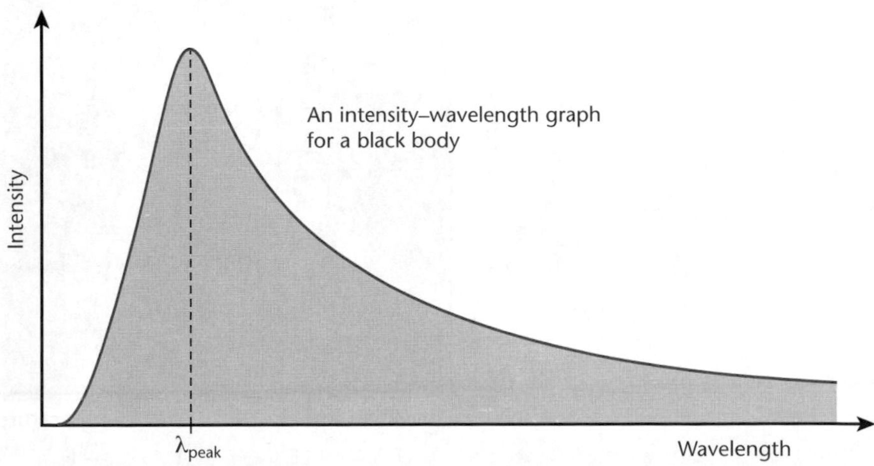

So, from the value of wavelength at which a star's spectrum peaks, we can find out the surface temperature of the star.

Intensity of emission

AQA B 1

Note that comparison of the intensity of light from a star that we see here on Earth with its intensity as known from its temperature gives us an indication of how far away it is, by application of the inverse square law, $I \alpha 1/R^2$.

Laboratory observations of the rate of emission of energy per unit area of a body, its emission intensity in Wm^{-2}, is also related to temperature. The relationship this time is given by **Stefan's law:**

intensity of emission, in Wm^{-2} = constant $\times T^4$

This is also useful, because it tells us the intensity with which a star radiates, without having to go there.

Luminosity of a star is the rate of emission of energy, measured in watts. It is the star's intensity of emission multiplied by its total surface area:

luminosity, in W = intensity of emission $\times$ star surface area

Star classification

AQA B 1

Colour is related to temperature – just as it is for a body heated on Earth. At relatively low temperatures, a body glows dull red and its colour shifts towards the blue end of the spectrum as it gets hotter.

One way to classify stars is simply by their temperature, as known by examining their spectra and applying Wien's law. This gives a sequence that runs from class O, the hottest 'blue' stars with surface temperatures of around 30 000 K, to 'red' class M stars, with surface temperature of about 3000 K.

The full sequence of temperature classifications is shown on the **Hertzsprung–Russell diagram** below. (Note that temperature decreases from left to right along the temperature axis of the diagram.)

The Hertzsprung–Russell diagram is a graph onto which stars can be plotted, according to their temperature and luminosity. Stars with similar positions are regarded as stars of the same kind, and classified accordingly, and given names according to their colour (as indicated by temperature as well as direct appearance) and according to their size (as indicated by their luminosity). Thus there are blue giants, red giants, white dwarfs and red dwarfs.

The Sun is plotted near the centre of the diagram and and is a yellow star.

As stars change, or evolve, their positions on the diagram change.

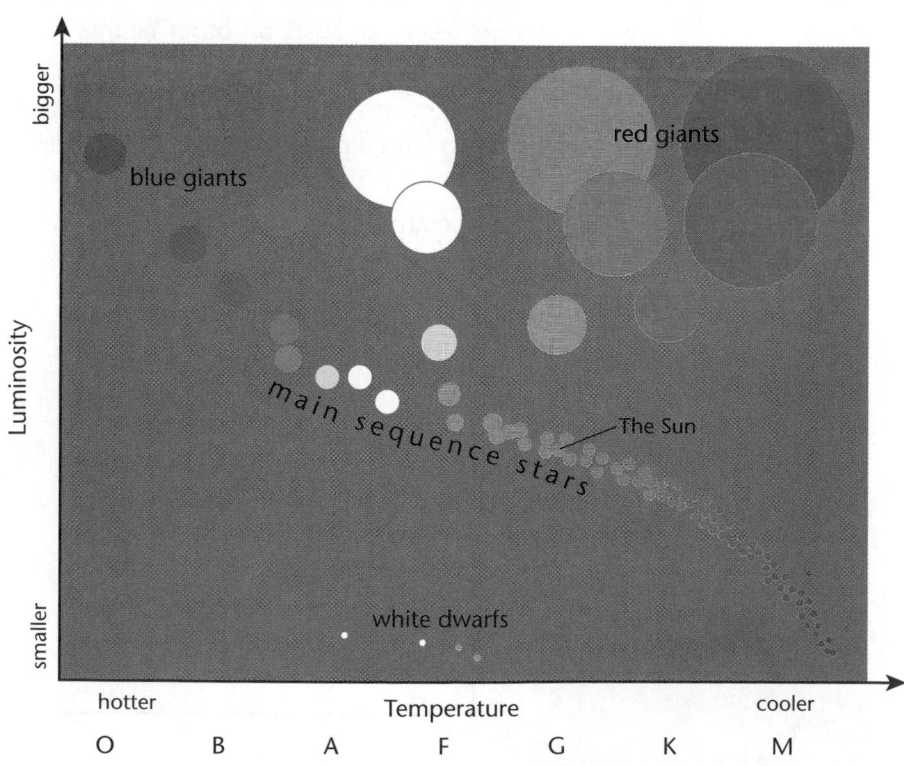

The Hertzsprung–Russell diagram – a visual classification of stars

The nuclear processes within stars

AQA B 1

Stars gain their energy by converting mass to energy. The conversion factor is large (the square of the speed of light c, as given by $E = mc^2$) so small amounts of mass make large amounts of energy available. The processes involve interactions on the nuclear scale, including:

$${}^{1}_{1}p + {}^{1}_{1}p \rightarrow {}^{2}_{1}d + {}^{0}_{1}e^{+} + \upsilon_e$$

$${}^{2}_{1}d + {}^{1}_{1}p \rightarrow {}^{3}_{2}He + \gamma$$

$${}^{3}_{2}He + {}^{3}_{2}He \rightarrow {}^{4}_{2}He + {}^{1}_{1}p + {}^{1}_{1}p$$

p = proton
d = deuterium

Overall:

$$4{}^{1}_{1}p \rightarrow {}^{4}_{2}He + 2{}^{0}_{1}e^{+} + 2\upsilon_e + 2\gamma$$

The process is a source of photons and neutrinos, which carry energy away from stars. Positrons are also produced, but these are antimatter and do not survive long since they are surrounded by matter.

Red shift and the expanding Universe

AQA B 1

Further study of absorption spectra of distant light sources reveals a very significant observation. Other galaxies are moving away from our own, and the further away they are, the faster they are moving. The space around us is getting bigger.

This knowledge comes from the Doppler effect. There is a shift in frequency of received light. The familiar patterns of absorption lines are shifted towards the red end of the spectrum, towards longer wavelength and lower frequency. It is this red shift that tells us that the Universe seems to be expanding. The size of the shift is given by:

$$\Delta f = f\, v/c$$

where f is the frequency of a dark absorption line in the laboratory, v is the relative speed of source and observer, which in this case, is called **speed of recession**, and c is the speed of light.

Discovery of the expansion of the Universe gave rise to the Big Bang theory. This is supported by other evidence, particularly by the existence of cosmic microwave background radiation. It is a theory that scientists are very much continuing to explore, looking for further support or for falsification.

The Big Bang idea supposes that the Universe – which means all of time and all of space – began from a single point, and went through phases of rapid expansion, or **inflation**, followed by the establishment of the four forces (gravitational, electric, strong and weak forces) and of particles like quarks, electrons, photons and neutrinos. A process of **nucleosynthesis** followed – the creation of protons and neutrons and of small nuclei, setting up a universal composition that we still see – a Universe of about 75% hydrogen and 24% helium.

For a long time, this soup was too hot for atoms to form – much of the energy of the Universe existed as photons, or radiation (of which the cosmic microwave background is the remains). Atoms did form eventually, though, and from initial random irregularities in their distribution the matter clumped together to make the first stars. Processes of change are still going on.

The Hubble law and the age of the Universe

AQA B 1

The statement that speed of recession is proportional to distance from us is called the **Hubble law**. The constant of proportionality is the Hubble constant.

speed of recession = Hubble constant × distance, or $v = Hd$

A megaparsec is a measure of distance, equal to about 3.26 million light years or about 3×10^{22} metres.

Speed of recession has been measured for many galaxies, in order to find out, as closely as possible, the value of the Hubble constant. The most reliable measurements available so far suggest that the value is $71 \pm 7\,\text{km}\,\text{s}^{-1}$ megaparsec^{-1}, which equates to $2.3 \times 10^{-18}\,\text{s}^{-1}$.

Quasars are objects with very large red shift, suggesting that they are a long way away. Since light takes time to travel the light we receive from such distant objects must have left them billions of years ago. This, in turn, suggests that they are objects formed in the early Universe. Quasars are now believed to be bodies of matter surrounding supermassive black holes in the centres of young galaxies.

The value of Hubble's constant provides an estimate of the age of the Universe. Since the Big Bang theory suggests that everything in the Universe was once at a single point, the speed with which any two bodies have been moving apart since the beginning of the Universe is given by:

$v = d/T$, where T is the total time available, or the age of the Universe

We can combine this with the equation above:

$Hd = d/T$

which becomes:

$T = 1/H$

So if H is $71\,\text{km}\,\text{s}^{-1}$ megaparsec^{-1}, then T is approximately 13.5 billion years. This is a crude estimate, based (for example) on an assumption that the Hubble constant has indeed always been constant, but it is the best we have.

Galaxies and dark matter

AQA B 1

After all our work in Physics, it seems that there is still plenty of work to do. Observations of galaxies, for example, tell us that there is far more matter in the Universe than we have been able to detect directly until now – and we do not know what it is! Since it emits no light, it is called **dark matter**.

Evidence for its existence comes from observing the rotation of galaxies. They behave exactly as if they were much bigger than the combined masses of their visible stars. The scientists who solve this riddle will make their names.

Progress check

1 State the nature of the evidence that can show:
  a the chemical composition of a star
  b the surface temperature of a star.

2 Sketch intensity–wavelength graphs (spectra) for a red giant and a white dwarf, on the same axes.

3 A few galaxies whose light is blue-shifted do exist. Blue shifted absorption lines are shifted towards the blue end of the visible spectrum.
  a Explain how blue shift happens.
  b Explain why such galaxies must be relatively close to our own.

1 a The absorption spectra
  b The peak on the spectrum is at a wavelength that is related to temperature by Wien's law.
2 red giant has higher peak at longer wavelength.
3 a The relative motion of source and observer is towards each other.
  b The galaxies have some motion superimposed on the overall expansion; for further galaxies relative motion is dominated by the expansion; since rate of recession is greater at greater distance.

Practice examination questions

1 The diagram shows some possible outcomes when alpha particles are fired at gold foil.

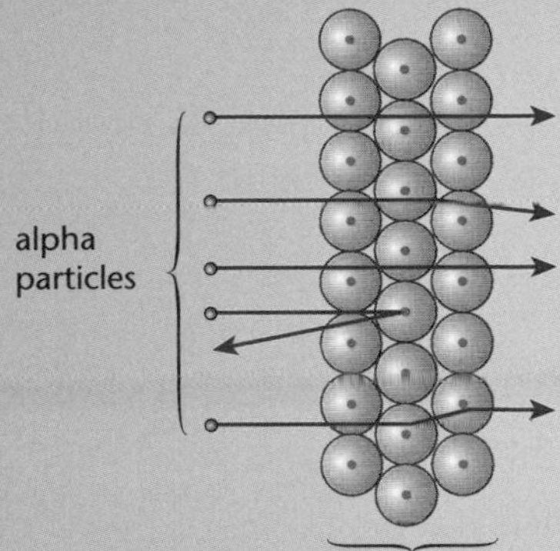

(a) Explain how these outcomes show that:

(i) the nucleus is positively charged [2]

(ii) most of the atom is empty space. [1]

(b) In **deep inelastic scattering** electrons are used to penetrate the nucleus.

(i) What is meant by 'deep inelastic scattering'? [2]

(ii) Explain why electrons are able to probe further into the nucleus than alpha particles can. [2]

(iii) Explain why the proton and the neutron are not fundamental particles. [1]

(iv) How does deep inelastic scattering provide evidence for quarks? [1]

(v) Describe the quark structure of a proton and a neutron. [2]

2 1.00 g of carbon is obtained from the wood of a living tree.

(a) Calculate the number of carbon atoms present in 1.00 g, assuming that all the carbon atoms are carbon-12.
The Avagadro constant, $N_A = 6.02 \times 10^{23}$ mol^{-1}.
The mass of 1 mole of carbon-12 is 12 g precisely. [2]

(b) Natural carbon consists of approximately:
99% carbon-12
1% carbon-13
1×10^{-10} % carbon-14.

(i) Explain whether the actual number of carbon atoms in a 1.00 g sample is likely to be significantly different from the answer to (a). [2]

(ii) Estimate the number of carbon-14 atoms present in the sample. [1]

3 For nuclei with up to 20 protons, a stable nucleus has approximately equal numbers of protons and neutrons.

(a) Explain how, for nuclei with less than 20 protons, the type of decay undergone by unstable nuclei depends on the neutron–proton ratio. [2]

(b) Describe how the decay of $^{13}_{6}C$ results in the formation of a more stable nucleus. [2]

(c) How does the neutron–proton ratio change for more massive stable nuclei? [2]

Practice examination questions (continued)

4 (a) How does the structure of a **meson** differ from that of a **baryon**? [2]

(b) The Feynman diagram illustrates the interaction between two protons.

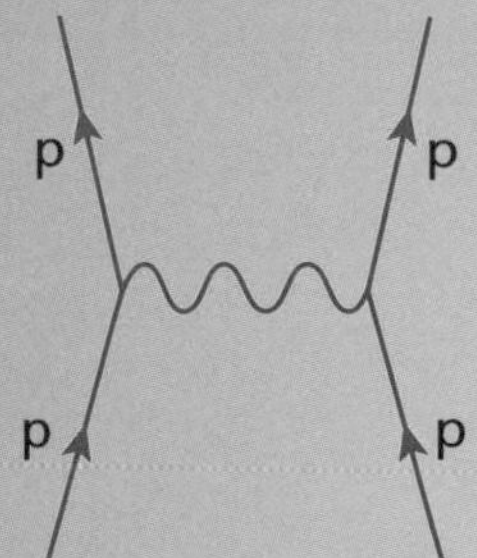

The exchange particle is a pion.

(i) Which pion is exchanged in this interaction? [1]

(ii) Explain why it is this pion rather than a different one. [2]

5 When electromagnetic radiation is incident on a clean surface of copper, *photoelectric emission* may take place. No emission occurs unless the photon energy is greater than or equal to the *work function*.

(a) Explain the meaning of the terms:

(i) photoelectric emission [2]

(ii) photon [2]

(iii) work function. [2]

(b) The work function of copper is 5.05 eV.

(i) Calculate the minimum frequency of radiation that causes photoelectric emission.
$e = 1.60 \times 10^{-19}$ C
$h = 6.63 \times 10^{-34}$ J s [2]

(ii) Calculate the maximum kinetic energy of the emitted electrons when electromagnetic radiation of frequency 1.80×10^{15} Hz is incident on the copper surface. [3]

Practice examination answers

Chapter 1 Force, motion and energy

1 The vector diagram is shown on the right [1].
The resultant velocity is 19 m s^{-1} [1]
at an angle of 18° E of N [1].

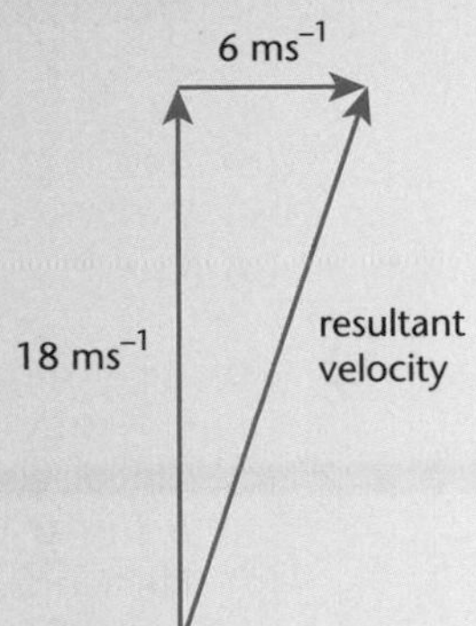

[Total: 3]

2 (a) 450 N × cos 50° [1] = 289 N [1]
(b) 289 N [1]
(c) The light fitting is in equilibrium [1] so the sum of the forces on it is zero [1].
(d) The vector diagram is shown on the right [1].
S = 345 N [1].

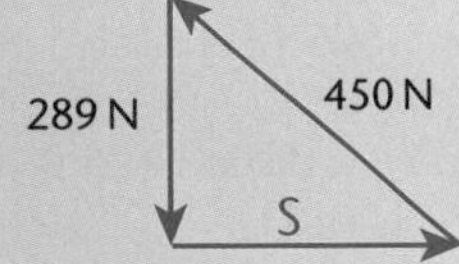

[Total: 7]

3 (a) Acceleration = change in velocity ÷ time [1] = 24 m s^{-1} ÷ 80 s [1] = 0.30 m s^{-2} [1].
(b) Between 230 and 290 s [1].
(c) (i) 20 s [1]
(ii) 120 s [1].
(d) 960 m + 1680 m + 720 m + 900 m + 1500 m + 900 m = 6660 m [2] (for all correct).
(e) 3360 m – 3300 m = 60 m [1]. [Total: 9]

4 (a) Use $s = ut + ½at^2$ [1] $t = \sqrt{(2 \times 0.39\ m \div 10\ m\ s^{-2})}$ [1] = 0.28 s [1].
(b) s = ut [1] = 35 m s^{-1} × 0.28 s = 9.8 m [1].
(c) u = s ÷ t [1] = 15 m ÷ 0.28 s = 54 m s^{-1} [1]. [Total: 7]

5 (a) t = Δv ÷ a [1] = 14 m s^{-1} ÷ 2 m s^{-2} = 7 s [1].
(b) distance = average speed × time [1] = 21 m s^{-1} × 7 s = 147 m [1]. [Total: 4]

6 (a) Use $v^2 = u^2 + 2\ as$ [1] $u^2 = -2 \times -10\ m\ s^{-2} \times 3.5\ m = 70\ m^2 s^{-2}$ [1] u = 8.4 m s^{-1} [1].
(b) Time travelling upwards, t = distance ÷ average speed = 3.5 m ÷ 4.2 m s^{-1} = 0.83 s [1]
Time in air = 1.66 s [1]. [Total: 5]

7 (a) Use $v^2 = u^2 + 2as$ [1] $a = v^2 \div 2s = (60\ m\ s^{-1})^2 \div 2 \times 1500\ m$ [1] = 1.2 m s^{-2} [1].
(b) F = ma [1] = 70 000 kg × 1.2 m s^{-2} [1] = 84 kN [1].
(c) As the aircraft gains speed the size of the resistive forces increases [1].
This causes the resultant force to decrease [1]. [Total: 8]

8 (a) (i) The pull of the trailer on the car [1]
(ii) 150 N backwards [1].
(b) (i) 130 N forwards [1]
(ii) a = F ÷ m [1] = 130 N ÷ 190 kg [1] = 0.68 m s^{-2} [1].
(c) (i) 0 [1]
(ii) 150 N backwards [1]. [Total: 8]

9 (a) (i) 40 kg × g = 400 N [1]
(ii) The child pulls the Earth [1].
(b) (i) 400 N × cos 66° [1] = 163 N [1]
(ii) a = F/m [1] = 73 N ÷ 40 kg [1] = 1.8 m s^{-2} [1].
(c) Use $v^2 = u^2 + 2as$ [1] $v^2 = 2 \times 1.8\ m\ s^{-2} \times 5.5\ m = 19.8\ m^2 s^{-2}$ [1] v = 4.4 m s^{-1} [1].
[Total: 10]

10 (a) The astronaut gains momentum in the opposite direction to the push on the spacecraft [1]. The spacecraft gains the same amount of momentum, in the direction of the astronaut's push [1].
(b) The change in momentum of the gas [1] is balanced by an equal and opposite change in momentum of the astronaut [1]. [Total: 4]

11 (a) W = F × s [1] = 30 000 N × 18 m [1] = 540 kJ [1].
(b) 540 kJ [1]
(c) P = W ÷ t = 540 kJ ÷ 24 s [1] = 22.5 kW [1].
(d) $P_{in} = P_{out}$ ÷ efficiency [1] = 22.5 kW ÷ 0.45 = 50 kW [1]. [Total: 8]

12 (a) $\Delta E_p = mg\Delta h$ [1] = 2.0×10^5 kg × 10 N kg^{-1} × 215 m [1] = 4.30×10^8 J [1].
(b) $E_k = ½mv^2 = 4.30 \times 10^8$ J [1] v = $\sqrt{(2 \times 4.30 \times 10^8 \text{ J} \div 2.0 \times 10^5 \text{ kg})}$ [1] = 65.6 m s^{-1} [1].
(c) Kinetic energy of water leaving turbines each second = $\frac{1}{2}mv^2 = \frac{1}{2} \times 2.0 \times 10^5$ kg × (8 m s^{-1})^2 = 6.4×10^6 J [1]. Maximum power input = 4.30×10^8 W – 6.4×10^6 W = 4.24×10^8 W [1].
(d) Efficiency = useful power out ÷ power in = 2.5×10^8 W ÷ 4.24×10^8 W [1] = 0.59 [1]. [Total: 10]

13 (a) Energy = $\frac{1}{2}Fx = \frac{1}{2}$ × 25 N × 0.15 m [1] = 1.875 J [1].
(b) (i) $\frac{1}{2}mv^2$ = 1.875 J [1] v^2 = 2 × 1.875 J ÷ 0.020 kg = 187.5 m^2 s^{-2} [1] v = 13.7 m s^{-1} [1]
(ii) $mg\Delta h$ = 1.875 J [1] Δh = 1.875 J ÷ 0.20 N [1] = 9.4 m [1]. [Total: 8]

14 (a) Moment = force × distance to pivot = 60 N × 0.45 m [1]
= 27 N m [1].
(b) F = 27 N m ÷ 0.06 m [1] = 450 N [1].
(c) A bigger moment is caused for the same applied force [1].
So the force acting on the branch is greater [1]. [Total: 6]

15 (a) Horizontal component = 4500 × cos 50 = 2893 N [1].
Vertical component = 4500 × cos 40 = 3447 N [1].
(b) pressure = normal force ÷ area [1] = 3447 N ÷ 0.09 m^2 [1] = 38300 Pa [1].
(c) To reduce the pressure on the ground [1] so that the girder does not penetrate the surface [1]. [Total: 7]

16 (a) A ductile material undergoes plastic deformation [1] but a brittle material breaks [1].
(b) (i) 340 MPa [1]
(ii) up to 300 MPa [1]
(iii) E = stress ÷ strain = 3.00×10^8 Pa ÷ 1.50×10^{-3} [1] = 2.00×10^{11} Pa [1].
(iv) The curve would have the same shape [1]; the values on the axes would be the same as the stress–strain graph is the same for all samples of the material [1].
[Total: 8]

Chapter 2 Electricity

1 (a) $R = \rho l/A$ [1].
(b) Resistivity is a property of a material [1]; the resistance of an object also depends on its dimensions [1].
(c) Cross-sectional area, $A = 2.0 \times 10^{-5}\ m^2$ [1] $R = 3.00 \times 10^{-5}\ \Omega\ m \times 1.0 \times 10^{-2}\ m \div 2.0 \times 10^{-5}\ m^2$ [1] $= 1.5 \times 10^{-2}\ \Omega$ [1]. [Total: 5]

2 (a) $1/R = 1/10\ \Omega + 1/10\ \Omega$ [1] $R = 5\ \Omega$ [1].
(b) $I = V/R$ [1] $= 6.0\ V \div 20\ \Omega$ [1] $= 0.30\ A$ [1].
(c) $0.30\ A \div 2$ [1] $= 0.15\ A$ [1].
(d) $V = IR = 0.15\ A \times 10\ \Omega$ [1] $= 1.5\ V$ [1]. [Total: 9]

3 (a) (i) The graph should show the resistance decreasing with increasing temperature [1] in a non-linear way [1]
(ii) The resistance of a metallic conductor increases with increasing temperature [1] linearly [1]
(iii) Ohm's law only applies to metallic conductors [1] at constant temperature [1].
(b) (i) Current in circuit $= 9.0\ V \div 600\ \Omega = 1.5 \times 10^{-2}\ A$ [1].
Thermistor p.d. $= 1.5 \times 10^{-2}\ A \times 500\ \Omega$ [1] $= 7.5\ V$ [1]
(ii) The p.d. across the thermistor decreases [1] as it has a smaller proportion of the circuit resistance [1]
(iii) Current in circuit $= 9.0\ V \div 175\ \Omega = 5.1 \times 10^{-2}\ A$ [1]. Resistor p.d. $= 5.1 \times 10^{-2}\ A \times 100\ \Omega$ [1] $= 5.1\ V$ [1]
(iv) Current in circuit $= 3.9\ V \div 300\ \Omega = 0.013\ A$ [1]. Resistor p.d. $= 5.1\ V$ [1]. Resistor value $= 5.1\ V \div 0.013\ A = 392\ \Omega$ [1]. [Total: 17]

4 (a) The current in the cell and the digital voltmeter is negligible [1] so there is no voltage drop due to internal resistance [1]. With the moving coil voltmeter there is a significant current in the cell; the voltage drop is the p.d. across the internal resistance [1].
(b) $R = V/I$ [1] $= 0.05\ V \div 1.55 \times 10^{-3}\ A$ [1] $= 32.3\ \Omega$ [1]. [Total: 6]

5 (a) Current in each lamp, $I = P/V = 6\ W \div 12\ V = 0.50\ A$. Current in battery $= 4 \times 0.50\ A = 2.0\ A$ [1].
(b) $\Delta q = I\Delta t$ [1] $= 2.0\ A \times 60\ s$ [1] $= 120\ C$ [1].
(c) $R = V/I$ [1] $= 12\ V \div 2.0\ A$ [1] $= 6.0\ \Omega$ [1]. [Total: 7]

6 (a) (i) $I^2 = P/R$ [1] $I = \sqrt{(10\ W \div 4.7\ \Omega)}$ [1] $= 1.46\ A$ [1]
(ii) $V = IR = 1.46\ A \times 4.7\ \Omega$ [1] $= 6.9\ V$ [1]
(iii) $l = AR/\rho$ [1] $= 2.0 \times 10^{-7}\ m^2 \times 4.7\ \Omega \div 5.0 \times 10^{-7}\ \Omega\ m$ [1] $= 1.88\ m$ [1].
(b) A smaller length of material is needed [1] for the same resistance [1]. [Total: 10]

7 (a) Potential divider [1].
(b) Apply a varying p.d. to component X [1].
(c) Circuit B varies the current and only allows investigation of the characteristics when the diode is conducting [1]. Circuit A allows the diode to be investigated when it is not conducting and when it is conducting [1].
(d) (i) The completed table is:

resistance /Ω	10.8	3.55	1.73	1.15	0.80

[Two marks for all correct.]
(ii) The graph is shown on the right.
Marks are awarded for:
Scales and labelling of axes [1]
Correct plotting [2]
Drawing a smooth curve [1]
(iii) 0.785 V [1]
(iv) The diode only conducts over a narrow range of voltages [1].

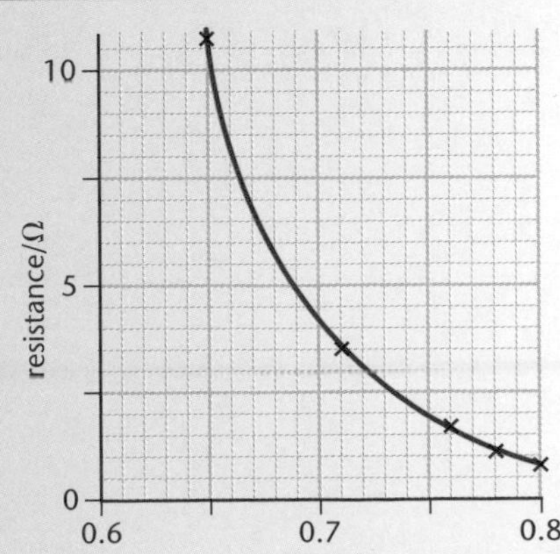

[Total: 12]

8 (a) $v = I/nAe$ [1] = 5 A ÷ (8.0 × 10^{28} m^{-3} × 1.25 × 10^{-6} m^2 × 1.60 × 10^{-19} C) [1] = 3.1 × 10^{-4} m s^{-1} [1].
(b) The cross-sectional area of the tungsten is smaller than that of the copper [1] and the concentration of free electrons is less [1].
(c) Heating is due to collisions between the free electrons and the metal ions [1]. The collisions are more frequent in the tungsten than in the copper [1]. [Total: 7]

9 (a) Effective resistance of the two parallel resistors = 4 Ω [1].
Circuit current, I = V/R = 3.0 V ÷ 7.5 Ω [1] = 0.40 A [1].
(b) V = IR = 0.40 A × 7.0 Ω [1] = 2.8 V [1]. [Total: 5]

Chapter 3 Waves, imaging and information

1 (a) In a transverse wave the vibrations are at right angles to the direction of wave travel [1].
Examples include any electromagnetic wave, e.g. radio, light [1].
In a longitudinal wave the vibrations are parallel to the direction of wave travel [1].
Examples include sound or any other compression wave [1].
(b) (i) In an unpolarised wave the vibrations are in all directions perpendicular to the direction of travel [1]. In a polarised wave the vibrations are in one direction only [1].
(ii) Sound is a longitudinal wave; there are no vibrations perpendicular to the direction of travel [1]. [Total: 7]

2 (a) The incident light, refracted light and normal are all in the same plane [1].
sin i ÷ sin r = constant [1].
(b) (i) λ = v ÷ f [1] = 3.00 × 10^8 m s^{-1} ÷ 6.00 × 10^{14} Hz [1] = 5.00 × 10^{-7} m [1]
(ii) n = c_{air} ÷ c_{water} = 3.00 × 10^8 m s^{-1} ÷ 2.25 × 10^8 m s^{-1} [1] = 1.33 [1]
(iii) sin r = sin i ÷ n = sin63° ÷ 1.33 [1] r = sin^{-1}0.668 = 42° [1]. [Total: 9]

3 (a) (i) 1.2 m [1]
(ii) f = 1/T = 1 ÷ 2.50 × 10^{-4} s [1] = 4000 Hz [1]
(iii) v = fλ [1] = 2000 Hz × 1.2 m [1] = 2400 m s^{-1} [1].
(b) (i) The wave profile does not move along the rope [1]
(ii) By superposition [1] of a wave and its reflection [1].
(c) (i) 0° [1]
(ii) 180° [1]. [Total: 11]

4 (a) (i) The waves are in phase [1] and interfere constructively [1]
(ii) The waves are out of phase [1] and interfere destructively [1].
(b) (i) λ = v ÷ f [1] = 330 m s^{-1} ÷ 2000 Hz = 0.165 m [1]
(ii) Separation of maxima, x = λD/a = 0.165 m × 5.0 m ÷ 0.75 m = 1.10 m [1].
CE is half this distance = 0.55 m [1]. [Total: 8]

5 (a) ${}_g n_p = {}_a n_p \div {}_a n_g$ [1] = 1.24 ÷ 1.58 = 0.785 [1].
(b) sin C = 0.785 [1] C = sin^{-1}0.785 = 52° [1].
(c) Light has further to travel along path B than along path A [1]. This causes a pulse to be elongated [1]. [Total: 6]

Chapter 4 Waves, particles and the Universe

1 (a) (i) Particles that approach a nucleus are deflected away from it [1]. So the nucleus must have the same sign of charge as the alpha particles [1]
(ii) Many particles pass through undeviated [1].
(b) (i) High energy electrons [1] are used to penetrate nucleons [1]
(ii) Electrons are not repelled by the nucleus [1] and they are not affected by the strong nuclear force [1]
(iii) The proton and neutron are made up of other particles [1]
(iv) The scattering of the electrons shows that there are dense regions of charge within a nucleon [1]
(v) A proton is uud [1]; a neutron is udd [1]. [Total: 11]

2 (a) 6.02×10^{23} mol^{-1} $\times$ 1/12 mol [1] = 5.02×10^{22} atoms [1].
(b) (i) No [1] since 99% of carbon is carbon-12 [1]
(ii) 5×10^{10} [1]. [Total: 5]

3 (a) If the neutron–proton ratio is less than 1 the nucleus undergoes β^+ decay [1]; if the neutron–proton ratio is greater than 1 the nucleus decays by β^- emission [1].
(b) When $^{13}_{6}C$ decays, the neutron–proton ratio changes from 1.17 [1] to 0.86, which is closer to 1 [1]
(c) It becomes greater than 1 [1] and reaches a value of 1.5 for massive nuclei [1]. [Total: 6]

4 (a) A meson consists of a quark and an antiquark [1]. A baryon is made up of three quarks [1].
(b) (i) π^0 [1].
(ii) Other pions have charge [1]. Exchange particle here must be neutral [1]. [Total: 5]

5
(a) (i) Emission of electrons [1] when a metal surface is illuminated with electromagnetic radiation [1]
(ii) A quantum of electromagnetic radiation [1] of a particular frequency [1]
(iii) The minimum energy [1] an electron needs to escape from a metal [1].
(b) (i) $f_0 = \phi \div h$ [1] = 5.05 eV $\times$ 1.60×10^{-19} C $\div$ 6.63×10^{-34} J s = 1.22×10^{15} Hz [1]
(ii) $(\frac{1}{2}mv^2)_{max} = hf - \phi$ [1] = 6.63×10^{-34} J s $\times$ 1.80×10^{15} Hz – (5.05 eV $\times$ 1.60×10^{-19} C) [1] = 3.85×10^{-19} J [1]. [Total: 11]

Notes

Notes

Notes

Notes

Notes

Notes

Index

Revise A2

AQA Physics

Graham Booth & David Brodie

Contents

Chapter 3 Thermal and particle physics

Chapter 4 Further physics

Chapter 5 Synthesis and How Science Works

AS/A2 Level Physics courses

AS and A2

All Physics A Level courses are in two parts, with three separate units or modules in each part. Most students will start by studying the AS (Advanced Subsidiary) course. Some will then go on to study the second part of the A Level course, called the A2. It is also possible to study the full A Level course, both AS and A2, in any order.

How will you be tested?

Assessment units

For AS Physics, assessment is in three units. For the full A Level, you will take a further three units that make up A2 Physics. AS Physics forms 50% of the assessment weighting for the full A Level.

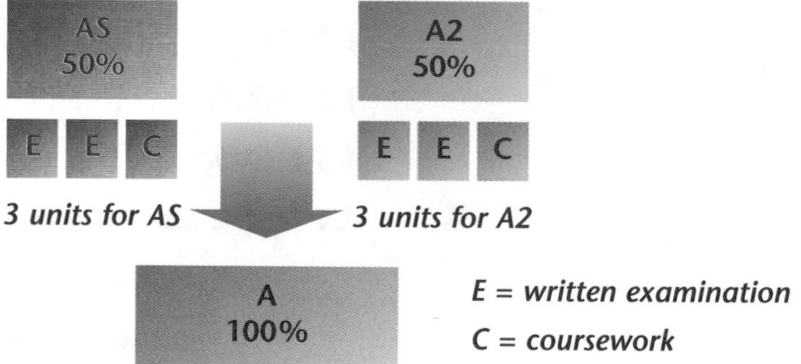

Units that are assessed by written examination can normally be taken in either January or June. Coursework units can be completed for assessment only in June. It is also possible to study the whole course before taking any of the unit tests. There is a lot of flexibility about when exams can be taken and the diagram below shows just some of the ways that the assessment units may be taken for AS and A Level Physics.

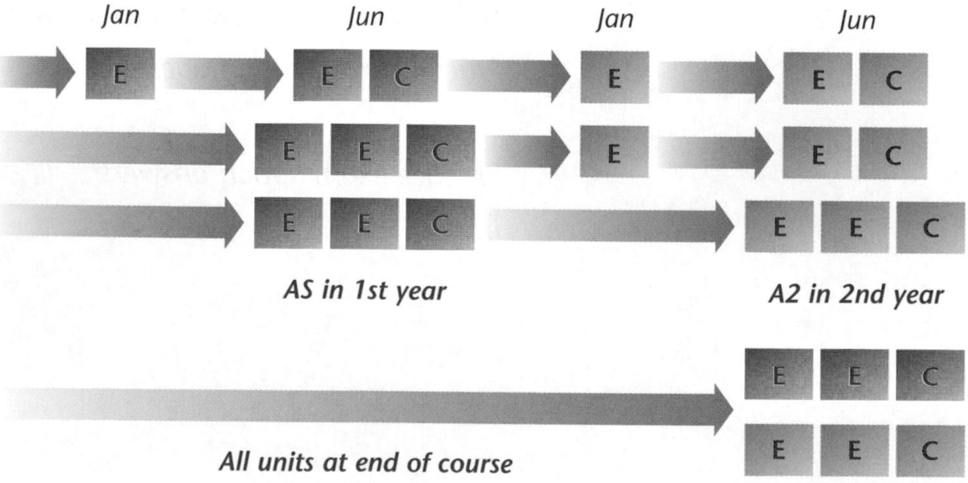

If you are disappointed with a unit result then you can re-sit the unit examination. The higher mark counts.

External assessment

For external assessment, written examinations are set and marked by teachers and lecturers from across the country, and overseen by the awarding bodies.

Internal assessment

Internal assessment is based on work that you do over a period of time, and is marked by your own teachers or lecturers. Some or all of your internally assessed work is based on practical and investigative skills and communication of your work.

Synoptic assessment

Some questions will require you to make and use connections within and between different areas of physics. They require a 'synoptic' or 'overview' approach. So, for example, you could be asked to compare energy storage by different systems such as stretched springs and charged capacitors, or to recognise and distinguish patterns in relationships such as those dealing with different kinds of rate of change.

What will you need to be able to do?

You will be tested by 'assessment objectives'. These are the skills and abilities that you should have acquired by studying the course. The assessment objectives for A2 Physics are shown below.

Knowledge and understanding of science and How Science Works

- recognising, recalling and showing understanding of scientific knowledge
- selection, organisation and communication of relevant information in a variety of forms.

Application of knowledge and understanding of science and How Science Works

- analysis and evaluation scientific knowledge and processes
- applying scientific knowledge and processes to unfamiliar situations including those related to issues
- assessing the validity, reliability and credibility of scientific information.

How Science Works

- demonstrating and describing ethical, safe and skilful practical techniques and processes
- selecting appropriate qualitative and quantitative methods
- making, recording and communicating reliable and valid observations and measurements with appropriate precision and accuracy
- analysis, interpretation, explanation and evaluation of the methodology, results and impact of their own and others' experimental and investigative activities in a variety of ways.

Specifications

AQA A Physics

ASSESSMENT UNIT	TOPIC, SECTION OR MODULE	CHAPTER REFERENCE	STUDIED	REVISED	PRACTICE QUESTIONS
4. Fields and Further Mechanics	*Further Mechanics*	*1.2, 1.3, 1.4, 1.5*			
	Gravitation	*2.1, 2.5*			
	Electric Fields	*2.2*			
	Capacitance	*2.3*			
	Magnetic Fields	*2.4, 2.5, 2.6*			
5. Nuclear Physics, Thermal Physics and an Optional Topic	*Radioactivity*	*3.3, 3.5*			
	Nuclear Energy	*3.4*			
	Thermal Physics	*3.1, 3.2*			
	Options: • *Astrophysics* • *Medical Physics* • *Applied Physics* • *Turning points in Physics*	 *4.1, 4.3, 4.4* *4.1, 4.2* *1.3, 3.1, 3.2* *1.6, 2.5, 3.5, 3.6*			
6. Investigative and Practical Skills, Internally assessed					

A2 assessment analysis

Unit 4	*1 h 45 min test*	*40%*
Unit 5	*1 h 45 min test*	*40%*
Unit 6	*Internally assessed practical skills and investigative skills*	*20%*

AQA B Physics

ASSESSMENT UNIT	TOPIC, SECTION OR MODULE	CHAPTER REFERENCE	STUDIED	REVISED	PRACTICE QUESTIONS
4. Physics Inside and Out	*Experiences Out of this World*	*1.1, 1.2, 1.3, 2.1, 2.5, 3.1 3.2, 3.6*			
	What Goes Around Comes Around	*1.3, 1.4, 1.5*			
	Imaging the Invisible	*1.5, 2.1, 2.4, 2.6, 3.5, 4.2*			
5. Energy Under the Microscope	*Matter Under the Microscope*	*3.1, 3.2*			
	Breaking Matter Down	*1.6, 2.2, 2.4, 2.5, 3.5*			
	Energy from the Nucleus	*2.3, 3.3, 3.4, 3.5, 4.2*			
6. Investigative and Practical Skills, Internally assessed					

A2 assessment analysis

Unit 4	*1 h 45 min test*	*40%*
Unit 5	*1 h 45 min test*	*40%*
Unit 6	*Internally assessed practical skills and investigative skills*	*20%*

Different types of questions in A2 examinations

For AQA A examinations Unit 4 consists of two sections. Section A provides 25 multiple choice questions whereas section B comprises of 4 or 5 structured questions requiring more extended responses. Unit 5 also has two sections both with 4 or 5 structured questions. Section B is devoted to one of the Option units. Synoptic assessment is addressed in all A2 units.

For AQA B both examinations are similar with between 5 and 8 long questions requiring extended responses. Synoptic assessment is addressed in both A2 units.

Short-answer questions

A short-answer question may test recall or it may test understanding by requiring you to undertake a short, one-stage calculation. Short-answer questions normally have space for the answers printed on the question paper. Here are some examples (the answers are shown in blue):

What is the relationship between electric current and charge flow?

Current = rate of flow of charge.

The current passing in a heater is 6 A when it operates from 240 V mains. Calculate the power of the heating element.

$P = I \times V = 6\,A \times 240\,V = 1440\,W$

Structured questions

Structured questions are in several parts. The parts are usually about a common context and they often become progressively more difficult and more demanding as you work your way through the question. They may start with simple recall, then test understanding of a familiar or an unfamiliar situation. The most difficult part of a structured question is usually at the end, where the candidate is sometimes asked to suggest a reason for a particular phenomenon or social implication.

When answering structured questions, do not feel that you have to complete one question before starting the next. The further you are into a question, the more difficult the marks are to obtain. If you run out of ideas, go on to the next question. Five minutes spent on the beginning of that question are likely to be much more fruitful than the same time spent racking your brains trying to think of an explanation for an unfamiliar phenomenon.

Here is an example of a structured question that becomes progressively more demanding.

(a) A car speeds up from $20\,m\,s^{-1}$ to $50\,m\,s^{-1}$ in 15 s.

Calculate the acceleration of the car.

acceleration = increase in velocity ÷ time taken
$= 30\,m\,s^{-1} \div 15\,s = 2\,m\,s^{-2}$

(b) The total mass of the car and contents is 950 kg.

Calculate the size of the unbalanced force required to cause this acceleration.

force = mass × acceleration
$= 950\,kg \times 2\,m\,s^{-2} = 1900\,N$

(c) Suggest why the size of the driving force acting on the car needs to be greater than the answer to (b).

The driving force also has to do work to overcome the resistive forces, e.g. air resistance and rolling resistance.

Extended answers

Questions requiring more **extended answers** will usually form part of structured questions. They will normally appear at the end of structured questions and be characterised by having at least three marks (and often more, typically five) allocated to the answers as well as several lines (up to ten) of answer space. These questions are also used to assess your abilities to communicate ideas and put together a logical argument.

The correct answers to extended questions are less well-defined than those for short-answer questions. Examiners may have a list of points for which credit is awarded up to the maximum for the question, or they may first of all judge the quality of your response as poor, satisfactory or good before allocating it a mark within a range that corresponds to that quality.

As an example of a question that requires an extended answer, a structured question on the use of solar energy could end with the following:

Suggest why very few buildings make use of solar energy in this country compared to countries in southern Europe. [5]

Points that the examiners might look for include:

- the energy from the Sun is unreliable due to cloud cover
- the intensity of the Sun's radiation is less in this country than in southern Europe due to the Earth's curvature
- more energy is absorbed by the atmosphere as the radiation has a greater depth of atmosphere to travel through
- fossil fuels are in abundant supply and relatively cheap
- the capital cost is high, giving a long payback time
- photo-voltaic cells have a low efficiency
- the energy is difficult to store for the times when it is needed the most

Full marks would be awarded for an argument that puts forward three or four of these points in a clear and logical way.

Free-response questions

Some test papers may make use of free-response and open-ended questions. These types of questions allow you to choose the context and to develop your own ideas. Examples could include 'Describe a laboratory method of determining *g*, the value of free-fall acceleration' and 'Outline the evidence that suggests that light has a wave-like behaviour'. When answering these types of questions it is important to plan your response and present your answer in a logical order.

Stretch and Challenge

Stretch and Challenge is a concept that is applied to the structured questions in Units 4 and 5 of the exam papers, for A2. In principle, it means that the sub-questions become progressively harder so as to challenge more able students and help differentiate between A and A* students.

Stretch and Challenge questions are designed to test a variety of different skills and your understanding of the material. They are likely to test your ability to make appropriate connections between different areas and apply your knowledge in unfamiliar contexts (as opposed to basic recall).

Exam technique

Continuity from AS to A2

If you have already studied AS units then you will find that the work at A2 follows on, developing ideas and skills further. This book provides all of the new material for A2 Physics, and it repeats some key points of examination technique and mathematical skills. However, A2 assessment can require knowledge of AS work, so for your preparation for tests you will need to look at this book and its AS companion.

What are examiners looking for?

Examiners use instructions to help you to decide the length and depth of your answer. If a question does not seem to make sense, you may have misread it – read it again!

State, define or list

This requires a short, concise answer, often recall of material that can be learnt by rote.

Explain, describe or discuss

Some reasoning or reference to theory is required, depending on the context.

Outline

This implies a short response, almost a list of sentences or bullet points.

Predict or deduce

You are not expected to answer by recall but by making a connection between pieces of information.

Suggest

You are expected to apply your general knowledge to a 'novel' situation, one which you have not directly studied during the A2 Physics course.

Calculate

This is used when a numerical answer is required. You should always use units in quantities and use significant figures with care. Look to see how many significant figures have been used for quantities in the question and give your answer to this degree of precision. If the question uses 3 (sig figs), then give your answer to 3 (sig figs) also.

Some dos and don'ts

Dos

Do answer the question

- No credit can be given for good Physics that is irrelevant to the question.

Do use the mark allocation to guide how much you write

- Two marks are awarded for two valid points – writing more will rarely gain more credit and could mean wasted time or even contradicting earlier valid points.

Do use diagrams, equations and tables in your responses

- Even in 'essay-type' questions, these offer an excellent way of communicating Physics. It is worth your while to practise drawing good clear and labelled sketches. Normally, you should do this without use of rulers or other drawing instruments.

Do write legibly

- An examiner cannot give marks if the answer cannot be read.

Do write using correct spelling and grammar. Structure longer essays carefully

- Marks are now awarded for the quality of your language in exams.

Don'ts

Don't fill up any blank space on a paper

- In structured questions, the number of dotted lines should guide the length of your answer.
- If you write too much, you waste time and may not finish the exam paper. You also risk contradicting yourself.

Don't write out the question again

- This wastes time. The marks are for the answer!

Don't contradict yourself

- The examiner cannot be expected to choose which answer is intended. You could lose a hard-earned mark.

Don't spend too much time on a part that you find difficult

- You may not have enough time to complete the exam. You can always return to a difficult calculation if you have time at the end of the exam.

What grade do you want?

Everyone would like to improve their grades but you will only manage this with a lot of hard work and determination. You should have a fair idea of your natural ability and likely grade in Physics and the hints below offer advice on improving that grade.

For a Grade E

You cannot afford to miss the easy marks. Even if you find Physics difficult to understand and would be happy with a Grade E, there are plenty of questions in which you can gain marks.

- You must memorise all definitions.
- You must practise exam questions to give yourself confidence that you do know some Physics. In exams, answer the parts of questions that you know first. You must not waste time on the difficult parts. You can always go back to these later.
- The areas of Physics that you find most difficult are going to be hard to score on in exams. Even in the difficult questions, there are still marks to be gained. Show your working in calculations because credit is given for a sound method. You can always gain some marks if you get part of the way towards the solution.

For a Grade C

You must have a reasonable grasp of Physics but you may have weaknesses in several areas and you will be unsure of some of the reasons for the Physics.

- Many Grade C candidates are just as good at answering questions as the Grade A students but holes and weaknesses often show up in just some topics.
- To improve, you will need to master your weaknesses and you must prepare thoroughly for the exam. You must become a better all-rounder.

For a Grade A

You will need to be a very good all-rounder.

- You must go into every exam knowing the work extremely well.
- You must be able to apply your knowledge to new, unfamiliar situations.
- You need to have practised many, many exam questions so that you are ready for the type of question that will appear.

The exams test all areas of the specification and any weaknesses in your Physics will be found out. There must be no holes in your knowledge and understanding. For a Grade A, you must be competent in all areas.

For a Grade A*

The Awarding bodies have now introduced an A* grade for A level. This follows on from the introduction of this grade at GCSE and is intended to be awarded to a relatively small number of the highest scoring candidates.

A* grades are awarded for the A level qualification only and not for the AS qualification or individual units.

What marks do you need?

The table below shows how your average mark is transferred into a grade.

average	*80%*	*70%*	*60%*	*50%*	*40%*
grade	*A*	*B*	*C*	*D*	*E*

To achieve an A* grade you must score over 80% on all six units combined (grade A) but also **90%** on the three **A2 units** combined. This is quite a difficult target!

Essential mathematics

This section describes some of the mathematical techniques that are needed in studying A2 Level Physics.

Quantities and units

Physical quantities are described by the appropriate words or symbols, for example the symbol R is used as shorthand for the value of a *resistance*. The quantity that the word or symbol represents has both a numerical value and a unit, e.g. 10.5 Ω. When writing data in a table or plotting a graph, only the numerical values are entered or plotted. For this reason headings used in tables and labels on graph axes are always written as (physical quantity)/(unit), where the slash represents division. When a physical quantity is divided by its unit, the result is the numerical value of the quantity.

Resistance is an example of a **derived** quantity and the ohm is a derived unit. This means that they are defined in terms of other quantities and units. All derived quantities and units can be expressed in terms of the seven **base** quantities and units of the SI, or International System of Units.

The quantities, their units and symbols are shown in the table. The candela is not used in AS or A2 Level Physics.

Quantity	Unit	Symbol
length	metre	m
mass	kilogram	kg
time	second	s
electric current	ampere	A
temperature difference	kelvin	K
amount of substance	mole	mol
luminous intensity	candela	cd

Equations

Physical quantities and homogeneous equations

The equations that you use in Physics are relationships between physical quantities. The value of a physical quantity includes both the numerical value and the unit it is measured in.

An equation must be **homogeneous**. That is, the units on each side of the equation must be the same.

For example, the equation:

4 cats + 5 dogs = 9 camels

is nonsense because it is not homogeneous.

Likewise:

4 A + 5 V = 9 Ω

is not homogeneous, and makes no sense.

However, the equations:

4 cats + 5 cats = 9 cats

and:

4 A + 5 A = 9 A

are homogeneous and correct.

Checking homogeneity in an equation is useful for:

- finding the units of a constant such as resistivity

- checking the possible correctness of an equation; if the units on each side are the same, the equation may be correct, but if they are different it is definitely wrong.

When to include units after values

A physicist will often write the equation:

$$4\,\text{A} \quad + \quad 5\,\text{A} \quad = \quad 9\,\text{A}$$

as:

$$4 \quad + \quad 5 \quad = \quad 9$$

total current = 9 A

That is, it is permissible to leave out the unit in working, providing that the unit is given with the answer.

The equals sign

The = sign is at the heart of Mathematics, and it is a good habit never to use it incorrectly. In Physics especially, **the = sign is telling us that the two quantities either side are physically identical**. They may not look the same in the equation, but in the observable world we would not be able to tell the difference between one and the other. We cannot, for example, tell the difference between (4 + 5) cats and 9 cats however long we stare at them. This is a rule that is so obvious that people often forget it.

We can tell the difference between (4 + 5 + 1) cats and 9 cats (if they keep still for long enough).

(4 + 5 + 1) cats ≠ 9 cats

The ≠ sign means **not** equal. An equals sign in this 'equation' would not be telling the truth.

We cannot tell the difference between (4 + 5 + 1) cats and (9 + 1) cats.

(4 + 5 + 1) cats = (9 + 1) cats

The equals sign is telling the truth.

This reveals another rule that is also quite simple, but people often find it hard because they forget that = signs can only ever be used to tell the truth about indistinguishability. The new rule is that **you cannot change one side of an equation without changing the other in the same way.**

Rearranging equations

Often equations need to be used in a different form from that in which they are given or remembered. The equation needs to be rearranged. The rules for rearranging are:

- both sides of the equation must be changed in exactly the same way
- add and subtract before multiplying and dividing, and finally deal with roots and powers.

For example, suppose that you know the equation $v^2 = u^2 + 2as$ (and you know that the = sign here is telling the truth), and you want to work out u.

First, subtract 2as from both sides.

$$v^2 - 2as = u^2$$

(Check that you feel happy that this = sign is being honest. It will be if you have done the same thing to both sides of the original equation.)

Second, 'square root' both sides of the equation.

$$\sqrt{(v^2 - 2as)} = u$$

The hard part is not changing the equation (provided you remember the rules) but knowing what changes to make to get from the original form to the one you want.

It's not rocket science, but it does take practice.

For simple equations such as V = IR there is an alternative method, using the 'magic triangle'.

Write the equation in to the triangle. Cover up the quantity you want to work out. The pattern in the triangle tells you the equation you want.

The alternative method is to apply the rules as above, and change both sides of the equation in the same way. So if you divide both sides by R, you get V/R = I.

For equations such as these, it makes sense to use whichever method works best for you.

Drawing graphs

Graphs have a number of uses in Physics:

- they give an immediate, visual display of the relationship between physical quantities
- they enable the values of quantities to be determined
- they can be used to support or disprove a hypothesis about the relationship between variables.

When plotting a graph, it is important to remember that values determined by experiment are not exact. Every measurement has a certain level of accuracy and a certain level of precision.

An accurate measurement is one that agrees with the 'true' value. (Of course, the only way to find out a value of a physical quantity is to measure it, or to calculate it from other measured values. So the 'true' value is an ideal, and may be impossible to know.)

A precise measurement is one that agrees closely with repeated measurements. Precise measurements generally allow values to be given with confidence to more decimal places.

Perfect accuracy and perfect precision are fundamentally impossible. For these reasons, having plotted experimental values on a grid, the graph line is drawn as the best straight line or smooth curve that represents the points. Where there are 'anomalous' results, i.e. points that do not fit the straight line or curve, these should always be checked. If in doubt, ignore them, but do add a note in your experimental work to explain why you have ignored them and suggest how any anomalous results could have arisen.

Graphs, gradients and rates of change

Quantities such as velocity, acceleration and power are defined in terms of a **rate of change** of another quantity with time. This rate of change can be determined by calculating the gradient of an appropriate graph. For example, *velocity* is the *rate of change of displacement with time*. Its value is represented by the gradient of a displacement–time graph. Different techniques are used to determine the gradient of a straight line and a smooth curve. For a straight line:

- determine the value of Δy, the change in the value of the quantity plotted on the y-axis, using the whole of the straight line part of the graph

A common error when determining the gradient of a graph is to work it out using the gridlines only, without reference to the scales on each axis.

- determine the corresponding value of Δx
- calculate the gradient as $\Delta y \div \Delta x$

For a smooth curve, the gradient is calculated by first drawing a tangent to the curve and then using the above method to determine the gradient of the tangent. To draw a tangent to a curve:

- mark the point on the curve where the gradient is to be determined
- use a pair of compasses to mark in two points on the curve, close to and equidistant from the point where the gradient is to be determined
- join these points with a ruler and extend the line beyond each point.

These techniques are illustrated in the diagrams below.

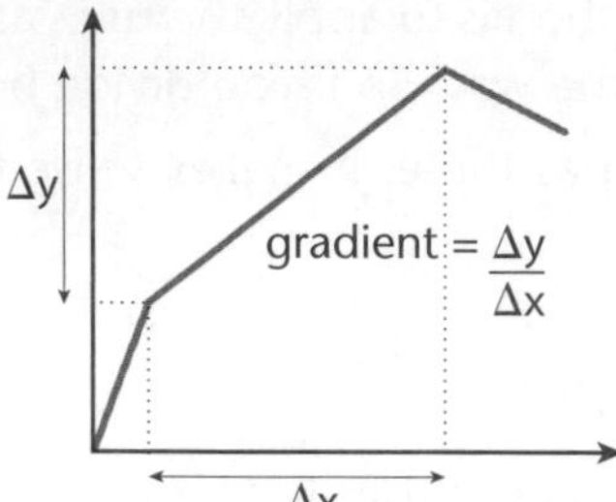

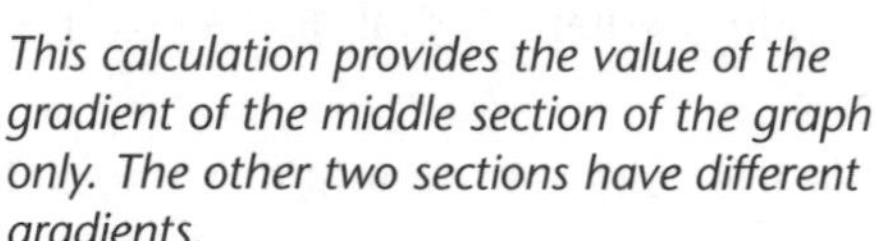

This calculation provides the value of the gradient of the middle section of the graph only. The other two sections have different gradients.

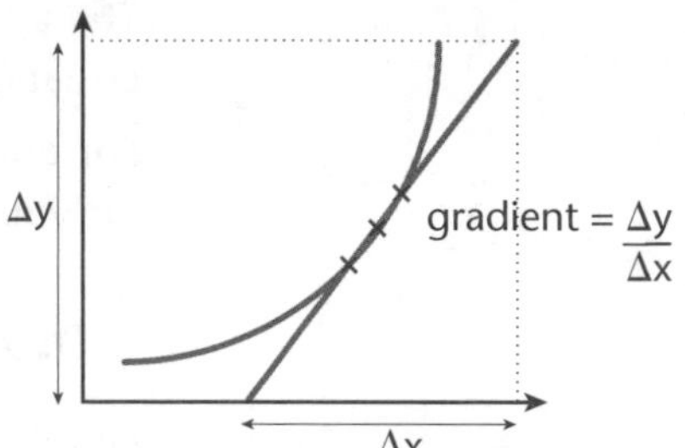

This method provides the value of the gradient at the middle of the three red marks. The graph has a continuously changing gradient that starts small and increases.

The area under a curve

The area between a graph line and the horizontal axis, often referred to as the 'area under the graph' can also yield useful information. This is the case when the product of the quantities plotted on the axes represents another physical quantity.

For example, on a *speed–time* graph this area represents the distance travelled. In the case of a straight line, the area can be calculated as that of the appropriate geometric figure. Where the graph line is curved, then the method of 'counting squares' is used.

- Count the number of complete squares between the graph line and the horizontal axis.
- Fractions of squares are counted as '1' if half the square or more is under the line, otherwise '0'.
- To work out the physical significance of each square, multiply together the quantities represented by one grid division on each axis.

These techniques are illustrated in the diagrams below:

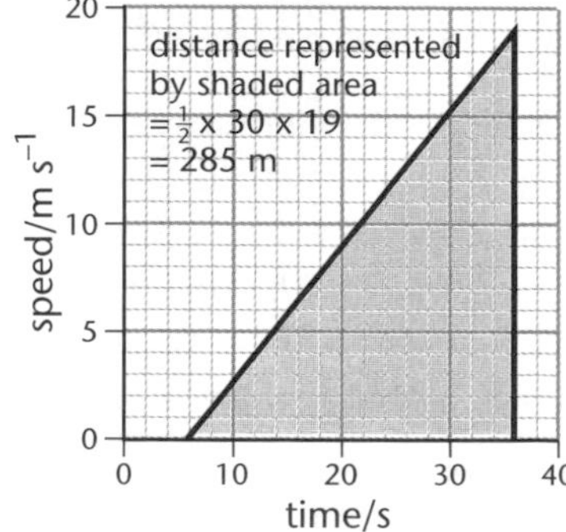

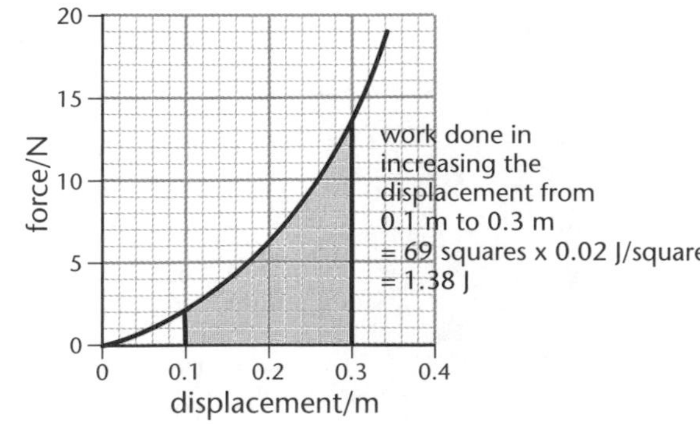

Equations from graphs

By plotting the values of two variable quantities on a suitable graph, it may be possible to determine the relationship between the variables. This is straightforward when the graph is a straight line, since all straight line graphs have an equation of the form $y = mx + c$, where m is the gradient of the graph and c is the value of y when x is zero, i.e. the intercept on the y-axis. The relationship between the variables is determined by finding the values of m and c.

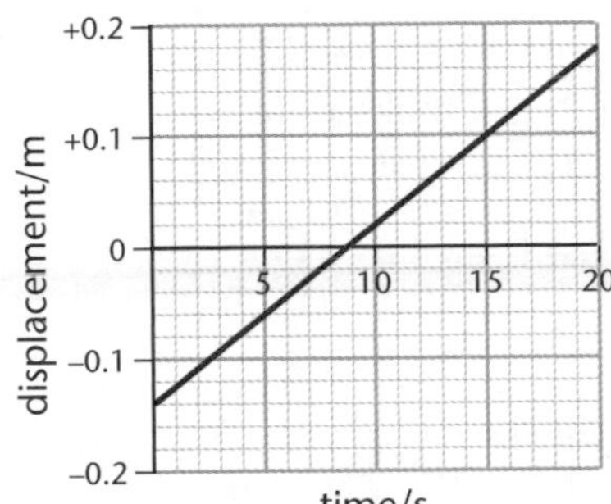

gradient = 0.32 ÷ 20
= $0.016\,m\,s^{-1}$
intercept on y-axis = –0.14

equation is s = 0.016t – 0.14

The straight line graph on the left shows how the displacement, s, of an object varies with time, t.

The equation that describes this motion is:

$$s = 0.016t - 0.14$$

(Note that no units are shown in the equation, for simplicity. The unit for displacement, s, is the metre and the unit for time, t, is the second. The physical value 0.016 has unit metre per second, and the physical value 0.14 is measured in metres.)

There is an important and special case of the equation $y = mx + c$. This is the case for which c is zero, so that $y = mx$. The straight line of this graph now passes through the graph's origin.

m is the gradient of the graph and is constant (since the graph is a single straight line). Note that different symbols can be used for the constant gradient in place of m. It is quite common, for example, to use k. That gives $y = kx$.

In Physics, we are often seeking simple patterns, and this is about as simple as patterns can be. If $y = mx$ or $y = kx$, whatever changes happen to x then y always changes by the same proportion.

Trigonometry and Pythagoras

In the right-angled triangle shown here, the sides are labelled o (opposite), a (adjacent) and h (hypotenuse). The relationships between the size of the angle θ and the lengths of these sides are shown on the left.

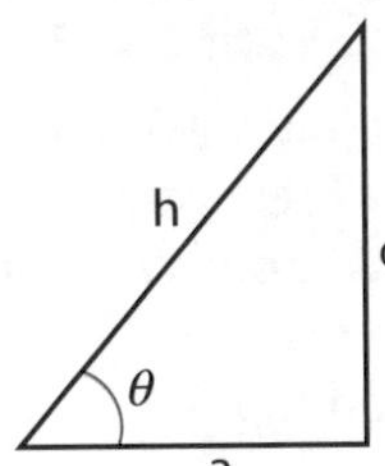

- $\sin\theta = o/h$
- $\cos\theta = a/h$
- $\tan\theta = o/a$

Pythagoras' theorem gives the relationship between the sides of a right-angled triangle: $h^2 = a^2 + o^2$

Multiples

For A2 Level Physics, you are expected to be familiar with the following multiples of units:

Name	Multiple	Symbol
pico-	10^{-12}	p
nano-	10^{-9}	n
micro-	10^{-6}	μ
milli-	10^{-3}	m
kilo-	10^{3}	k
mega-	10^{6}	M
giga-	10^{9}	G

For example, the symbol MHz means 1×10^{6} Hz and mN means 1×10^{-3} N.

Four steps to successful revision

Step 1: Understand

- Mark up the text if necessary – underline, highlight and make notes.
- Re-read each paragraph slowly.

GO TO STEP 2

Step 2: Summarise

- Now make your own revision note summary:
  What is the main idea, theme or concept to be learnt?
  What are the main points? How does the logic develop?
  Ask questions: Why? How? What next?
- Use bullet points, mind maps, patterned notes.
- Link ideas with mnemonics, mind maps, crazy stories.
- Note the title and date of the revision notes
  (e.g. Physics: Electricity, 3rd March).
- Organise your notes carefully and keep them in a file.

This is now in short-term memory. You will forget 80% of it if you do not go to Step 3.
GO TO STEP 3, but first take a 10 minute break.

Step 3: Memorise

- Take 25 minute learning 'bites' with 5 minute breaks.
- After each 5 minute break test yourself:
  Cover the original revision note summary
  Write down the main points
  Speak out loud (record on tape)
  Tell someone else
  Repeat many times.

The material is well on its way to long-term memory.
You will forget 40% if you do not do Step 4. GO TO STEP 4

Step 4: Track/Review

- Create a Revision Diary (one A4 page per day).
- Make a revision plan for the topic, e.g. 1 day later, 1 week later, 1 month later.
- Record your revision in your Revision Diary, e.g.
  Physics: Electricity, 3rd March 25 minutes
  Physics: Electricity, 5th March 15 minutes
  Physics: Electricity, 3rd April 15 minutes
  ... revisit each topic at monthly intervals.

Chapter 1

Mechanics

The following topics are covered in this chapter:

- *Newton's first and third laws*
- *Momentum*
- *Motion in a circle*
- *Oscillations, resonance and damping*
- *Simple harmonic motion*
- *Relativity*

1.1 Newton's first and third laws

LEARNING SUMMARY

After studying this section you should be able to:

- *state Newton's first and third laws of motion*
- *apply the first law to situations where the forces acting on an object are balanced and unbalanced*
- *apply the third law to the forces acting on two objects that interact*

Newton's laws of motion

With his three laws of motion, Newton aimed to describe and predict the movement of every object in the Universe. Like many other 'laws' in physics, they give the right answers much of the time. The paths of planets, moons and satellites are all worked out using Newton's laws.

This section looks at formal statements of the first and third law and how they apply to everyday situations.

The first law

4

KEY POINT

Newton's first law states that:
An object maintains its state of rest or uniform motion unless there is a resultant, or unbalanced, force acting on it.

The phrase 'uniform motion' means moving in a straight line at constant speed, i.e. moving at constant velocity.

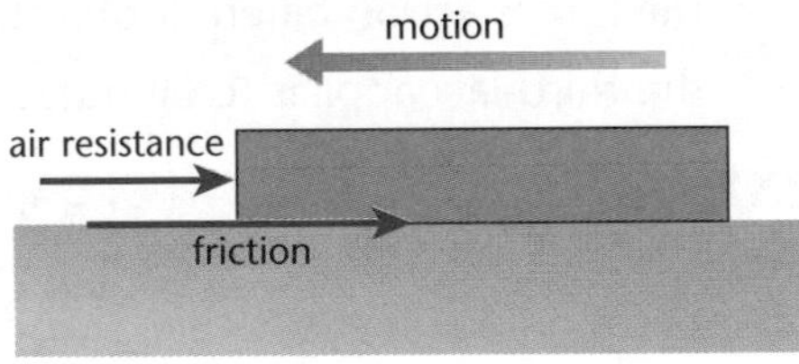

friction and air resistance create a resultant force on this sliding object

At first, this seems contrary to everyday experience. Give an object a push and it slows down before coming to rest. This is not 'continuing in a state of uniform motion'. Newton realised that in this case the unseen **resistive forces** of **friction** and **air resistance** together act in opposition to the motion; as there is no longer a driving force after the object has been pushed, there is a resultant backwards-directed force acting on it.

Key point from AS

- **When the forces on an object are balanced, the vector diagram is a closed figure.**
  Revise AS section 1.1

Motion without resistive forces is difficult to achieve on Earth. There is always air resistance or friction from a surface that a moving object rests on. The nearest that we can get to modelling motion with no resistive forces is to study motion on ice or an air track.

The converse of Newton's first law is also true:

> if an object is at rest or moving at constant velocity the resultant force on it is zero

This means that where there are two forces acting on an object that satisfies these conditions, they must be equal in size and opposite in direction. If there are three or more forces acting then the vector diagram is a closed figure.

One example of the first law is a vehicle moving at constant velocity; the **driving force** and **resistive force** are equal in size and act in opposite directions.

A normal reaction force acts on all four wheels; the arrow shown on the diagram represents the sum of these forces.

When net force is zero, as here, then Newton's first law tells us that the motion must be unchanging and that the body cannot be accelerating.

The diagram shows an example of a situation where three forces sum to zero; it illustrates a vehicle parked on a hill at rest and a vector triangle that shows the forces.

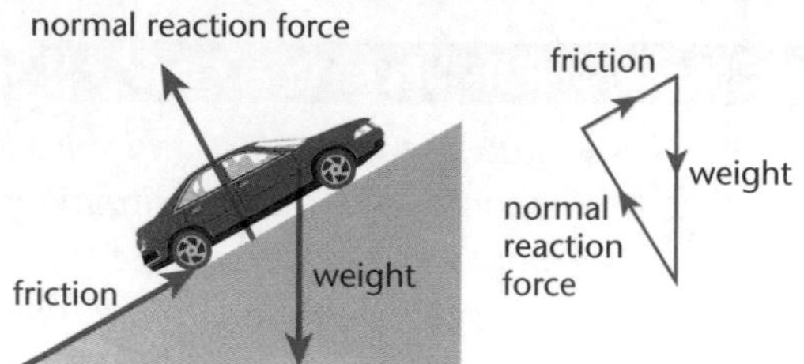

The third law

AQA B 4

This is both the simplest in its statement and the most misunderstood of the three laws. The statement given here is not a direct translation of the original, but it helps to remove some of the misunderstanding.

KEY POINT

Newton's third law can be stated as:

If object A exerts a force on object B, then B exerts a force of the same type, equal in size and opposite in direction on A.

The laws were originally written in Latin. The third law is directly translated as 'to every action there is an equal and opposite reaction'. This was widely misinterpreted as meaning that two equal and opposite forces act on the same object, resulting in zero acceleration.

According to the third law, forces do not exist individually but in pairs. However, it is important to remember that:

- the forces are of the same type, i.e. both gravitational or electrical
- the forces act on different objects
- the third law applies to all situations.

Some examples

AQA B 4

Application of Newton's third law can lead to some surprising results. If you step off a wall the Earth pulls you down towards the ground. According to the third law, you also pull the Earth up with an equal-sized force. So why doesn't the Earth accelerate upwards to meet you instead of the other way round? The answer is it does, but if you apply $F = ma$ to work out the Earth's acceleration, it turns out to be minimal.

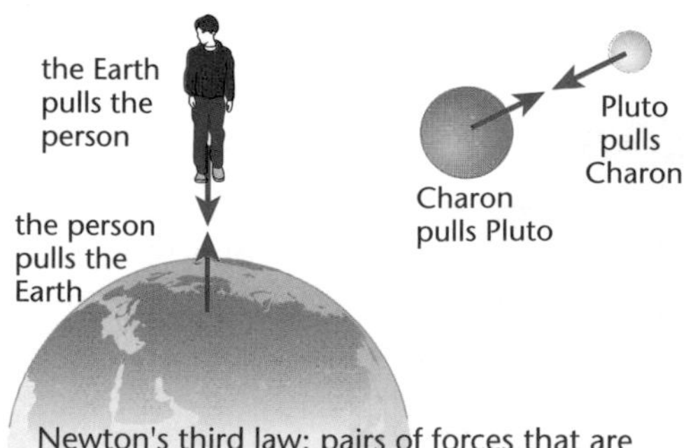

Newton's third law: pairs of forces that are equal in size and opposite in direction

Key point from AS

- **force = mass × acceleration**

Revise AS section 1.3

This type of situation is where the greatest misunderstanding of Newton's third law occurs. Remember, third law pairs of forces act on **different** objects.

Planets pull moons and, according to the third law, moons pull planets with an equal-sized force. Why does the moon go round the planet instead of the other way round?

Again, the answer is it does. In fact, they both rotate around a common centre of mass. In the case of the Earth and its Moon, the centre of mass is relatively close to the centre of the Earth so it is a reasonable approximation to say that the Moon orbits the Earth.

The diagram shows another pair of forces that are equal in size and opposite in direction. This is an application of the **first** law – the vase is in a state of rest so there is no resultant force acting on it.

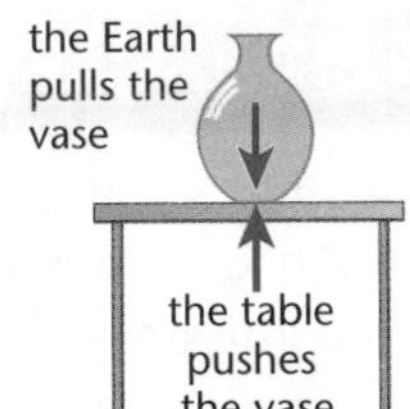

These forces are equal in size and opposite in direction acting on a single object – an application of Newton's first law, not third.

Forces acting on bodies in lifts

AQA B 4

When a body is accelerating, there *must* be a net (unbalanced) force acting on it. When it is not accelerating (whether stationary or with a constant velocity that is not zero), then the net force *must* be zero.

Note the close link between force and acceleration – always in the same direction, and when one is zero then so is the other.

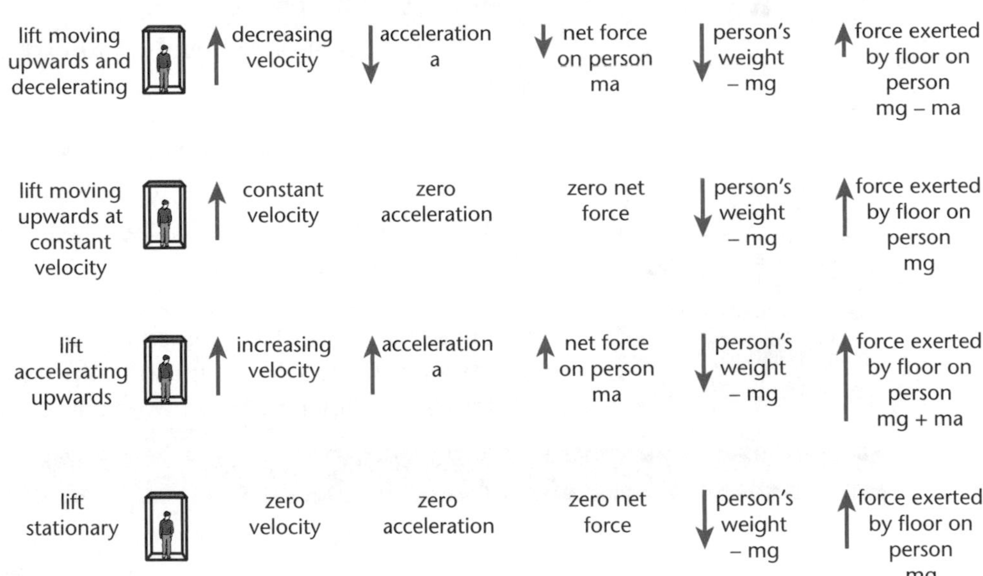

Note that the net force acting on the person is the resultant of their weight and the upwards force that the floor exerts on them.

Progress check

1 Explain why a driving force is needed to maintain motion on the surface of the Earth.

2 Suggest why the driving force needed to maintain motion on the surface of the Moon is less than for the same vehicle on the surface of the Earth.

3 A foot kicks a ball. What are the forces that make up the pair in the sense of the third law?

1 The driving force is needed to balance the resistive forces that oppose motion.
2 There is no air resistance on the Moon.
3 The ball pushes the foot.

1.2 Momentum

After studying this section you should be able to

LEARNING SUMMARY

- *apply the principle of conservation of momentum to collisions in one dimension*
- *state the relationship between the change of momentum of an object and the resultant force on it*
- *explain the difference between an elastic and an inelastic collision*

Momentum

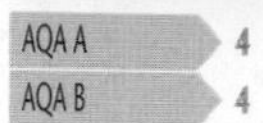

Which is more effective at demolishing a brick wall, a 1 kg iron ball moving at 50 m s^{-1} or a 1000 kg iron ball moving at 1 m s^{-1}? Both the mass and the velocity need to be taken into account to answer this question.

A collision between two objects need not involve physical contact. Imagine two positive charges approaching each other; the repulsive forces could cause them to reverse in direction without actually touching.

No matter how hard you throw a tennis ball at a garden wall you will not knock the wall down; nor will you be very successful if you use a tennis ball to play ten-pin bowling!

Newton realised that what happens to a moving object involved in a collision depends on two factors:

- the mass of the object
- the velocity of the object.

He used the concept of **momentum** to explain the results of collisions between objects.

KEY POINT

momentum = mass × velocity

$$p = m \times v$$

Momentum is a vector quantity, the direction being the same as that of the velocity. It is measured in N s or kg m s^{-1}; these units are equivalent.

A ten-pin bowling ball has a mass of several kilogrammes, so even at low speeds it has much more momentum than a tennis ball (mass 0.06 kg) travelling as fast as you can throw it.

Everything that moves has momentum and can exert a force on anything that it interacts with. This applies as much to the light that is hitting you at the moment as it does to a collision between two vehicles.

Conservation of momentum

When a force causes an object to change its velocity, there is also a change in its momentum. When two objects interact or collide, they exert equal and opposite forces on each other and so the momentum of each one changes. This is illustrated in the diagram, where the blue ball loses momentum and the black ball gains momentum.

The phrase 'equal and opposite' is used as a shorthand way of writing 'equal in size and opposite in direction'.

Like the forces they exert on each other during the collision, the changes in momentum of the balls are *equal in size and opposite in direction.*

It follows that:

- the combined momentum of the balls is the same before and after they collide.

> Since p = mv and kinetic energy is given by
> $E_k = \frac{1}{2}mv^2$,
> the two quantities are related by the formula
> $E_k = \frac{p^2}{2m}$

This is an example of the **principle of conservation of momentum.**

> **KEY POINT**
>
> The principle of conservation of momentum states that:
> When two or more objects interact the total momentum remains constant provided that there is no external resultant force.

In the context of the colliding balls, an external resultant force could be due to friction or something hitting one of the balls. Either of these would result in a change in the total momentum.

Different types of interaction

Objects moving in opposite directions

When two objects approach each other and 'collide' head-on, the result of the collision depends on whether they stick together or whether one or both rebound.

The diagram shows two 'vehicles' approaching each other on an air track. The vehicles join together and move as one.

> To make two air track vehicles stick together, one has some Plasticine stuck to it and the other is fitted with a pin that sticks into the Plasticine.

0.40 m s^{-1} →   0.30 m s^{-1} ←   velocity = ?

0.30 kg   0.20 kg   0.50 kg

before   **after**

The principle of conservation of momentum can be used to work out the combined velocity, v, after the collision.

Taking velocity from left to right as being positive:

total momentum before collision =

$$(0.3 \text{ kg} \times +0.4 \text{ m s}^{-1}) + (0.2 \text{ kg} \times -0.3 \text{ m s}^{-1}) = 0.06 \text{ N s}.$$

This must equal the momentum after the collision, so $0.5 \text{ kg} \times v = 0.06$ N s,

i.e. $v = 0.12$ m s^{-1}. As this is positive, it follows that the 'vehicle' is moving from left to right.

Rebound

When a light object collides with a heavier one, the light object may rebound, reversing the direction of its momentum. The diagram shows an example of such a collision.

> A rebound collision can be modelled on an air track by fitting the vehicles with repelling magnets.

> The figures here are rounded. The 3.3 is a rounded value of 3⅓.

Application of the principle of conservation of momentum in this case gives:

$$(2 \text{ kg} \times 10 \text{ m s}^{-1}) + (10 \text{ kg} \times 0 \text{ m s}^{-1}) = (2 \text{ kg} \times v) + (10 \text{ kg} \times 3.3 \text{ m s}^{-1})$$

So $v = -6.7$ m s^{-1}. The significance of the negative sign is that the green ball is now travelling from right to left.

Recoil

When firefighters use water to put out a fire, it often takes two people to hold the hose. This is due to the recoil as water is forced out of the nozzle at high speed.

In some situations two objects are stationary before they interact, having a combined momentum of zero. After they interact, both objects move off but the combined momentum is still zero. Examples of this include:

- a stationary nucleus undergoes radioactive decay, giving off an alpha particle
- a person steps off a boat onto the quayside
- two ice skaters stand facing each other; then one pushes the other
- two air track 'vehicles' fitted with repelling magnets are held together and then released.

In each case, both objects start moving from rest. For the combined momentum to remain zero, each object must gain the same amount of momentum but these momentum gains must be in *opposite directions*. Such an interaction is sometimes called an **explosion**.

In recoil situations, each object gains the same amount of momentum, but in opposite directions. It follows that the one with the smaller mass ends up with the greater speed.

The ice skater who does the pushing **recoils** as an equal and opposite force is exerted by the one who is pushed.

In the example shown in the diagram above, an 80 kg man pushes a 50 kg child. Their combined momentum is zero both before and after this interaction.

Taking velocity from left to right as positive, this means that:

$$(80\text{ kg} \times -0.6\text{ m s}^{-1}) + (50\text{ kg} \times v) = 0$$

So v, the velocity of the child after the interaction, is 0.96 m s^{-1} to the right.

Elastic or inelastic

AQA A 4
AQA B 4

In all three examples above, both momentum and energy are conserved. This is always the case. A collision where the total **kinetic** energy before and after the interaction is the same is called an **elastic** collision. Where there is a gain or loss of kinetic energy as a result of the interaction the collision is called **inelastic**.

It is important to emphasise here that total energy is **always** conserved.

In an elastic collision: momentum, kinetic energy and total energy are conserved.

In an inelastic collision: momentum and total energy are conserved, but the amount of kinetic energy changes.

Momentum and the second law

AQA A 4
AQA B 4

Newton's second law establishes the relationship between the resultant force acting on an object and the change of momentum that it causes. An important result that follows from the second law and the definition of the newton is F = m × a.

The expression $\Delta p/\Delta t$ can be read as: change in momentum ÷ time taken for the change.

KEY POINT

Newton's second law states that:

The rate of change of momentum of an object is proportional to the resultant force acting on it and acts in the direction of the resultant force.

$$\Delta p/\Delta t \propto F$$

The definition of the size of the newton fixes the proportionality constant at one. This enables the second law to be written as:

Force = rate of change of momentum

$F = \Delta p/\Delta t$

Jet and rocket engines use the same principle; hot gases are ejected from the back of the engine to provide a force in the forwards direction. The difference is that jet engines are designed to operate within the Earth's atmosphere, so they can take oxygen from their surroundings.

In this form, the second law is useful for working out the force in situations such as jet and rocket propulsion where the change in momentum each second can be calculated easily. Note that the change in momentum, Δp, is called **impulse**.

A rocket carries its own oxygen supply, so that it can fire the engines when it is travelling in a vacuum. A spacecraft travelling where it is not affected by any gravitational fields or resistive forces maintains a constant velocity, so it only needs to fire the rocket engines to **change** speed or direction.

The useful materials carried by a rocket are its payload. Rockets must use a lot of fuel to lift a payload above the atmosphere. They have a low payload:fuel ratio.

The principle of **rocket propulsion** can be seen by blowing up a balloon and letting it go; the air squashed out of the neck of the balloon gains momentum as it leaves, causing the balloon and the air that remains inside to gain momentum **at the same rate but in the opposite direction**. The diagram shows the principle of rocket propulsion.

The unit kg m s^{-2} is equivalent to the N, so a rate of change of momentum of 4 500 000 kg m s^{-2} is another way of stating: a force of 4 500 000 N.

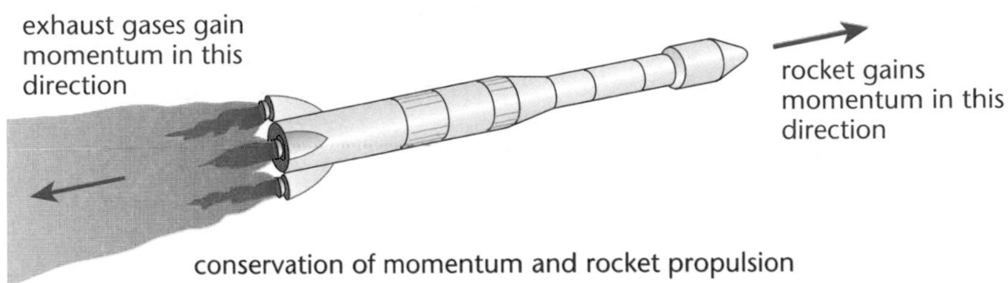

conservation of momentum and rocket propulsion

The equation $F = \dfrac{\Delta p}{\Delta t}$ is the same as the equation F = ma. The latter equation shows that for a body of fixed mass, the force and acceleration vary in fixed proportion.

In this example, the exhaust gases are emitted at a rate of 3000 kg s^{-1}. The speed of the exhaust gases relative to the rocket is 1500 m s^{-1}. The rate of change of momentum of these gases is therefore 4 500 000 kg m s^{-2}, i.e. 4 500 000 kg m s^{-1} each second. The size of the force on both the gases and the rocket is 4 500 000 N.

Progress check

1 Two vehicles fitted with repelling magnets are held together on an air track. The masses of the vehicles are 0.10 kg and 0.15 kg.

The vehicles are released. The 0.10 kg vehicle moves to the left with a speed of 0.24 m s^{-1}. Calculate the velocity of the heavier vehicle.

2 A ball of mass 0.020 kg hits a wall at a speed of 12.5 m s^{-1} and rebounds at a speed of 10.0 m s^{-1}.

a Calculate the change in momentum of the ball.

b If the ball is in contact with the wall for 0.010 s, calculate the size of the force between the ball and the wall.

c Explain whether this collision is elastic or inelastic.

1 0.16 m s^{-1} to the right

2 a 0.45 N s in the direction of the rebound

b 45 N

c Inelastic. The ball has less kinetic energy after rebound than before it hits the wall.

1.3 Motion in a circle

After studying this section you should be able to:

- *use the radian to measure angular displacement*
- *describe the forces on an object moving in a circle*
- *recall and use the expressions for centripetal force and centripetal acceleration*
- *understand rotational motion through moments of inertia*

LEARNING SUMMARY

Measuring angles

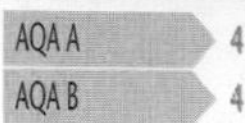

When an object travels through a complete circle it moves through an angle of 360°, although the distance it travels depends on the radius of the circle. When a train or motor vehicle travels round a bend, the outer wheels have further to travel than the inner wheels. Similarly, when a compact disc is being played all points on the disc move through the same angle in any given time, but the linear speed at any point depends on its distance from the centre.

An alternative unit to the degree for measuring angular displacement is the **radian**, defined as the arc length divided by the radius of the circle. This is shown in the diagram.

Angle θ is 1 radian when $x = r$.

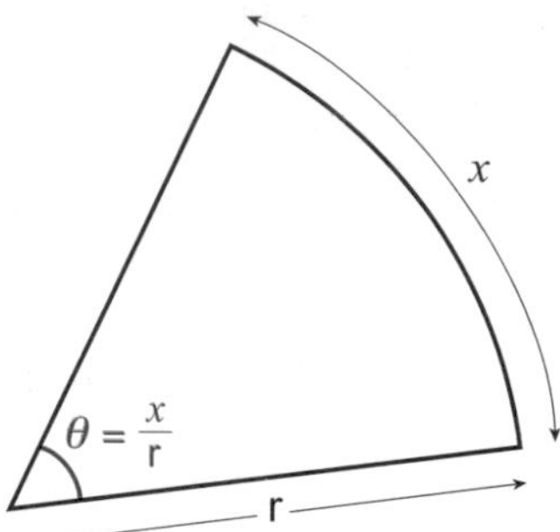

Measuring an angle in radians

To convert an angle in degrees to one in radians, multiply by π and divide by 180.

It follows that movement through one complete circle, or 360°, is equal to a displacement of θ = circumference ÷ radius = $2\pi r \div r = 2\pi$ radians.

Speed in a circle

When an object moves in a circular path its velocity is always at right angles to a radial line, a line joining it to the centre of the circle.

Two objects, such as points on a CD, at different distances from the centre each complete the same number of revolutions in any specified time but they have different speeds. They have the same **angular velocity** but different linear velocities.

Angular velocity is a vector quantity that can have one of two directions: clockwise or anticlockwise.

> Angular velocity, ω, is defined as:
>
> *the rate of change of angular displacement with time.*
>
> It is represented by the gradient of an angular displacement–time graph.
>
> average angular velocity = angular displacement ÷ time
>
> $$\omega = \Delta\theta/\Delta t$$
>
> where $\Delta\theta$ is the angular displacement in time Δt.
>
> KEY POINT

The relationship between angular velocity and linear velocity for an object moving in a circle of radius r is:

$$v = r\omega.$$

The **time period** for one revolution, T, is equal to the angular displacement (2π radians) divided by the angular velocity:

$$T = 2\pi/\omega.$$

The frequency of a circular motion, f, is equal to 1/T:

$$f = 1/T = \omega/2\pi.$$

KEY POINT

$$\omega = 2\pi f$$

Centripetal acceleration

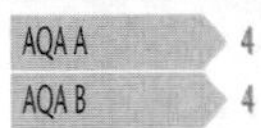

Any object that changes its velocity is accelerating. Since velocity involves both speed and direction, a change in either of these quantities is an acceleration. An object moving at a constant speed in a circle is therefore accelerating as it is continually changing its direction. The vector diagram below shows the change in velocity, Δv, for an object following a circular path. This change in direction is always towards the centre of the circle.

Note that Δv is the change in velocity from v_1 to v_2, so $v_1 + \Delta v = v_2$

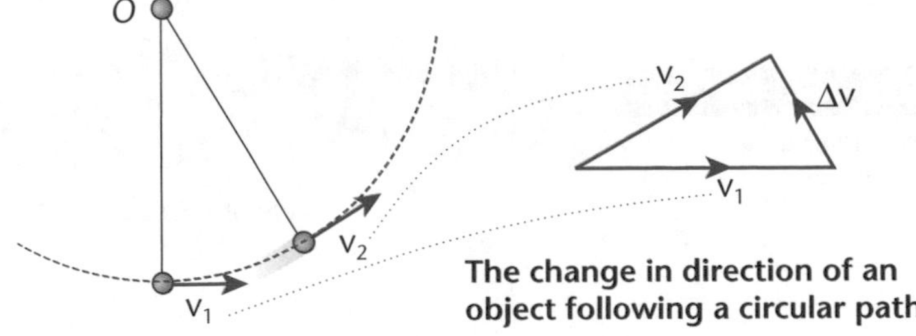

The change in direction of an object following a circular path

The acceleration of an object moving in a circle:

- is called **centripetal acceleration**
- is directed towards the centre of the circle.

KEY POINT

Centripetal acceleration, a, is the acceleration of an object moving in a circular path.

$$a = v^2/r = r\omega^2$$

where v is the linear velocity, ω is the angular velocity and r is the radius of the circle.

The unbalanced force

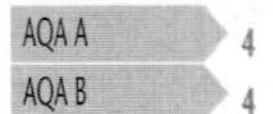

To make an object accelerate, there has to be an unbalanced force acting in the same direction as the acceleration (see Newton's first law). In the case of circular motion the unbalanced force is called the **centripetal force**.

It is important to understand that the centripetal force is the resultant of all the forces acting on an object. A common misconception is that it is an extra force that appears when an object moves in a circular path.

When a train travels around a bend, the centripetal force comes from the push of the outer rail on the wheel. For a road vehicle such as a car, bus or bicycle it is the push of the road on the tyres; if there is insufficient friction due to ice or a slippery surface, the vehicle does not complete the turn and may leave the road.

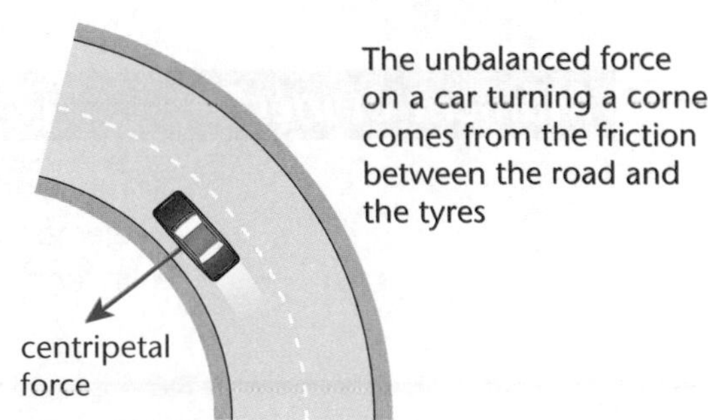

The unbalanced force on a car turning a corner comes from the friction between the road and the tyres

The size of the unbalanced force required to maintain circular motion can be calculated using F = ma:

> An object moving at constant speed in a circular path requires an unbalanced force towards the centre of the circle.
>
> centripetal force, $F = mv^2/r = mr\omega^2$
>
> KEY POINT

The theme park ride

AQA A 4
AQA B 4

You feel the apparent change in weight when you travel in a lift. As the lift accelerates upwards you feel heavier because the normal contact force is greater than your weight.

Some theme park rides involve motion in a vertical circle. Part of the thrill of being on one of these rides is the apparent change in weight as you travel round the circle. This is because the force that you feel as you sit or stand is not your weight, but the normal contact force pushing up. If the forces on you are balanced, this force is equal in size to your weight, but it can become bigger or smaller if the forces are unbalanced, causing you to feel heavier or lighter.

Key points from AS

- **Vector diagrams**
  Revise AS section 1.1

The diagram below shows the forces acting on a person moving in a vertical circle when they are at the top and bottom of the circle.

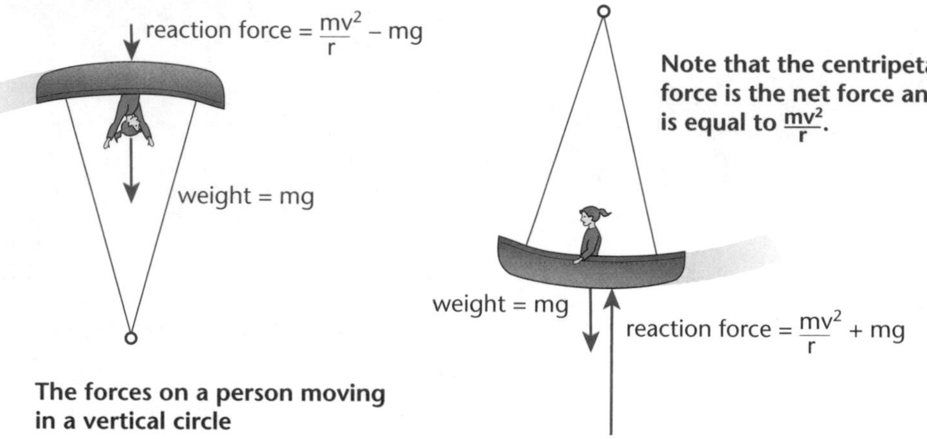

The forces on a person moving in a vertical circle

Check that in each case the resultant of the forces is equal to mv^2/r towards the centre of the circle.

At the bottom of the circle:

- The person feels heavier than usual as the size of the normal reaction force is equal to the person's weight plus the centripetal force required to maintain circular motion.

At the top of the circle:

- The person may feel lighter than usual as the person's weight is providing part of the centripetal force required to maintain circular motion. The rest comes from the normal reaction force, which is less than the person's weight if $mv^2/r < 2mg$.
- If the speed of the ride is such that the required centripetal force is the same as the person's weight then he or she feels 'weightless' as the normal reaction force is zero.

Another possibility is that the speed of the ride is such that the required centripetal force is less than the person's weight. In this case a safety harness is required to stop the person from falling out.

Moment of inertia

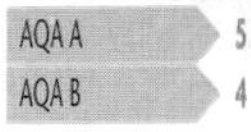

Moment of inertia is a measure of the resistance of a body to rotational (angular) acceleration. As such, it is comparable to mass of a body as a measure of resistance to linear acceleration.

$$\text{Note that mass, (m)} = \frac{\text{force}}{\text{linear acceleration}} = \frac{F}{a}$$

$$\text{and moment of inertia, (I)} = \frac{\text{torque}}{\text{angular acceleration}} = \frac{T}{\alpha}$$

where torque (T) = force (F) × perpendicular distance from force to axis of rotation (r) and angular acceleration (α) = $\Delta\omega/\Delta t$

The moment of inertia (I) depends on the mass (and distribution of mass) of the body (m) and its distance from the axis of rotation (r). $I = mr^2$. Note that the unit of moment of inertia is kg m^2

For a mass m in circular motion, r is simply the radius of orbit.

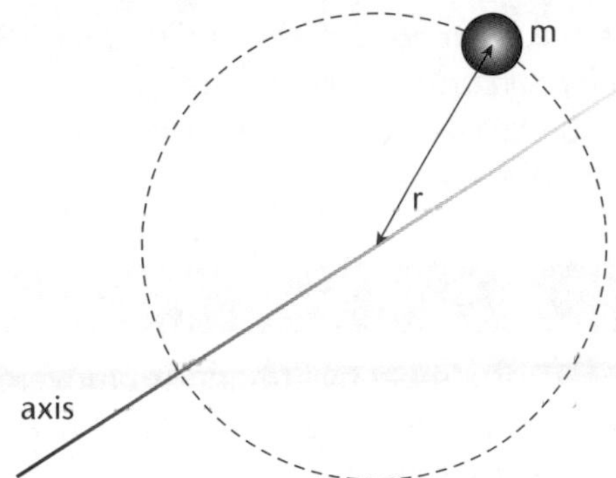

For a symmetrical cylinder rotating about a central axis, $I = \frac{1}{2}mr^2$

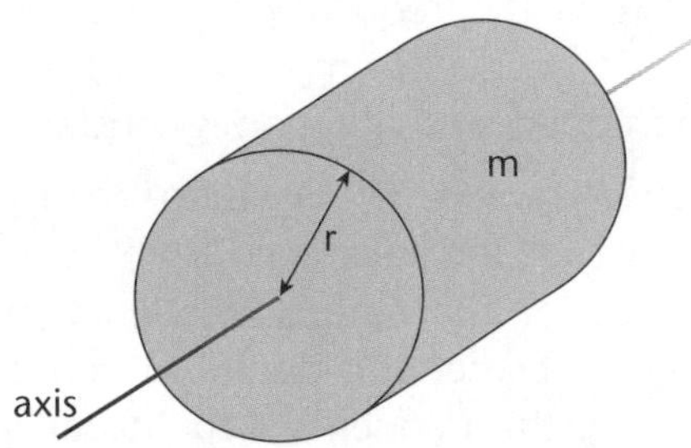

The expressions for the moments of inertia of different shaped bodies will be provided in an examination.

The equations of motion of linear dynamics have an exact analogy when dealing with rotating bodies subject to a uniform angular acceleration. The table shows the relationship between linear motion and rotational motion,

Linear motion	*Rotational motion*
$v = u + at$	$\omega_2 = \omega_1 + \alpha t$
$s = ut + \frac{1}{2}at^2$	$\theta = \omega_1 t + \frac{1}{2}\alpha t^2$
$v^2 = u^2 + 2as$	$\omega_2^2 = \omega_1^2 + 2\alpha\theta$
$s = \frac{1}{2}(u + v)t$	$\theta = \frac{1}{2}(\omega_1 + \omega_2)t$

All rotating bodies conserve angular momentum just as bodies in linear motion conserve linear momentum. Hence,

Angular momentum (L) $= I\omega$

Rotating bodies also possess angular kinetic energy and this is given by:

Angular kinetic energy $= \frac{1}{2}I\omega^2$

The power (rate of doing work) is given by the equation, $P = T\omega$

Progress check

1 a Convert into radians:
i 90°
ii 720°
b An object is displaced through an angle of 8π radians. Explain why it has returned to its original position.

1 a i $\pi/2$ **ii** 4π **b** It has travelled four complete revolutions.

1.4 Oscillations, resonance and damping

After studying this section you should be able to:

LEARNING SUMMARY

- *explain the difference between a free and a forced vibration*
- *describe resonance and how it occurs*
- *distinguish between a damped oscillation and an undamped one and give examples of each*

Oscillations

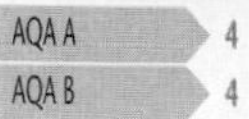

An **oscillation** is a repetitive to-and-fro movement. Loudspeaker cones oscillate, as do swings in a children's playground. However, there is an important difference between these oscillations.

The terms 'oscillation' and 'vibration' have the same meaning. They are both used to describe a regular to-and-fro movement.

The loudspeaker cone is an example of a **forced vibration**; the cone is forced to vibrate at the frequency of the current that passes in the coil. The amplitude of vibration depends on the size of the current. When reproducing speech or music, the frequency of vibration is constantly changing.

A child's swing, like a string on a piano or a guitar is an example of a **free vibration**. Once displaced from its normal position, it vibrates at its own **natural frequency**. The natural frequency of vibration of an object is the frequency it vibrates at when it is displaced from its normal, or rest, position and released. For a swing, this frequency depends only on the length of the chain or rope; in the case of a stringed instrument it is also affected by the tension and mass per unit length of the string.

The vibration of particles in a solid is a free vibration. Like a swing, the energy of the particles changes from kinetic to potential (stored) energy.

- In a free vibration there is a constant interchange between potential (stored) energy and kinetic energy.
- A swing has zero kinetic energy and maximum gravitational potential energy when at its greatest displacement; the kinetic energy is at a maximum and the gravitational potential energy at a minimum when the displacement is zero. This is shown in the diagram.

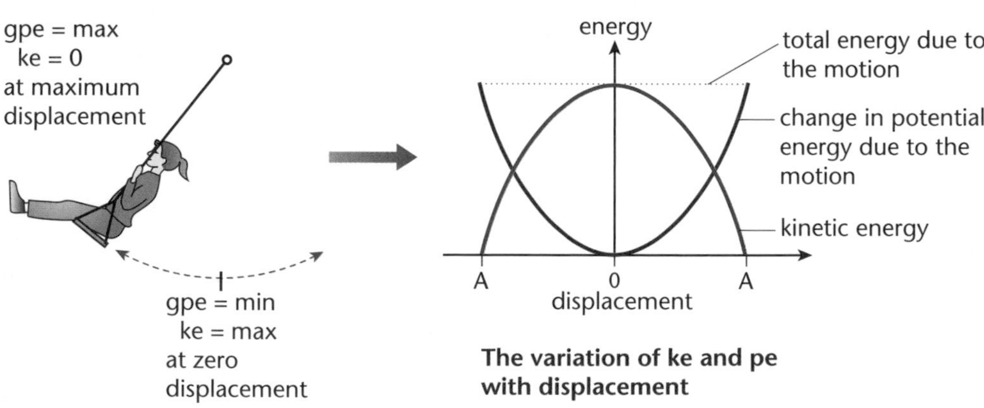

The variation of ke and pe with displacement

total energy = $E_k + E_p$

Resonance

AQA A 4
AQA B 4

After a swing is given a push, it vibrates at its natural frequency. To make it swing higher, subsequent pushes need to coincide with the vibrations of the swing. This is an example of **resonance**, the large amplitude oscillation that occurs when the frequency of a forced vibration is equal to the natural frequency of the vibrating object.

There are many everyday examples of resonance. You may have noticed windows vibrating due to the sound from a bus waiting at a stop. Electrical resonance is used to tune in a radio or television to a particular station.

The diagram shows how a vibration generator can be used to force a string to vibrate at any frequency, and the large amplitude vibration that occurs when the driving frequency coincides with the natural frequency of the string, also called its **resonant frequency**. At this frequency the string is able to absorb sufficient energy from the vibration generator to replace the energy lost due to resistive forces.

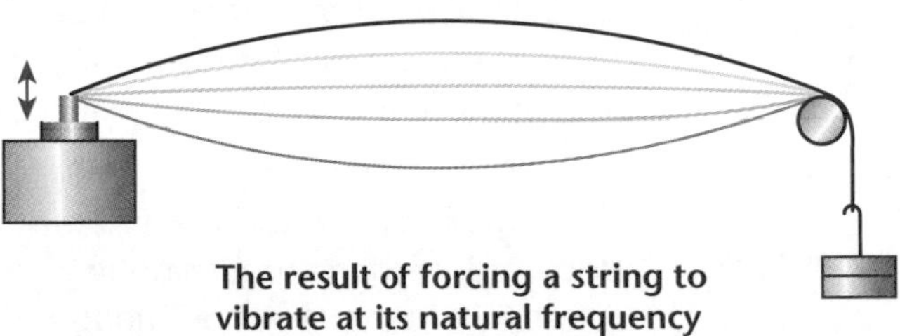

The result of forcing a string to vibrate at its natural frequency

If the frequency of the forced oscillation is increased slowly from zero, the amplitude of vibration of the string starts to increase at a frequency below the resonant frequency. It has its maximum value at this frequency and it then decreases again as the frequency continues to increase. The variation of amplitude around the resonant frequency is shown below.

Vibration generator frequency

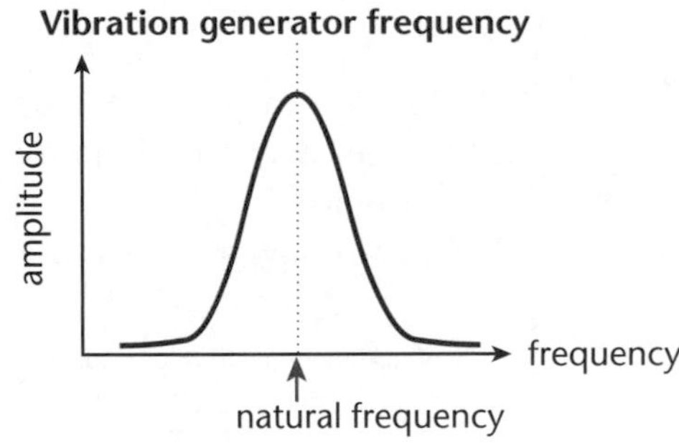

The sharpness of the peak depends on the resistive forces acting. The greater the resistive force, the smaller the amplitude of vibration.

These are some practical examples of resonance.

- Microwave cookers heat food by causing water molecules in the food to resonate, absorbing the energy from the microwaves.
- Magnetic resonance imaging (MRI) is used in medicine to produce a detailed picture of the inside of the body by making the atomic nuclei resonate.
- Resonance in mechanical systems can be a nuisance or a danger. When the frequency of the engine vibrations in a vehicle matches the natural frequency of the exhaust system the result is a rattle as the exhaust system vibrates against the body of the vehicle.
- Resonance of parts of aircraft can cause failure due to excessive stress and several helicopter crashes have been attributed to resonance of the pilot's eyeballs resulting in the pilot being unable to see overhead power lines.

Damping

4
4

Just as for linear motion, all mechanical oscillations are subject to resistive forces. The effect of resistive forces in removing energy from a vibrating object is known as **damping**.

A string or mass on a spring vibrating in air is **lightly damped** as the main resistive force is air resistance. The effect of light damping is to gradually reduce the energy, and therefore amplitude, of the vibrating object.

A car mechanic tests the shock absorbers by pushing down on the car bonnet. If the car oscillates, the shock absorbers need replacing.

The body of a motor vehicle is connected to the wheels by springs. When the vehicle goes over a bump in the road the springs compress, giving the passengers a smoother ride. If motor vehicles were lightly damped they would continue to oscillate after going over a bump. Shock absorbers increase the resistive force, so that when the body of the vehicle has been displaced it returns to its original position without oscillating. A motor vehicle is **critically damped** so that it returns to the normal position in the minimum time to avoid vibrations that could cause 'car-sickness' and possibly loss of control by the driver.

Very large resistive forces result in **heavy damping**. Imagine a mass on a spring suspended so that the mass is in a viscous liquid such as syrup. If the mass is displaced, the force opposing its movement is very large and it takes a long time to return to the normal position.

The effect of damping is to transfer the energy from the vibrating object to heat the surroundings.

One effect of damping is to decrease the effects of resonance. The greater the damping, the smaller the amplitude of vibration at the resonant frequency. This is very important in mechanical structures such as bridges which can resonate due to the effects of wind. Resonance caused the suspension bridge built over the Tacoma Narrows, in America, to collapse in 1939. Since then, engineers have designed suspension bridges with shock absorbers built into the suspension to absorb the energy and prevent excessive vibration.

Progress check

1 Describe the difference between a free vibration and a forced vibration.
2 Explain why a washing machine sometimes vibrates violently when the drum is spinning.
3 Explain how a free vibration is affected by the amount of damping.

1 A free vibration occurs when an object is displaced and it vibrates at its natural frequency.
A forced vibration can be at any frequency and is caused by an external oscillating force.
2 The washing machine has a natural frequency of vibration. If the drum rotates at this frequency it causes the machine to resonate.
3 Damping removes energy from a vibrating object. Increasing the damping reduces the amplitude of vibration and may stop any vibration from occurring.

1.5 Simple harmonic motion

After studying this section you should be able to:

- explain what is meant by simple harmonic motion
- apply the equations of simple harmonic motion
- sketch graphs showing the change in displacement, velocity and acceleration in simple harmonic motion

LEARNING SUMMARY

What is simple harmonic motion?

As the name implies, it is the simplest kind of oscillatory motion. One set of equations can be used to describe and predict the movement of any object whose motion is simple harmonic.

The motion of a vibrating object is **simple harmonic** if:

- its acceleration is proportional to its displacement
- its acceleration and displacement are in opposite directions.

The resultant force on an object in simple harmonic motion is a restoring force since it is always directed towards the equilibrium position that the object has been displaced from.

The second bullet point means that the acceleration, and therefore the resultant force, always acts towards the equilibrium position, where the displacement is zero.

Common examples of simple harmonic motion (often abbreviated to shm) include the oscillations of a simple pendulum (provided the amplitude is small, equivalent to an angular displacement of less than about 10°) and those of a mass suspended vertically on a spring.

The diagram shows how accelerations are related to displacements, x, from equilibrium positions.

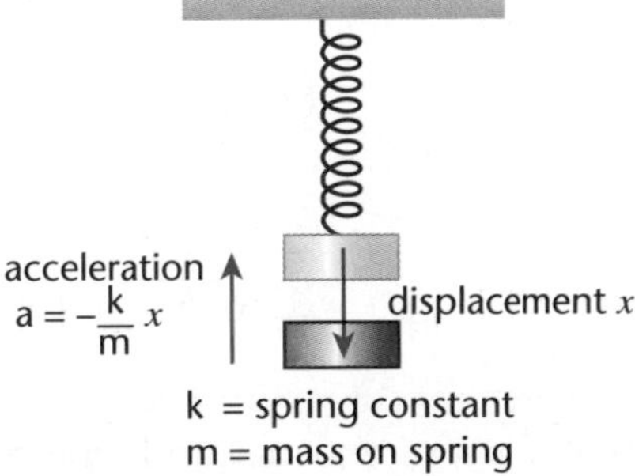

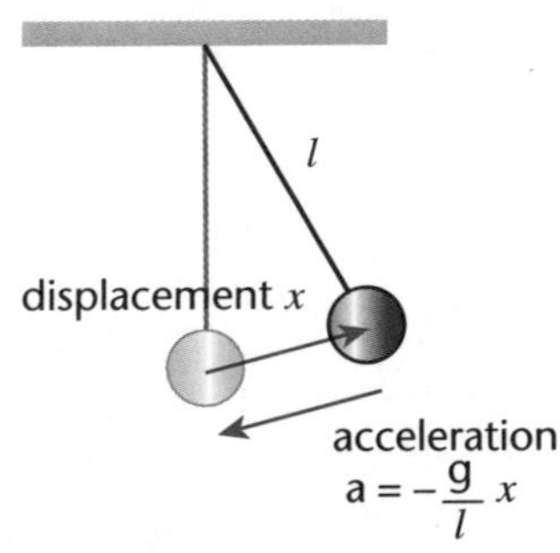

k, the spring constant, is defined as the force per unit extension $k = F/x$

- the numerical value of the acceleration is equal to a constant multiplied by the displacement, showing that acceleration is proportional to displacement
- the negative value of the acceleration shows that it is in the opposite direction to the displacement.

Frequency of simple harmonic motion

KEY POINT

The acceleration of any object whose motion is simple harmonic is related to the frequency of vibration by the equation:

$$a = -(2\pi f)^2 x$$

where a is the acceleration, f is the frequency and x is the displacement.

Since $(2\pi f)^2$ is always positive, a and x have the opposite signs so they are always in opposite directions.

Simple harmonic motion is like circular motion in that it repeats over and over again – it is cyclic. The mathematics of circular motion is therefore useful for describing and predicting simple harmonic motion. As for circular motion, the 'angular velocity' of simple harmonic motion is written as ω (omega), and is equal to $2\pi f$. So the above equation can also be written as:

$$a = -\omega^2 x$$

The remainder of this section uses $2\pi f$.

A graph of acceleration against displacement is therefore a straight line through the origin, as shown in the diagram. The gradient of the graph has a negative value and is numerically equal to $(2\pi f)^2$.

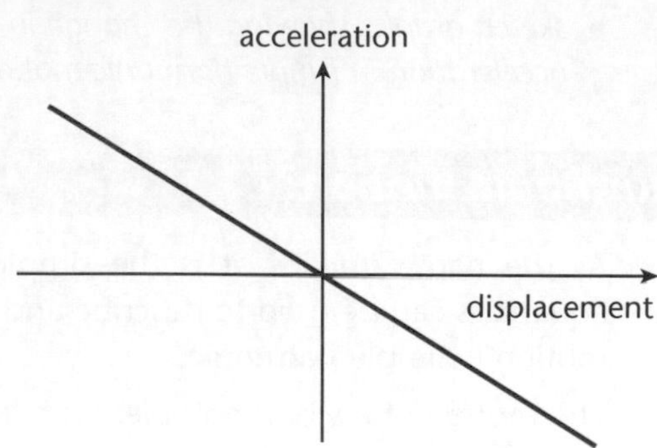

The constants in the equations for the acceleration of the mass on a spring (constant = k/m) and the pendulum bob (constant = g/l) give useful information about the frequency and time period of the oscillation.

For a mass on a spring:

$$(2\pi f)^2 = k/m$$

So:

> The relationship between the frequency, f, and the time period, T, is $f = l/T$ so each of these quantities is the reciprocal of the other.

KEY POINT

$$f = \frac{1}{2\pi}\sqrt{\frac{k}{m}} \text{ and } T = 2\pi\sqrt{\frac{m}{k}}$$

For a simple pendulum:

$$(2\pi f)^2 = g/l$$

So:

KEY POINT

$$f = \frac{1}{2\pi}\sqrt{\frac{g}{l}} \text{ and } T = 2\pi\sqrt{\frac{l}{g}}$$

These equations for frequency and time period show that:

> The spring constant, k, is a measure of the stiffness of a spring.

- increasing the mass on a spring causes the frequency to decrease and the time period to increase
- for a given mass, the stiffer the spring the higher the frequency of oscillation
- the frequency and time period of a pendulum do not depend on the mass of the pendulum bob, as m does not appear in the equation
- increasing the length of a pendulum causes the frequency to decrease and the time period to increase.
- sets of measurements of T or f against l will provide an accurate determination of g.

Displacement and time

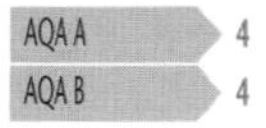

The displacement of an object in simple harmonic motion varies sinusoidally with time. This means that a displacement–time graph has the shape of a sine or cosine curve, and the displacement at any time, t, can be written in terms of a sine or cosine function.

The symbol x_0 is also used as an alternative to A to represent amplitude.

KEY POINT

The displacement of an object in simple harmonic motion can be calculated using either of the equations:

$$x = A \sin 2\pi ft$$

$$x = A \cos 2\pi ft$$

where A is the amplitude of the motion.

The graphs below show the variation of displacement with time for a simple harmonic motion.

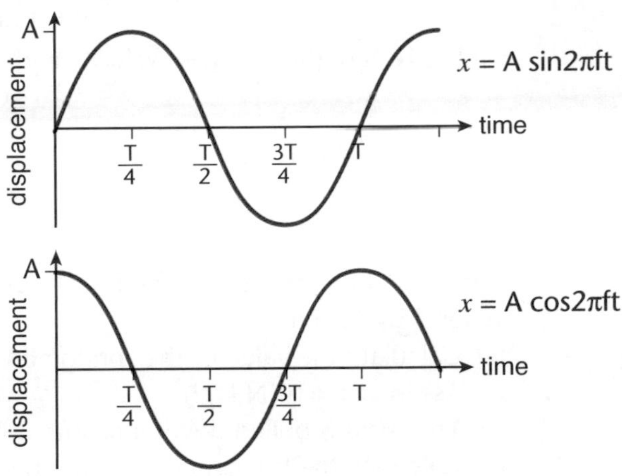

Sine and cosine are identical functions with a 90°, or $\pi/2$ radians, phase difference.

Which equation and graph apply to a simple harmonic motion depends on the displacement when the time is zero.

- If the displacement is zero at time t = 0, then the sine applies.
- If the displacement is at the maximum at time t = 0, then the cosine applies.

Displacement, velocity and acceleration

AQA A 4
AQA B 4

These bullet points apply to any motion.

Displacement, velocity and acceleration are linked graphically because:

- the gradient of a displacement–time graph represents velocity
- the gradient of a velocity–time graph represents acceleration.

The following diagram shows the variation of all three variables for a simple harmonic motion where the variation of the displacement with time is represented by a sine function.

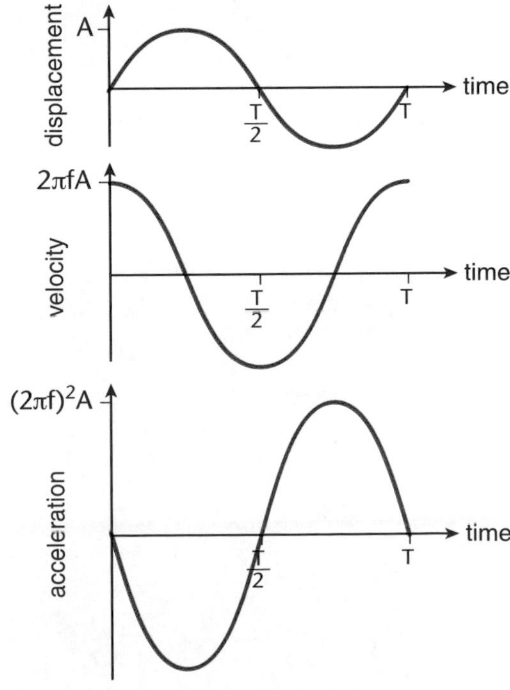

Note that the gradient of the displacement–time graph equals velocity. For example, where the gradient of the displacement–time graph is zero then the velocity is zero. Similarly, the gradient of the velocity–time graph is equal to acceleration. So, for example, when the gradient of the velocity–time graph is large and positive, then the acceleration is large and positive.

These graphs show that:

> These statements about the acceleration follow from the definition of simple harmonic motion.

- when the displacement is zero the velocity is at a maximum and the acceleration is zero
- when the displacement has its maximum value the velocity is zero and the acceleration is a maximum in the opposite direction.

Knowledge of the frequency and amplitude of a simple harmonic motion enables the maximum speed and the velocity at any displacement to be calculated.

> The ± symbol is used because at any value of the displacement the velocity could be in either direction.

KEY POINT

The maximum speed of an object in simple harmonic motion is:

$$v = \pm 2\pi f A$$

The relationship between velocity and displacement is:

$$v = \pm 2\pi f \sqrt{(A^2 - x^2)}$$

Progress check

When a 100 g mass is attached to the bottom of a vertical spring it causes it to extend by 5.0 cm.

a Calculate the value of the spring constant, k.
Assume g = 10 N kg^{-1}.
The mass is pulled down a further 3 cm and released.

b Calculate the frequency and time period of the resulting oscillation.

c i What is the value of the amplitude of the oscillation?
ii Calculate the maximum speed of the mass.

a 20 N m^{-1}
b 2.25 Hz and 0.44 s
c i 3 cm
ii 0.42 m s^{-1}

1.6 Relativity

After studying this section you should be able to:

- *recall the postulates of special relativity*
- *calculate the amount of time dilation and length contraction when a moving object is observed*
- *explain why there is a maximum speed to which an object can be accelerated*

LEARNING SUMMARY

Does time pass at a constant rate?

In everyday life, we sense and judge speed relative to the surface of the Earth or vehicles that we travel in. If you are in a car travelling at 25 m s^{-1} and another car travelling at the same speed is approaching you, then its speed relative to your vehicle is 50 m s^{-1}. If, on the other hand, it is in front of you and travelling in the same direction, its speed relative to you is zero.

Each car has its own **frame of reference** and a pedestrian on the roadside has a different frame of reference again. Measurements of the same motion from different frames of reference can produce different answers. **Special relativity** considers only frames of reference that are not accelerating.

> The laws of physics apply equally in all non-accelerating frames of reference.
> This is the first postulate (assumption) of Einstein's theory of special relativity.
>
> KEY POINT

In the diagram, the driver of the red car sees the blue car approaching at 50 m s^{-1} but, from the frame of reference of the stationary observer, the blue car is approaching at 30 m s^{-1}.

What about the light that passes between the two vehicles? Are the different measurements of its speed by the drivers and pedestrian affected by their motion? For anything else the answer would be 'yes', but for light the answer is 'no'.

Light is special – it does not behave like objects such as cars. Measurements of its speed in air or in a vacuum all give the same answer, whether or not there is relative motion of the observer and the source of light.

> The speed of light in a vacuum always has the same value in any frame of reference, whether or not the light originated in another frame of reference.
> This is the second postulate of Einstein's theory.
>
> KEY POINT

The invariance of the speed of light was verified experimentally by Michelson and Morley using a very sensitive interferometer. An interferometer is a device used to measure wavelength extremely accurately by the production of interference fringes. For absolute motion to be confirmed there would be no shift in the interference fringes and despite extensive and exhaustive measurements over a prolonged period of time (different days, different times of the year) no shift was ever observed. The conclusion reached meant that the speed of light has the same value for all observers whether they are moving or not.

The invariance in the speed of light leads to the conclusion that neither time nor distance have absolute values. For most people this is a surprising conclusion – the idea that there is no single universal time but that measurement of time in other frames of reference can vary when there is relative motion is an idea that takes some getting used to. Each of the drivers of the cars in the diagram would notice that time passes ever so slightly more slowly in the other vehicle. This effect is known as **time dilation**, and it is an effect that increases as relative speed increases. We do not normally notice the effects of time dilation because we move at speeds very much less than the speed of light.

Far more muons reach the surface of the Earth than is predicted by the half-life of 2 microseconds. Muons created in cosmic ray showers move through the atmosphere at speeds which are within 1% of that of the speed of light. Due to time dilation effects the apparent half-life of a muon increases to around 60 microseconds as measured by an observer on Earth.

As well as having different measurements of time, the car drivers in the diagram have different measurements of distance. Each driver sees the other car as being shorter than its driver sees it; this is known as **length contraction**. Length contraction only occurs in the direction of motion.

> **KEY POINT**
>
> The factor by which time is dilated and length is contracted is $\sqrt{1-\frac{v^2}{c^2}}$,
>
> where v is the speed of the object relative to the observer and c is the speed of light.

It follows that time t_0 and length l_0 measured within a frame of reference are observed by someone moving relative to that frame to have values:

- $t = \dfrac{t_0}{\sqrt{\left(1 - \frac{v^2}{c^2}\right)}}$
- $l = l_0\sqrt{\left(1 - \frac{v^2}{c^2}\right)}$

How is mass affected?

A consequence of length contraction is that as an object speeds up under the action of a constant force, its apparent acceleration decreases. This is matched by an apparent increase in mass.

For everyday objects moving at everyday speeds, there is no significant increase in mass.

> **KEY POINT**
>
> The apparent mass, m, is given by the expression
>
> $$m = \frac{m_0}{\sqrt{(1 - v^2/c^2)}}$$
>
> where m_0 is the **rest mass**, the mass of the object measured within its frame of reference.

This expression shows that the mass of an object tends towards infinity as its speed approaches the speed of light. It is therefore impossible to accelerate an object beyond the speed of light – the maximum possible speed.

Progress check

1 A muon has a half-life of 2 microseconds, measured within its frame of reference. Calculate the observed half-life when the muon moves at 0.99c.

2 Explain why a moving spacecraft appears shorter but not narrower to an external observer.

1 14 microseconds

2 Length contraction only occurs in the direction of motion.

Sample question and model answer

In this question, take the value of g, free-fall acceleration, to be 10 m s^{-2}.

A toy vehicle, mass 60 g, is at rest at the foot of a slope.

A second vehicle, mass 90 g, which is moving at a speed of 1.20 m s^{-1} hits the back of the stationary vehicle and they stick together.

When objects stick together, there is always some energy transfer from kinetic energy to internal energy, as some deformation occurs.

(a) (i) What type of collision occurs? [1]

Inelastic — 1 mark

However, total energy is always conserved, no matter what type of collision.

(ii) Which of the following quantities are conserved in this collision? [1]

kinetic energy momentum total energy

Momentum and total energy. — 1 mark

Always write down the formula that you intend to use, and each step in the working. Otherwise, a simple error such as using the wrong mass results in no marks.

(b) (i) Calculate the momentum of the larger vehicle before the collision. [3]

momentum = mass × velocity — 1 mark

= 0.090 kg × 1.20 m s^{-1} — 1 mark

= 0.108 kg m s^{-1} — 1 mark

The first mark here is for correct transposition of the equation.

The second mark is for using the correct values.

(ii) Calculate the combined speed of the vehicles after the collision. [3]

velocity = momentum ÷ mass — 1 mark

= 0.108 kg m s^{-1} ÷ 0.150 kg — 1 mark

= 0.72 m s^{-1} — 1 mark

It is not essential to include the units at each stage of the calculation, but it is good practice and you MUST include the correct unit with your answer to each part.

(c) Calculate the vertical height of the vehicles' centre of mass when they come to rest on the slope. [5]

As the vehicles rise up the slope, kinetic energy is transferred to gravitational potential energy — 1 mark

$\frac{1}{2}mv^2 = mg\Delta h$ — 1 mark

$\Delta h = v^2 \div 2g$ — 1 mark

$= (0.72 \text{ m s}^{-1})^2 \div (2 \times 10 \text{ m s}^{-2})$ — 1 mark

$= 0.026$ m — 1 mark

Practice examination questions

Throughout this section, take the value of g, free fall acceleration, to be 10 m s^{-2}.

1

A child of mass 40 kg sits on a swing. An adult pulls the swing back, raising the child's centre of mass through a vertical height of 0.60 m. This is shown in the diagram.

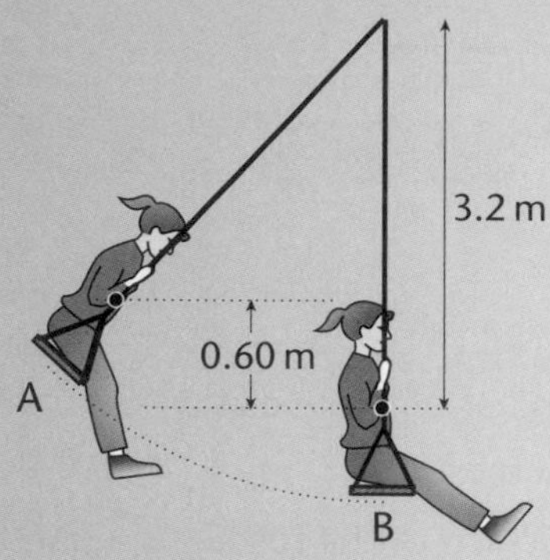

(a) Calculate the increase in gravitational potential energy when the child moves from position B to position A. [3]

(b) The child is released from position A.
Calculate her speed as she passes through position B. [3]

(c) Explain why there must be a resultant (unbalanced) force on the child as she swings through position B and state the direction of this force. [2]

(d) Calculate the size of the resultant force that acts on the child as she swings through position B. [3]

(e) One force that acts on the child at position B is the upward push of the swing.
Write a description of the other force that acts on the child. [2]

(f) Write down the value of the upward push of the swing when the child is:
(i) stationary at B.
(ii) moving through B. [2]

2

The diagram shows two 'vehicles' on an air track approaching each other. After impact they stick together.

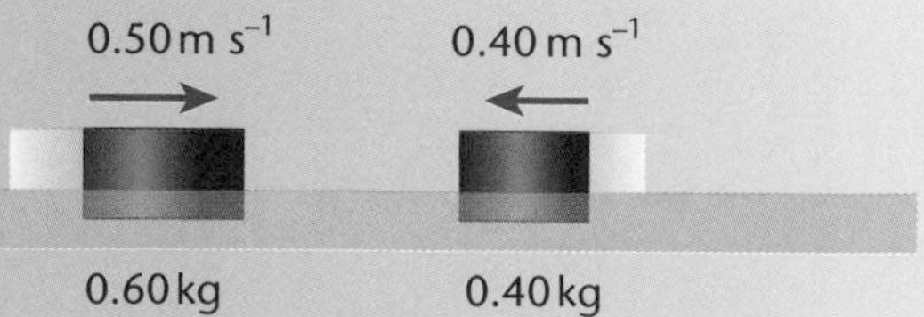

Practice examination questions (continued)

(a) Calculate the total momentum of the vehicles before the collision. [3]

(b) Calculate the velocity of the vehicles after the collision. State the direction of motion. [3]

(c) By performing appropriate calculations, explain whether the collision is elastic or inelastic. [3]

3

A communications satellite in a geostationary orbit (orbit time = 24 hours) goes round the Earth at a distance of 4.24×10^7 m from the centre of the Earth.

(a) Calculate the orbital speed of the communications satellite. [3]

(b) Calculate the centripetal acceleration of the satellite and state its direction. [4]

(c) What force causes this acceleration? [2]

4

A tennis ball has a mass of 60 g. A ball travelling at a speed of 15 m s^{-1} hits a player's racket and rebounds with a speed of 24 m s^{-1}. The ball is in contact with the racket for a time of 12 ms.

(a) Calculate the change in momentum of the ball. [3]

(b) Calculate the average force exerted by the racket on the ball. [4]

(c) Calculate the average force exerted by the ball on the racket. [1]

(d) Describe the energy transfers that take place while the ball is in contact with the racket. [3]

5

The diagram shows a tractor being used to pull a log.

The tension in the steel rope is 600 N.

(a) Calculate the horizontal component of the forwards force on the log. [2]

(b) What other horizontal force acts on the log and what is the size of this force? [2]

(c) Calculate the work done on the log when it is pulled a distance of 250 m. [3]

(d) What happens to the energy used to pull the log? [2]

Practice examination questions (continued)

6

A car of mass 800 kg travels around a circular corner of radius 100 m at a speed of 14 m s^{-1}.

(a) Calculate the centripetal acceleration of the car. [3]

(b) Calculate the size of the force needed to cause this acceleration. [3]

(c) (i) Write a description of the force that causes this acceleration. [2]

(ii) What is the other force that makes up the pair of forces in the sense of Newton's third law? [2]

(d) Suggest why, in wet or icy conditions, cars sometimes leave the road when travelling round a bend. [2]

7

A jet aircraft has a mass of 500 000 kg when fully laden.
It accelerates from rest to a take-off speed of 60 m s^{-1} in 20 s.

(a) Calculate the momentum of the aircraft as it leaves the ground. [2]

(b) Calculate the mean force required to impart this momentum to the aircraft. [2]

(c) Suggest why the force produced by the engines during take-off is greater than the answer to (b). [2]

(d) How does the principle of conservation of momentum apply to this event? [2]

8

In a hydroelectric power station, water falls at the rate of 3.2×10^5 kg s^{-1} through a vertical height of 180 m before entering a turbine.

(a) Calculate the loss in gravitational potential energy of the water each second. [3]

(b) Assuming that all this energy is transferred to kinetic energy, calculate the speed of the water as it enters the turbines. [3]

(c) The water leaves the turbines at a speed of 3 m s^{-1}. Calculate the maximum energy transfer to the turbine each second. [2]

9

(a) Explain the meaning of the expression *natural frequency of vibration*. [2]

(b) In what circumstances does an object resonate? [2]

(c) The speed of sound in air is 340 m s^{-1}.

Sound waves with a wavelength of 0.15 m cause a glass tumbler to vibrate at its natural frequency.

Calculate the natural frequency of vibration of the tumbler. [3]

Practice examination questions (continued)

10

A mass is suspended on a spring. After it is given a small displacement its equation of motion is:

$$a = -kx/m$$

(a) State the meaning of the symbols used in this equation. [4]

(b) The diagram shows how the displacement of the mass changes during one cycle of oscillation.

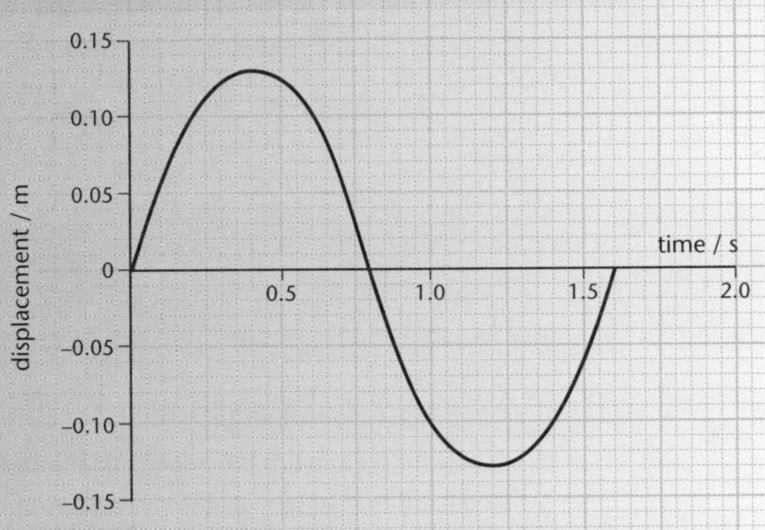

Calculate:

(i) the amplitude of the oscillation [1]

(ii) the frequency of the oscillation. [2]

(c) The mass on the spring is 0.50 kg. Calculate the spring constant. [3]

(d) (i) Calculate the maximum kinetic energy of the mass. [3]

(ii) At what point on its motion does the mass have its maximum kinetic energy? [1]

Practice examination questions (continued)

11

The graph shows how the displacement of a pendulum bob changes over one cycle of its motion.

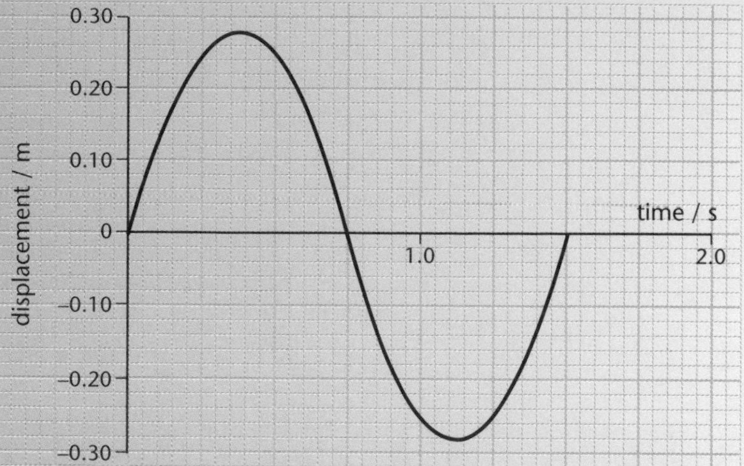

The equation that represents the variation of displacement with time is $x = A \sin 2\pi ft$.

(a) Write down the values of A and f. [2]

(b) Calculate the maximum velocity of the pendulum bob. [1]

(c) Calculate the velocity of the pendulum bob when its displacement is 0.20 m. [2]

Chapter 2

Fields

The following topics are covered in this chapter:

- *Gravitational fields*
- *Electric fields*
- *Capacitors and exponential change*
- *Magnetic fields*
- *Circular orbits*
- *Electromagnetic induction*

2.1 Gravitational fields

After studying this section you should be able to:

LEARNING SUMMARY

- *understand the concept of a field*
- *recall and use the relationship that describes the gravitational force between two masses*
- *describe the Earth's gravitational field and explain how the field strength varies with distance from the centre of the Earth*
- *understand gravitational field strength and gravitational potential as properties of points in the field*

Fields

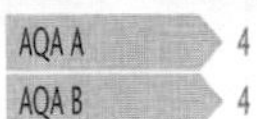

A field is a region of space where forces are exerted on objects with certain properties. In this and subsequent sections three types of field are considered:

- **gravitational fields** affect anything that has mass
- **electric fields** affect anything that has charge
- **magnetic fields** affect moving charge such as the electrons in the atoms of permanent magnets and electric currents.

These three types of field have many similar properties and some important differences. There are key definitions and concepts that are common to all three types of field.

Gravitational fields

Newton realised that all objects with mass attract each other. This seems surprising, since any two objects placed close together on a desktop do not immediately move together. The attractive force between them is tiny, and very much smaller than the frictional forces that oppose their motion.

The mass of the Earth is about 6×10^{24} kg.

Gravitational attractive forces between two objects only affect their motion when at least one of the objects is very massive. This explains why we are aware of the force that attracts us and other objects towards the Earth – the Earth is very massive.

The diagram represents the Earth's gravitational field. The lines show the direction of the force that acts on a mass that is within the field. Such lines are called field lines or lines of force.

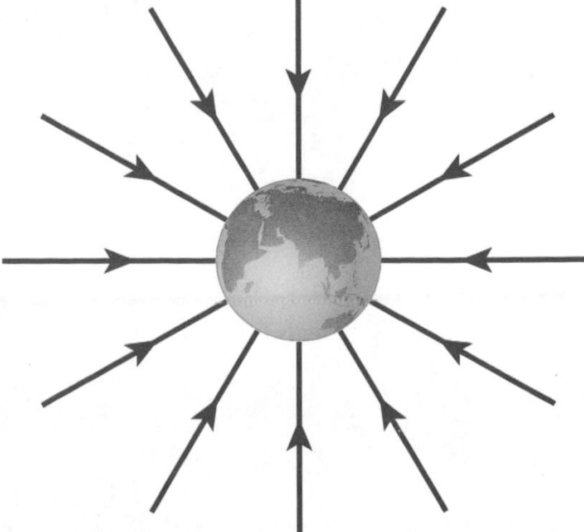

This diagram shows that:

- gravitational forces are always attractive – the Earth cannot repel any objects
- the Earth's gravitational pull acts towards the centre of the Earth
- the Earth's gravitational field is radial; the field lines become less concentrated with increasing distance from the Earth.

The less concentrated the field lines, the smaller the force. The force exerted on an object in a gravitational field depends on its position. If the **gravitational field strength** at any point is known, then the size of the force on a particular mass can be calculated.

Gravitational field strength is a vector quantity: its direction is towards the object that causes the field.

Always use the given value for g. Candidates can lose marks for using $10\,m\,s^{-2}$ if the formula sheet gives $g = 9.81\,m\,s^{-2}$.

KEY POINT

The gravitational field strength (g) at any point in a gravitational field is *the force per unit mass* at that point:

$$g = F/m$$

Close to the Earth's surface, g has the value of $9.81\ N\ kg^{-1}$, though the value of $10\ N\ kg^{-1}$ is often used in calculations.

Universal gravitation

AQA A 4
AQA B 4

In studying gravitation, Newton concluded that the gravitational attractive force that exists between any two masses:

- is proportional to each of the masses
- is inversely proportional to the square of their distances apart.

A point mass is one that has a radial field, like that of the Earth.

KEY POINT

Newton's law of gravitation describes the gravitational force between two point masses. It can be written as:

$$F = \frac{Gm_1m_2}{r^2}$$

where G is the universal gravitational constant and has the value

$$6.7 \times 10^{-11}\ N\ m^2\ kg^{-2}$$

and m_1 and m_2 are the values of the masses, and r is the separation of the centres of mass.

Remember that two objects attract *each other* with equal-sized forces acting in opposite directions as expected from Newton's third law.

Although the Earth is a large object, on the scale of the Universe it can be considered to be a point mass. The gravitational field strength at its centre is zero, since attractive forces pull equally in all directions. Beyond the surface of the Earth, the gravitational force on an object decreases with increasing distance. When the distance is measured from the centre of the Earth, the size of the force follows an **inverse square law**; doubling the distance from the centre of the Earth decreases the force to one quarter of the original value. The variation of force with distance from the centre of the Earth is shown in the diagram.

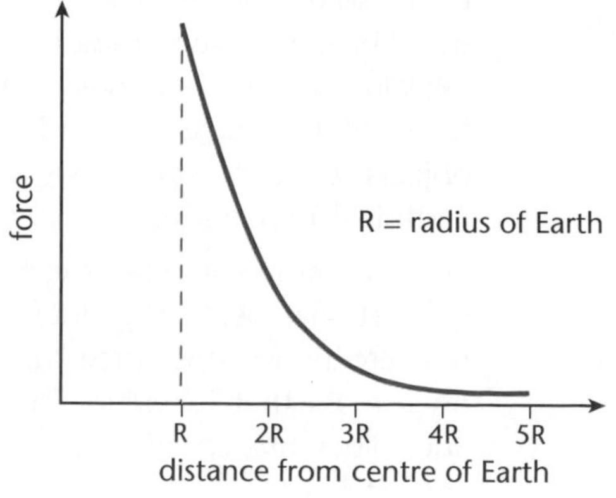

g and G

AQA A 4
AQA B 4

Newton's law of gravitation can be used to work out the value of the force between any two objects. It can also be used to calculate the strength of the gravitational field due to a spherical mass such as the Earth or the Sun.

A negative sign is sometimes used on the right-hand side of this equation, following the convention that attractive forces are given negative values and repulsive forces positive values.

KEY POINT

A small object, mass m, placed within the gravitational field of the Earth, mass M, experiences a force, F, given by

$$F = \frac{GMm}{r^2}$$

where r is the separation of the centres of mass of the object and the Earth.

It follows from the definition of gravitational field strength as the *force per unit mass* that the field strength at a point, g, is related to the mass of the Earth by the expression:

KEY POINT

$$g = \frac{F}{m} = \frac{GM}{r^2}$$

Gravitational field strength is a property of any point in a field. It can be given a value whether or not a mass is placed at that point. Like gravitational force, beyond the surface of the Earth the value of g follows an inverse square law. A graph of g against distance from the centre of the Earth has the same shape as that shown in the graph on page 46.

The radius of the Earth is about 6.4×10^6 m, so you would have to go much higher than aircraft-flying height for g to change by 1%.

Because the inverse square law applies to values of g when the distance is measured from the centre of the Earth, there is little change in its value close to the Earth's surface. Even when flying in an aircraft at a height of 10 000 m, the change in distance from the centre of the Earth is minimal, so there is no noticeable change in g.

The same symbol, g, is used to represent:

- gravitational field strength
- free-fall acceleration.

Gravitational field strength, g, is defined as the force per unit mass, g = F/m. From Newton's second law and the definition of the newton, free-fall acceleration, g, is also equal to the gravitational force per unit mass. The units of gravitational field strength, N kg^{-1}, and free-fall acceleration, m s^{-2}, are also equivalent.

Potential and potential energy

AQA A 4
AQA B 4

When an object changes its position relative to the Earth, there is a change in potential energy given by $\Delta E_p = mg\Delta h$. Δh is the change in height. It is not possible to place an absolute value on the potential energy of any object when h is measured relative to the surface of the Earth. Two similar objects placed at the top and bottom of a hill have different values of potential energy, but relative to the ground immediately around them the potential energy is zero for both objects, see diagram on the next page.

Absolute values of potential energy are measured relative to infinity. In this context, infinity means 'at a distance from the Earth where its gravitational field strength is so small as to be negligible'.

The cars would only have zero potential energy if an infinite distance from the Earth or any other object. Using the ground as the starting point we can only calculate *changes* in potential energy.

The car at the top of the hill has more potential energy than the one at the bottom, but relative to local ground level they both have zero.

Work has to be done to move an object from within the Earth's gravitational field to infinity.

Using infinity as the reference point:

- all objects at infinity have the same amount of potential energy, zero
- any object closer than infinity has a negative amount of potential energy, since it would need to acquire energy in order to reach infinity and have zero energy.

Just as every point in a gravitational field has a value of gravitational field strength, whether or not there is any body at that point, every point has a **gravitational potential**.

KEY POINT

The gravitational potential at a point in a gravitational field is the potential energy per unit mass placed at that point, measured relative to infinity.

So if the potential at any point in a field is known, the potential energy of a mass placed at that point can be calculated by multiplying the potential by the mass.

Calculating potential and potential energy

AQA A 4
AQA B 4

When an object is within the gravitational field of a planet, it has a negative amount of potential energy measured relative to infinity. The amount of potential energy depends on:

- the mass of the object
- the mass of the planet
- the distance between the centres of mass of the object and the planet.

The centre of mass of a planet is normally taken to be at its centre.

KEY POINT

The gravitational potential energy measured relative to infinity of a mass, m, placed within the gravitational field of a spherical mass M can be calculated using:

$$E_p = -\frac{GMm}{r}$$

where r is the distance between the centres of mass and G is the universal gravitational constant.

Gravitational potential energy is measured in joules (J).

To escape from a planet's surface a body must gain an amount of energy given by,

$E_p = GMm/r$

This can be done if the body has enough kinetic energy.

Kinetic energy required = $\Delta E_p = GMm/r$

$\frac{1}{2} mv^2 = GMm/r$

And so

$v = \sqrt{2GM/r}$

Since gravitational potential is the gravitational potential energy per unit mass placed at a point in a field, it follows that:

KEY POINT

Gravitational potential, V, is given by the relationship:

$$V = \frac{E_p}{m} = -\frac{GM}{r}$$

Gravitational potential is measured in J kg^{-1}.

The relationship between potential and potential energy is similar to that between gravitational force and gravitational field strength:

- potential and field strength are properties of a point in a field
- potential energy and force are the corresponding quantities for of a mass placed within a field.

Improving on an approximation

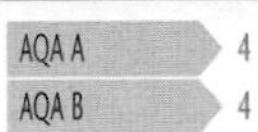

Note that the formula used for calculating changes in potential energy when a body changes its height above the surface of the Earth (or other body) is an approximation. That is,

$$\Delta E_p = mg\Delta h$$

makes an assumption that g is constant, which it is not. For small changes in height the equation works, since changes in g are then small. However, for large changes in height it is necessary to take account of the changes in g.

At lower height, the potential energy of a body-planet system can be written as:

$$E_{p1} = -GMm/r_1$$

And at greater height this becomes,

$$E_{p2} = -GMm/r_2$$

The change in potential energy in going from the lower height to the higher is then:

$$\Delta E_p = -GMm/r_2 - -GMm/r_1 = -GMm(1/r_1 - 1/r_2)$$

Equipotential surfaces

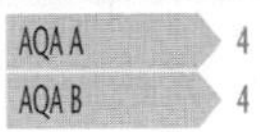

The potential energy of a satellite in a circular orbit around the Earth remains constant provided that its distance from the centre of the Earth does not change. To move to a higher or lower orbit the satellite must gain or lose potential energy. The satellite travels along an **equipotential surface**, the spherical shape consisting of points all at the same potential.

For a satellite in an elliptical orbit, there is an interchange between kinetic and potential energy as it travels around the Earth.

The diagram shows the spacing of equipotential surfaces around the Earth. The surfaces are drawn at equal differences of potential.

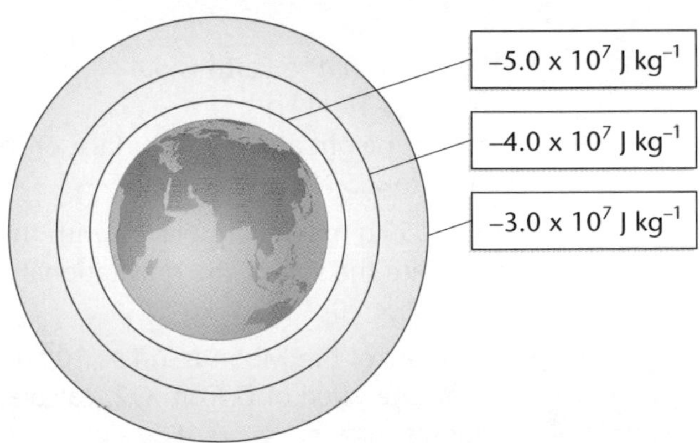

The diagram shows that:

- equipotential surfaces around a spherical mass are also spherical
- the spacing between the equipotential surfaces increases with increasing distance from the centre of the Earth.

To move to a lower orbit, a rocket can lose energy by firing the rocket engines 'backwards' so that the exhaust gases are expelled in the direction of motion.

For a **satellite** to move from an orbit where the potential is -4.0×10^7 J kg^{-1} to one where the potential is -3.0×10^7 J kg^{-1}, it needs to gain 1.0×10^7 J of gravitational potential energy for each kilogram of satellite. It does this by firing the rocket engines, transferring energy from its fuel supply.

Potential gradient

AQA A 4
AQA B 4

The graph shows that the rate of change of potential with distance, the potential gradient, decreases with increasing distance from the Earth.

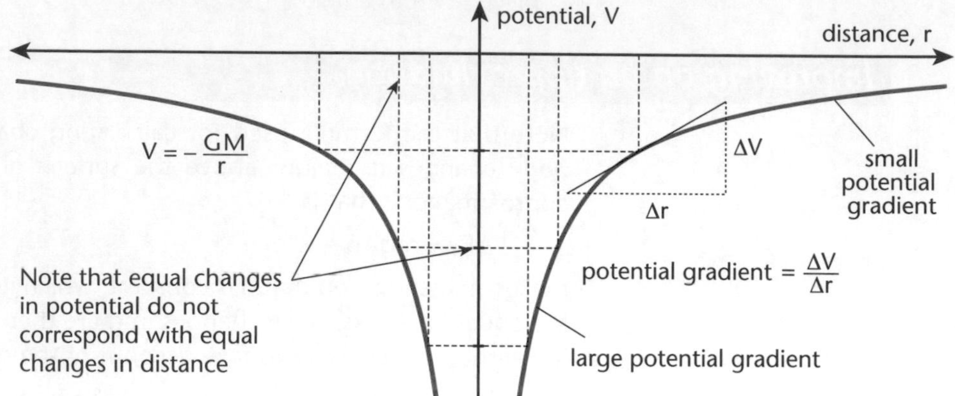

As the potential gradient decreases, so does the gravitational field strength.

The concept of potential gradient is similar to that of the gradient of a hill or slope. The steeper the slope, the greater the acceleration of an object free to move down it.

KEY POINT

The relationship between gravitational field strength and potential gradient is:

$$g = -\text{(potential gradient) or } g = -\frac{\Delta V}{\Delta r}$$

where ΔV is the change in potential over a small distance Δr.

In order for a spacecraft to escape from the Earth's gravitational potential well its total energy must be greater than 0, i.e. $E_k + E_p > 0$.
Since $E_k = \frac{1}{2}mv^2$ and $E_p = mV = -\frac{GmM}{r}$ the escape velocity is given by

$$v = \sqrt{\frac{2GM}{r}}$$

Progress check

1 The radius of the Earth is 6.4×10^6 m and the gravitational field strength at its surface is 10 N kg^{-1}.
At what height above the surface of the Earth is the gravitational field strength equal to 2.5 N kg^{-1}?

2 Two 2.5 kg masses are placed with their centres 10 cm apart.
Calculate the size of the gravitational attractive force between them.
$G = 6.7 \times 10^{-11}$ N m^2 kg^{-2}.

3 The mass of the Moon is 7.4×10^{22} kg and its radius is 1.7×10^6 m.
Using the value of G from Q2, calculate the value of free-fall acceleration at the Moon's surface.

1 1.28×10^7 m
2 4.2×10^{-8} N
3 1.7 m s^{-2}

2.2 Electric fields

After studying this section you should be able to:

- *describe the electric field due to a point charge and between two charged parallel plates*
- *calculate the force on a charge in an electric field*
- *compare gravitational and electric fields*
- *understand electric field strength and electric potential as properties of points in the field*

LEARNING SUMMARY

Charging up

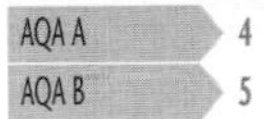

Electrostatic phenomena are due to the forces between charged objects. Similar charges repel and opposite charges attract.

Transfer of charge between two objects can happen when they slide relative to each other. It is caused by electrons leaving one surface and joining the other. This results in objects having one of two types of charge:

- an object that gains electrons has a negative charge
- an object that loses electrons has a positive charge.

In many cases any imbalance of charge on an object is removed by movement of electrons to or from the ground, but if at least one of the objects is a good insulator charge can build up.

A balloon is easily charged by rubbing but it is not possible to charge a hand-held metal rod since it is immediately discharged by exchange of electrons with the person holding it.

The smallest unit of charge is that carried by an electron. All quantities of charge must be a whole number multiple of $e = -1.6 \times 10^{-19}$ C.

The electric field

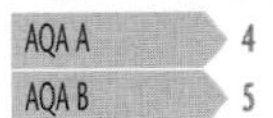

Unlike gravitational fields, where there are only attractive forces, electric fields can give rise to attraction or repulsion of objects that are charged. When drawing field lines that represent the forces due to a charged object, the arrows show the direction of the force on a positive charge.

The diagram shows the electric fields due to a 'point charge' and between a pair of oppositely charged parallel plates.

A pair of parallel plates become oppositely charged when connected to the positive and negative terminals of a d.c. supply.

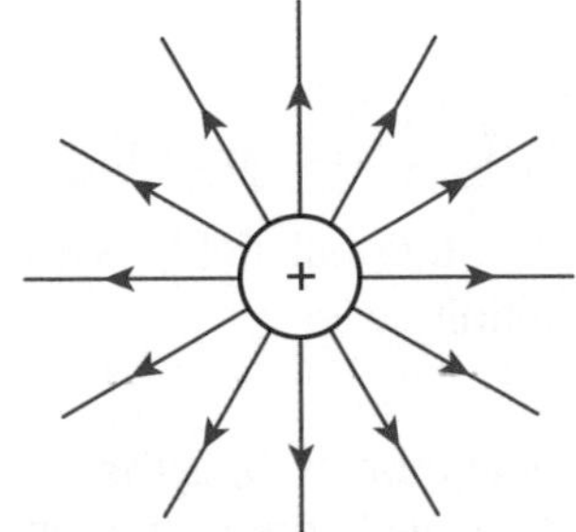

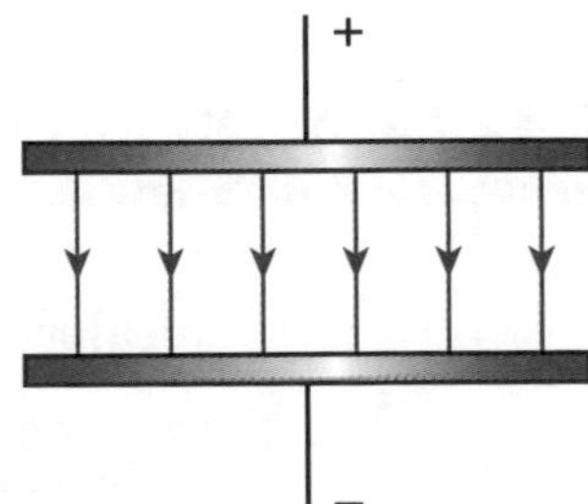

The field due to the point charge is radial, it decreases in strength with increasing distance from the charge. That between the parallel plates is uniform, it maintains a constant strength at all points between the plates.

The size of the force between two point charges is:

- proportional to each of the charges
- inversely proportional to the square of their distances apart.

Compare this equation with that for gravitational force.

Permittivity is a measure of the extent to which the medium reinforces the electric field. Water has a high permittivity due to its molecules being polarised.

KEY POINT

Coulomb's law states that the force between two point charges is given by

$$F = \frac{kQ_1 Q_2}{r^2}$$

where Q_1 and Q_2 represent the values of the charges, r is their distance apart and k has the value, in air or a vacuum, of $\frac{1}{4\pi\varepsilon_0} = 9.0 \times 10^9$ N m^2 C^{-2}.

The constant ε_0 is called the permittivity of free space.

Field strength

AQA A 4
AQA B 5

Electric field strength of a point:

- is defined *as the force per unit positive charge that would act if a small charge were placed at that point*
- is measured in N C^{-1}.

The test charge has to be small enough to have no effect on the field.

Coulomb's law can be used to express the field strength due to a point charge Q. Since the force between a charge Q and a small charge q placed within the field of Q is given by $F = \frac{kQq}{r^2}$, it follows that:

Compare the equations for electric field strength and gravitational field strength.

KEY POINT

The electric field strength, *E*, due to a point charge *Q* is given by the expression:

$$E = \frac{F}{q} = \frac{kQ}{r^2}$$

In a radial field, the field strength follows an inverse square law. This can be seen by the way in which the field lines spread out from a point charge. In a uniform field, like the one between two oppositely charged parallel plates, the field lines maintain a constant separation. The value of the electric field strength in a uniform field is the same at all points.

This gives an alternative unit for electric field strength, V m^{-1}, which is equivalent to the N C^{-1}.

KEY POINT

The electric field strength between two oppositely charged parallel plates is given by the expression:

$$E = \frac{V}{d}$$

where V is the potential difference between the plates and d is the separation of the plates.

Potential in a radial field

AQA A 4
AQA B 5

Like potential in a gravitational field, absolute potential in an electric field is measured relative to infinity.

Unlike the potential in the gravitational field of a point mass, the electric potential in the field of a point charge is positive, since work has to be done to move a positive charge from infinity to any point in the field.

KEY POINT

The electric potential at a point, V, is the work done per unit positive charge in bringing a small charge from infinity to that point.

In a radial field, $V = \frac{1}{4\pi\varepsilon_0}\frac{Q}{r}$

The unit of electric potential is the joule per coulomb, also known as a volt.

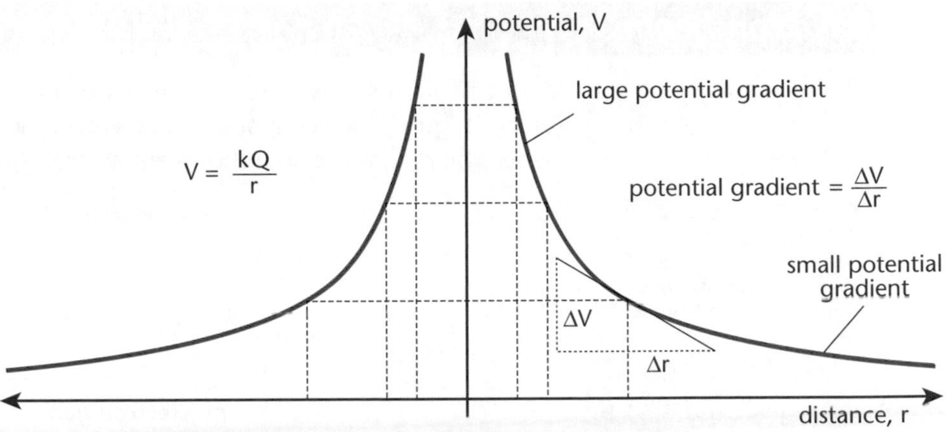

Potential and potential difference

AQA A 4
AQA B 5

In a uniform field the potential changes by equal amounts for equal changes in distance.

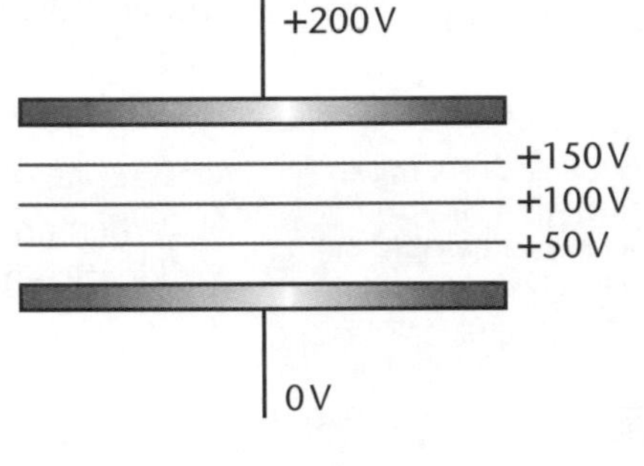

The diagram here shows the variation in potential between two oppositely charged plates.

In this example:

- potential has been measured relative to the lower plate, which has been given the value 0
- the equipotential lines are parallel to the plates
- an equipotential surface, joining points all at the same potential, is simply a surface drawn parallel to the plates.

Potential could be measured relative to the upper plate, in which case the potentials would have negative values.

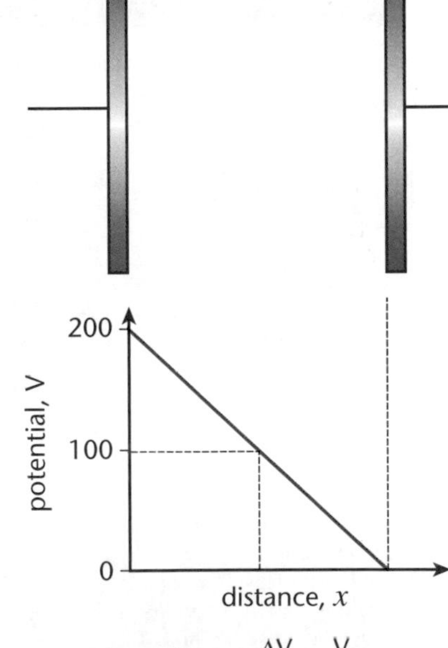

potential gradient = $\frac{\Delta V}{\Delta x} = \frac{V}{d}$

where d is the separation of the plates and V is the potential difference between them (200 V in this diagram)

The potential difference between any two points is just that – a difference in potential between them. The potential difference between the parallel plates represents the energy transfer per coulomb when charge moves between them. A charge, q, moving between the plates would gain or lose energy Vq.

KEY POINT

The potential of a point is work that would need to be done per unit charge if a small positive charge were moved to the point from outside the field. The potential difference between two points is simply the work that must be done per unit charge in moving charge between the two points. Potential difference, just like potential, is measured in joules per coulomb or volts.

Electron beams

AQA A 4
AQA B 5

Cathode ray oscilloscopes use beams of electrons. The beam is produced by an electron gun, a device that accelerates and focuses the electrons given off by a hot wire. The diagram below shows an electron gun.

The filament is the cathode, or negative terminal. The anode is usually cylindrical and is connected to the positive terminal of the high voltage supply.

Note that there is a potential difference (a difference in potential) between the anode and the cathode. Here, this difference in potential is marked as a 'high voltage'.

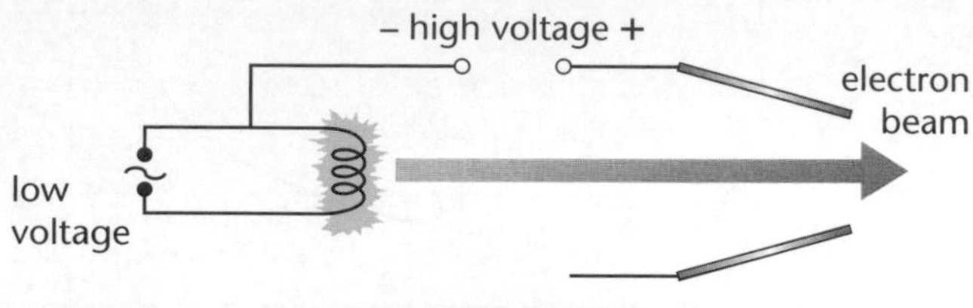

An electron gun

The low voltage supply heats the filament, causing it to emit electrons. The electrons are accelerated by the high voltage supply, gaining kinetic energy as they move towards the positively charged anode. Since the potential difference between the cathode and the anode represents the energy transfer per coulomb of charge, the kinetic energy of the accelerated electrons can be calculated.

When the electron is accelerated between two points which have a difference of potential of 1 volt between them, then the energy it gains is 1.6×10^{-19}J, which is also called one electron-volt, or 1eV.

KEY POINT

When an electron is accelerated through a potential difference, V, it gains kinetic energy

$$\frac{1}{2}m_e v^2 = eV$$

where m_e is the mass of an electron and e is the electron charge.

Deflecting a beam of electrons

In a cathode ray oscilloscope, vertical deflection of the electron beam is achieved by passing the beam between a pair of oppositely charged parallel plates. The effect of this can be seen by studying the path of a beam of electrons in a deflection tube.

When passing between the plates, the electrons have a constant speed in the direction parallel to the plates. Perpendicular to the plates, the force on each electron is equal to Ee (where E is the electric field strength), so they accelerate in this direction. The consequent motion of the electrons is similar to that of a body such as a stone projected horizontally on the Earth; in each case the result is a parabolic path, shown in the diagram.

The path of the electrons before entering the field and after leaving the field is a straight line as there is no resultant force acting on them.

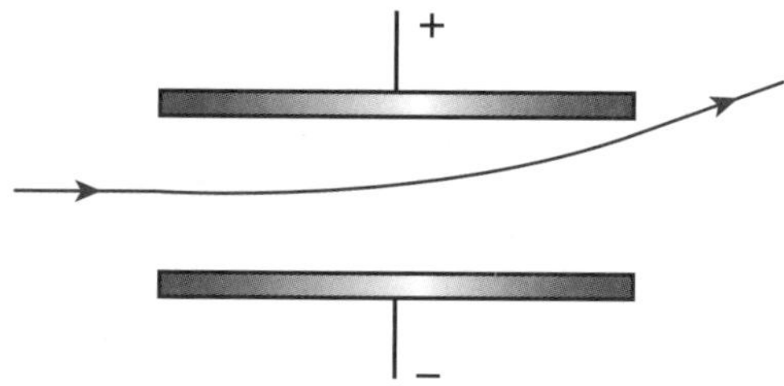

When moving between the plates:

- the electrons travel equal distances in successive equal time intervals in the direction parallel to the plates
- the electrons travel increasing distances in successive equal time intervals in the direction perpendicular to the plates.

Comparing electric and gravitational fields

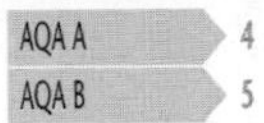

There are similarities and differences between electric and gravitational fields:

- electric field strength is defined as force per unit charge, gravitational field strength is defined as force per unit mass
- electric potential and gravitational potential are defined in similar ways
- the electric field due to a point charge is similar to the gravitational field of a point mass and the inverse square law applies in both cases
- electric fields can attract or repel charged objects, gravitational fields can only attract masses.
- the values of G (the universal gravitational constant) and $1/4\pi\varepsilon_0$ where ε_0 is the permittivity of a vacuum) are a very different, so gravitational forces and electric forces dominate at different scales in nature.

Progress check

1 Calculate the electric field strength due to a point charge of 3.0 μC at a distance of 0.10 m from the charge.

$$\frac{1}{4\pi\varepsilon_0} = 9.0 \times 10^9 \text{ N m}^2 \text{ C}^{-2}$$

2 The potential difference between two parallel plates is 300 V. They are placed 0.15 m apart.

a Calculate the value of the electric field strength between the plates.

b Calculate the size of the force on an electron, charge -1.6×10^{-19} C, placed midway between the plates.

c Explain how the size of the force on the electron varies as it moves from the negative plate to the positive plate.

1 2.7×10^6 N C^{-1}

2 a 2.0×10^3 N C^{-1}

b 3.2×10^{-16} N

c It stays the same as the field strength does not vary.

2.3 Capacitors and exponential change

After studying this section you should be able to:

- *describe the action of a capacitor and calculate the charge stored*
- *relate the energy stored in a capacitor to a graph of charge against voltage*
- *explain the significance of the time constant of a circuit that contains a capacitor and a resistor*

LEARNING SUMMARY

The action of a capacitor

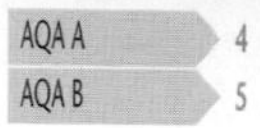

Capacitors store charge and energy. They have many applications, including smoothing varying direct currents, electronic timing circuits and powering the memory to store information in calculators when they are switched off.

A capacitor consists of two parallel conducting plates separated by an insulator. When it is connected to a voltage supply, charge flows onto the capacitor plates until the potential difference across them is the same as that of the supply. The charge flow and the final charge on each plate is shown in the diagram.

When a capacitor is charging, charge flows in all parts of the circuit except between the plates.

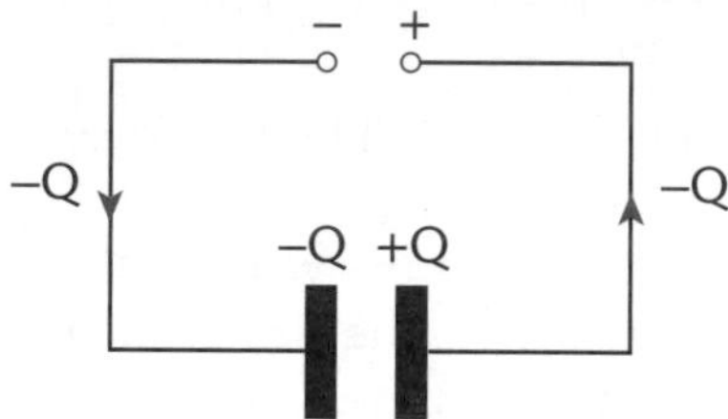

As the capacitor charges:

- charge –Q flows onto the plate connected to the negative terminal of the supply
- charge –Q flows off the plate connected to the positive terminal of the supply, leaving it with charge +Q
- the capacitor plates always have the same quantity of charge, but of the opposite sign
- no charge flows between the plates of the capacitor.

Capacitance

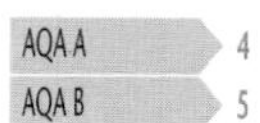

The capacitor shown in the diagram above is said to store charge Q, meaning that this is the amount of charge on each plate.

When a capacitor is charged, the amount of charge stored depends on:

- the voltage across the capacitor
- its capacitance, i.e. the greater the capacitance, the more charge is stored at a given voltage.

As the capacitor plates have equal amounts of charge of the opposite sign, the total charge is actually zero.
However, because the charges are separated they have energy and can do work when they are brought together.

> The **capacitance** of a capacitor, C, is defined as:
>
> $$C = \frac{Q}{V}$$
>
> Where Q is the charge stored when the voltage across the capacitor is V.
> Capacitance is measured in farads (F).
> 1 farad is the capacitance of a capacitor that stores 1 C of charge when the p.d. across it is 1 V.
>
> KEY POINT

One farad is a very large value of capacitance. Common values of capacitance are usually measured in picofarads (1 pF = 1.0×10^{-12} F) and microfarads (1 μF = 1.0×10^{-6} F).

The energy stored in a capacitor

AQA A 4
AQA B 5

Energy is needed from a power supply or other source to charge a capacitor.

A charged capacitor can supply the energy needed to maintain the memory in a calculator or the current in a circuit when the supply voltage is too low.

The amount of energy stored in a capacitor depends on:

- the amount of charge on the capacitor plates
- the voltage required to place this charge on the capacitor plates, i.e. the capacitance of the capacitor.

The graph below shows how the voltage across the plates of a capacitor depends on the charge stored.

When a charge ΔQ is added to a capacitor at a potential difference V, the work done is ΔQV. The total work done in charging a capacitor is $\Sigma \Delta QV$.

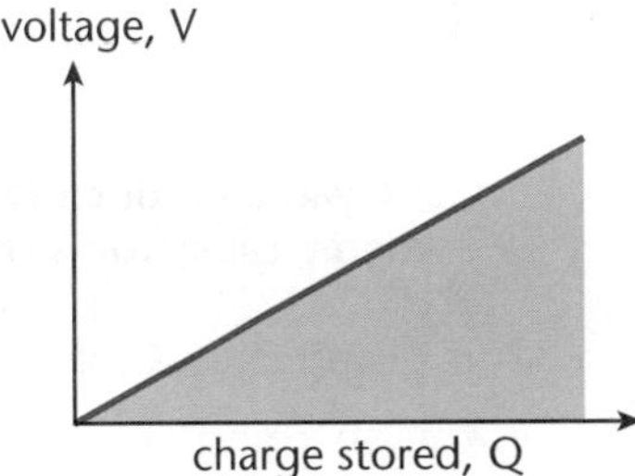

The shaded area between the graph line and the charge axis represents the energy stored in the capacitor.

> **KEY POINT**
>
> The energy, E, stored in a capacitor is given by the expression
>
> $$E = \frac{1}{2}QV = \frac{1}{2}CV^2$$
>
> where Q is the charge stored on a capacitor of capacitance C when the voltage across it is V.

Charging and discharging a capacitor

AQA A 4
AQA B 5

When a capacitor is charged by connecting it directly to a power supply, there is very little resistance in the circuit and the capacitor seems to charge instantaneously. This is because the process occurs over a very short time interval. Placing a resistor in the charging circuit reduces the rate of flow of charge (current) and slows the process down. The greater the values of resistance and capacitance, the longer it takes for the capacitor to charge.

Having a resistor in the circuit means that extra work has to be done to charge the capacitor, as there is always an energy transfer to heat when charge flows through a resistor.

The diagram below shows how the current changes with time when a capacitor is charging through a resistor.

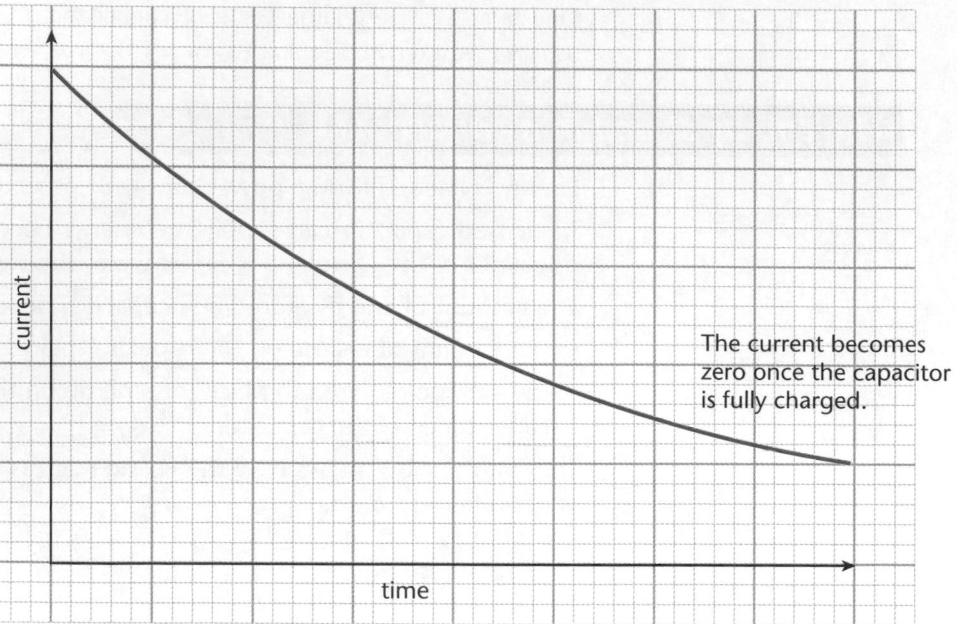

The variation of current with time when a capacitor is charging

> ### Key points from AS
> - **The relationship between resistance, current and voltage.**
> *Revise AS section 2.2*

As the p.d. across the capacitor rises, that across the resistor falls, reducing the current.

This graph shows that:

- the charging current falls as the charge on the capacitor, and the voltage across the capacitor, rise
- the charging current decreases by the same proportion in equal time intervals.

Whenever the rate of change of any quantity, such as current, is proportional to the quantity itself, then the quantity changes exponentially.

The second bullet point shows that the change in the current follows the same pattern as the activity of a radioactive isotope. This is an example of an **exponential change**. The charging current decreases exponentially.

The graph shown above can be used to work out the amount of charge that flows onto the capacitor by estimating the area between the graph line and the time axis. Since *current = rate of flow of charge* it follows that:

> **KEY POINT**
> On a graph of current against time, the area between the graph line and the time axis represents the charge flow.

> ### Key points from AS
> - **The relationship between current and charge flow.**
> *Revise AS section 2.1*

To calculate the charge flow:

- estimate the number of whole squares between the graph line and the time axis
- multiply this by the 'charge value' of each square, obtained by calculating $\Delta Q \times \Delta t$ for a single square.

The time constant

When a capacitor is charging or discharging, the amount of charge on the capacitor changes exponentially. The graphs in the diagram show how the charge on a capacitor changes with time when it is charging and discharging.

Graphs showing the change of voltage with time are the same shape. Since $V = \frac{Q}{C}$, it follows that the only difference between a charge–time graph and a voltage–time graph is the label and scale on the *y*-axis.

These graphs show the charge on the capacitor approaching a final value, zero in the case of the capacitor discharging.

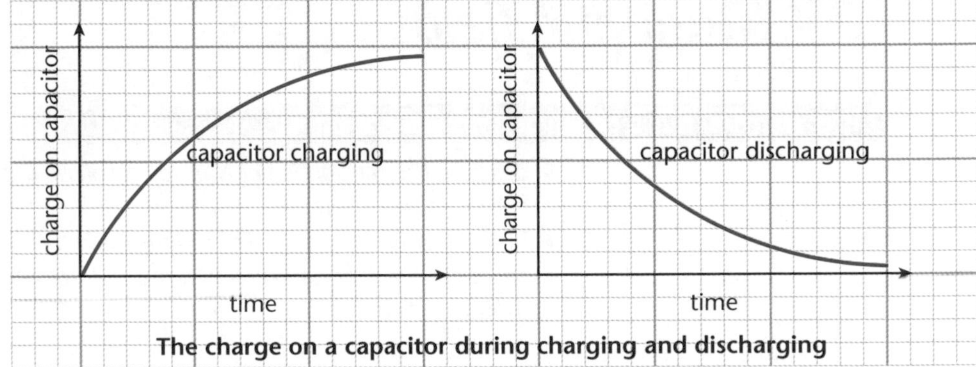

The charge on a capacitor during charging and discharging

The rate at which the charge on a capacitor changes depends on the **time constant** of the charging or discharging circuit.

> **KEY POINT**
>
> The time constant, t, of a capacitor charge or discharge circuit is the product of the resistance and the capacitance:
>
> $$t = RC$$
>
> t is measured in s.

The greater the values of R and C the longer the charge or discharge process takes. Knowledge of the values of R and C enables the amount of charge on a capacitor to be calculated at any time after the capacitor has started to charge or discharge. This is useful in timing circuits, where a switch is triggered once the charge, and therefore p.d., has reached a certain value.

e is a constant that has a value of 2.72 to 2 d.p. It is the base of the natural log and exponential functions.

The exponential function e is used to calculate the charge remaining on a capacitor that is discharging.

> **KEY POINT**
>
> The charge, Q, on a capacitor of capacitance C, remaining time t after starting to discharge is given by the expression
>
> $$Q = Q_0 e^{-t/RC}$$
>
> where Q_0 is the initial charge on the capacitor and RC is the time constant.

This expression shows that when t is equal to RC, i.e. after one time constant has elapsed, the charge remaining is equal to $Q_0 e^{-1}$, or $\frac{Q_0}{e}$

Progress check

1 The charge on a capacitor is 3.06×10^{-4} C when the p.d. across it is 6.5 V. Calculate the capacitance of the capacitor.

2 A capacitor has an intitial charge of 5.6×10^{-3} C. What is the charge remaining in the capacitor after 100s if the time constant is 45s?

3 A 50 μF capacitor is charged to a p.d. of 360 V. Calculate the energy stored in the capacitor.

1 4.71×10^{-5} F.
2 6.07×10^{-4} C.
3 3.24 J

2.4 Magnetic fields

After studying this section you should be able to:

- *describe the magnetic fields inside a solenoid and around a wire when current passes in them*
- *understand the meaning of magnetic field strength*
- *calculate the size and direction of the force that acts on a current in a magnetic field*

LEARNING SUMMARY

Magnetic fields

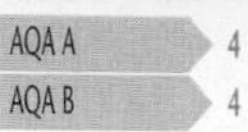

Magnetic fields exist around *moving* charged bodies. In a permanent magnet the moving charges are within atoms that are aligned. In electric currents in wires the moving charges are electrons.

Likewise, *moving* charged bodies experience force when they are in magnetic fields. Charged bodies do not experience magnetic force if there is no relative motion relative to the field.

Although the field lines are often curves, the force at any point acts in a straight line.

Like other fields, magnetic fields are represented by lines with arrows. The arrows show the direction of the force at any point in the field. The convention when drawing magnetic field lines is that the arrows show the direction of the force that would be exerted on the N-seeking pole of a permanent magnet placed at that point.

The diagrams show the field patterns around a bar magnet and between two different arrangements of pairs of magnets. Where equal-sized forces act in opposite directions, the result is a **neutral point**. At a neutral point, the resultant magnetic force is zero.

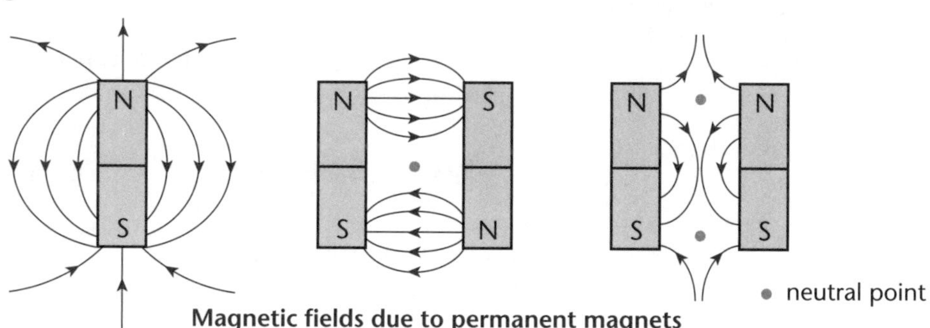

Magnetic fields due to permanent magnets

Magnetic field strength

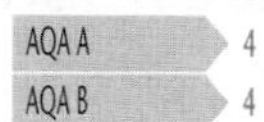

In electric and gravitational fields, the strength of the field is defined in terms of the force per unit mass or charge. In magnetic fields the situation is a little more complicated, and both charge and its velocity relative to the field are involved. The size of a force on charge q moving with velocity v perpendicular to the field is given by $F = Bqv$. If we write $B = F/qv$ then this is analogous to $g = F/m$ in gravitational fields and $E = F/q$ in electric fields. B could then be called magnetic field strength, but in fact is normally called **magnetic flux density**. The unit of flux density is the tesla, T.

Magnetic field strength and the force on a wire

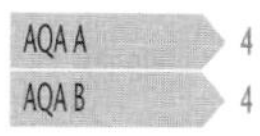

A current-carrying wire holds moving charges. In a magnetic field it experiences a force provided that it is not parallel to the field. 'Parallel to the field' means parallel to the field lines. The force has its maximum value when the current is

The force on any electric current that is parallel to a magnetic field is zero – there is no force.

perpendicular to the field. **Fleming's left hand rule** shows the direction of the force on a current of positively charged bodies (or of conventional current) that has a component which is perpendicular to a magnetic field. The diagram, left, illustrates Fleming's rule and its application to a simple **motor**.

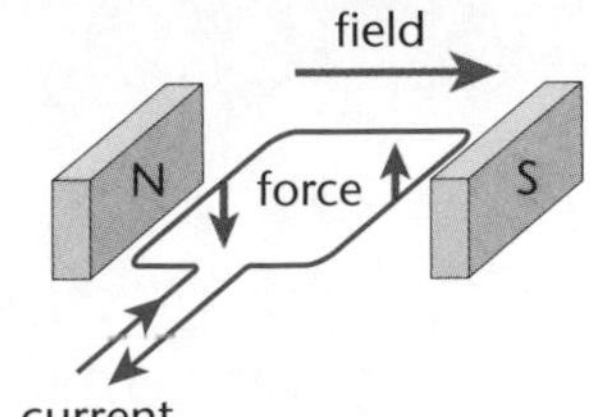

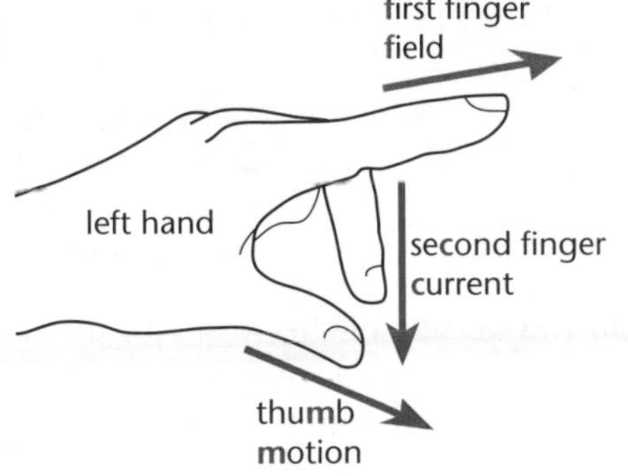

1 point your first finger in the direction of the magnetic field

2 point your second finger in the direction of the current

3 your thumb points in the direction of the resulting **motion**

Remember when using Fleming's rule that the current direction is conventional current, taken to be from + to –. The conventional current due to an electron flow is in the opposite direction to that of the electron movement. This arises because electric current was studied and conventions were established before electrons were discovered.

The size of the force on a current-carrying conductor in a magnetic field depends on:

- the size of the current
- the length of conductor in the field
- the orientation of the conductor relative to the field
- the strength of the magnetic field or flux density, B.

With the current and magnetic field directions shown in the diagram below the force is into the paper.

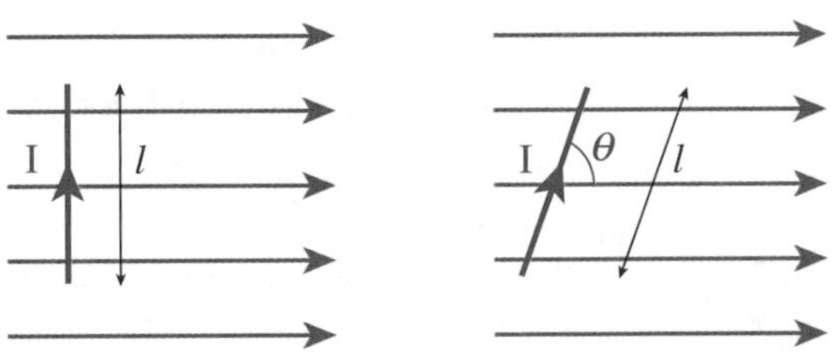

KEY POINT

The force on a current-carrying conductor in a magnetic field is given by the expressions:

when the conductor is perpendicular to the field lines:

$$F = BIl$$

when the angle between the conductor and the field lines is θ,

$$F = BIl\sin\theta$$

Fields due to currents

4
4

The diagram shows the magnetic field around a wire when a current passes in it.

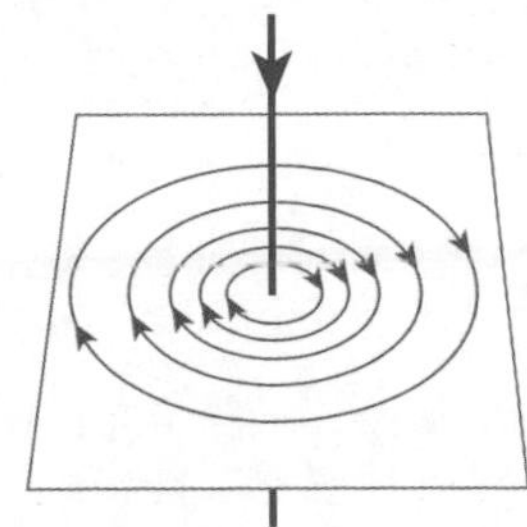

The magnetic field pattern is a set of concentric circles around the wire. The change in the spacing of the circles shows that the magnetic field strength decreases with increasing distance from the wire.

The magnetic field pattern due to a current passing in a **solenoid** is shown in the diagram below.

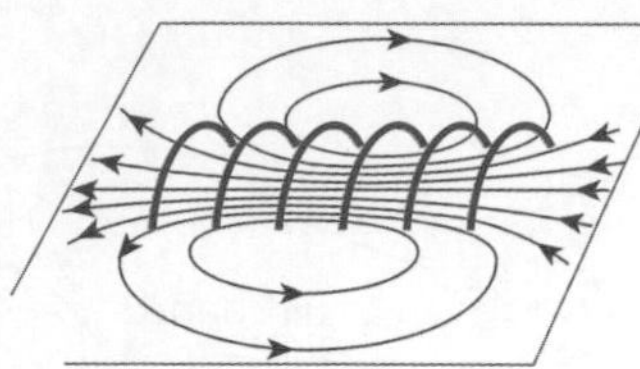

On the outside of the solenoid, the field resembles that of a bar magnet. Inside, away from the ends of the solenoid, the parallel field lines show that the magnetic field strength is uniform.

The strength of this uniform field:

- is independent of the diameter of the solenoid
- depends on the number of turns of wire per metre of length, n.

Progress check

1 Two parallel wires each carry a current in the same direction.
Use Fleming's rule to determine whether the force between them is attractive or repulsive.

2 The magnetic field strength between two permanent magnets is 5.6×10^{-3} T.
A wire carrying a current of 3.5 A is placed at right angles to their magnetic field. The length of wire within the field is 0.1 m.
Calculate the size of the force on the wire.

3 An electron, charge -1.6×10^{-19} C, moves at a velocity of 2.4×10^{7} m s^{-1} perpendicular to a magnetic field of strength 4.5×10^{-2} T.
Calculate the size of the force on the electron.

1 Attractive.
2 1.96×10^{-3} N
3 1.73×10^{-13} N

2.5 Circular orbits

LEARNING SUMMARY

After studying this section you should be able to:

- *describe the forces acting on planets, moons and satellites*
- *explain how charged particles are accelerated in a cyclotron*
- *understand how charged particles behave in electric and magnetic fields*

Movement in the Solar System

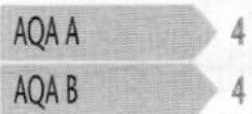

Mercury has a highly elliptical orbit. The other planets follow paths that are very close to being circles.

All planetary movement in the **Solar System** is anticlockwise, when viewed from above the North Pole. The further a **planet** is from the Sun, the slower the speed in its orbit. Although the orbits of the planets are ellipses, for most planets they are so close to circles that our understanding of circular motion can be applied.

Planets can then be considered to be:

- moving at constant speed in a circle around the Sun
- accelerating towards the Sun with centripetal acceleration v^2/r.

There are no resistive forces since the planets move through a vacuum. The only forces acting on them are gravitational. Gravitational attraction between a planet and the Sun provides the unbalanced force required to cause the centripetal acceleration.

This diagram shows that the gravitational force on a planet acts at its centre of mass and is directed towards the Sun's centre of mass.

An astronaut in a spacecraft experiences apparent weightlessness, even though he or she has not escaped from the Earth's gravitational field, and there is still a gravitational force pulling them down. The same experience is felt by a person inside a lift that has broken free and is tumbling down the lift shaft. The gravitational force is making these people accelerate towards the Earth. Neither of them could exert their usual force on bathroom scales that they (might!) happen to have with them. The scales will read zero.

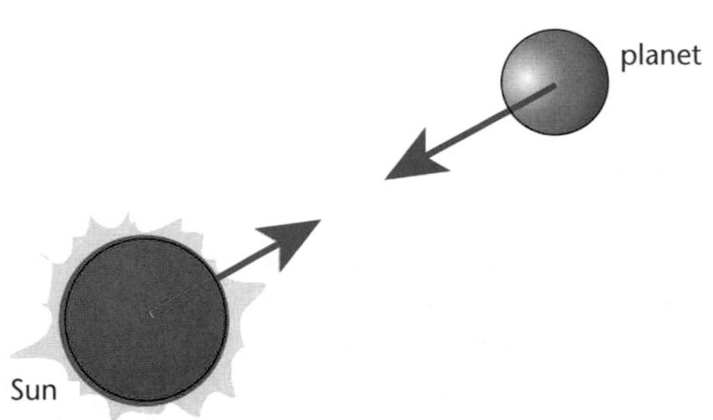

The force on the planet is:

- equal in size and opposite in direction to that on the Sun
- at right angles to its direction of motion
- the unbalanced, centripetal force required to maintain circular motion.

By equating the gravitational force to centripetal force mv^2/r, it emerges that the orbital speed depends only on the orbital radius and not on the mass of the planet.

Asteroids in the asteroid belt, between Mars and Jupiter, have a wide range of masses but similar orbit times.

KEY POINT

The centripetal force required to keep a planet in a circular orbit is the gravitational force between the planet and the Sun:

$$\frac{M_p v^2}{r} = \frac{GM_s M_p}{r^2}$$

where M_s is the mass of the Sun and M_p is the mass of the planet.

$$v^2 r = GM_s$$

The orbital period T of a planet is also related to the radius of its orbit by $\frac{T^2}{r^3}$ = constant for all planets in the Solar System (known as Kepler's third law).

Artificial satellites

AQA A 4
AQA B 4

The relationship between the orbital speed and radius of a planet can be applied to the orbit of a **satellite** around the Earth by replacing the mass of the Sun, M_S, with that of the Earth, M_E. This enables the speed of a satellite to be calculated at any orbital radius.

The relationship also applies to the Earth's natural satellite, the Moon.

Some communications satellites occupy **geo-synchronous** orbits (also called geostationary orbits). A satellite in a geo-synchronous orbit:

- orbits above the equator
- remains in the same position relative to the Earth's surface
- has an orbit time of 24 hours.

The radius of a geo-synchronous orbit can be calculated from $v^2r = GM_E$, v can be written as $2\pi r/t$ to work out the value of r.

Circular orbits in magnetic fields

AQA A 4
AQA B 5

When a charged particle moves at right angles to a magnetic field, the magnetic force on the particle is perpendicular to both its direction of motion and the magnetic field. This can result in circular motion.

The diagram shows the path and the force on an electron moving in a magnetic field directed into the paper.

When applying Fleming's rule to electrons, remember that the direction of the current is opposite to that of the electrons' motion, since electrons are negatively charged.
So for this diagram the second finger should point from right to left. The first finger represents the field and points down into the page. The thumb represents the direction of the magnetic force.

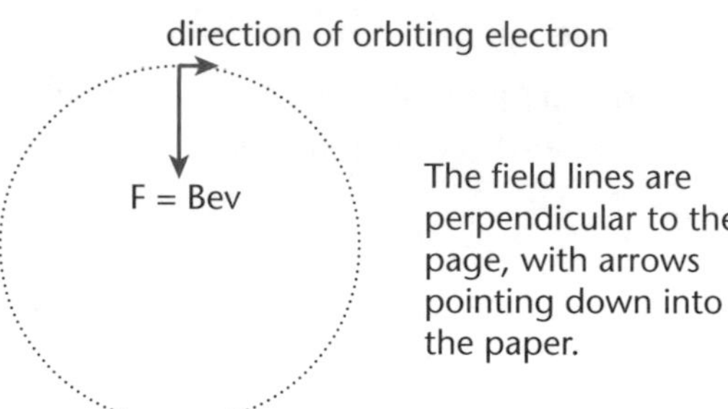

The electron follows a circular path, the magnetic force being the unbalanced force required to cause acceleration towards the centre of the circle. The radius of the circular path is proportional to the speed of the electron.

For an electron, Q = e, so the relationship is $Be = \frac{mv}{r}$.

KEY POINT

When a charge *Q* moves in a circular path in a magnetic field of strength *B*:

$$BQv = \frac{mv^2}{r}$$

so

$$BQ = \frac{mv}{r}$$

The cyclotron

AQA A 4
AQA B 5

A **cyclotron** uses a magnetic field to force charged particles to move in a 'circular' path, and an electric field to accelerate them as they travel around the circle. As the charged particles accelerate, the increase in speed results in an increase in the radius of the circle, so they spiral outwards.

Particles accelerated in a cyclotron are used to probe atomic nuclei and for treating some cancers.

A cyclotron consists of two D-shaped halves called **dees**. A magnetic field acting at right angles to the plane of the dees causes a beam of charged particles to follow a circular path. Particles such as protons and alpha particles are both suitable for use in cyclotrons.

The diagram shows the path of protons produced at the centre of the cyclotron.

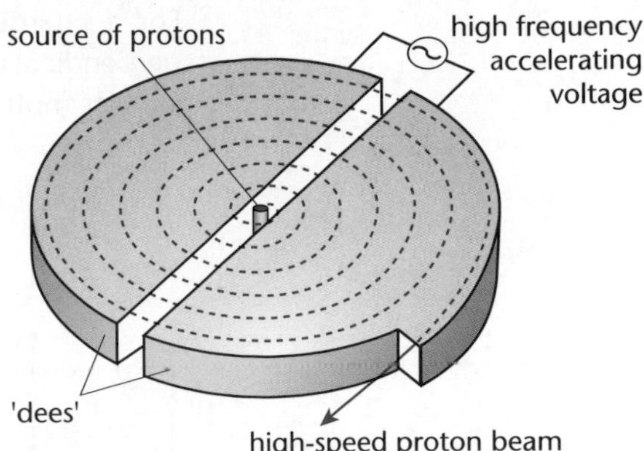

In a cyclotron:

- the near circular pathways are a result of a magnetic field acting perpendicularly to the cyclotron and the pathways
- the beam of charged particles is accelerated as it passes from one dee into the other
- this occurs because of the alternating electric field which changes polarity so that it attracts the particles as they enter a dee
- the frequency of the alternating voltage must be equal to the frequency of rotation of the particles
- the radius of orbit increases as the particles accelerate.

If the frequency of the accelerating voltage is fixed, each orbit takes the same time. With an increase in the radius of successive orbits, the particles travel increasing distances in a given time period i.e. faster.

The frequency of rotation of the charged particles in a cyclotron matches that of the accelerating voltage. The value of the magnetic field strength must be adjusted to also match this frequency.

KEY POINT

The frequency of rotation of a charged particle in a cyclotron, f, is related to the magnetic field strength, B, by the expression:

$$f = \frac{BQ}{2\pi m}$$

where Q is the charge on a particle of mass m.

Balanced electric and magnetic fields

AQA A 5
AQA B 5

Orbital motion of one charge around another is possible in radial electric fields.

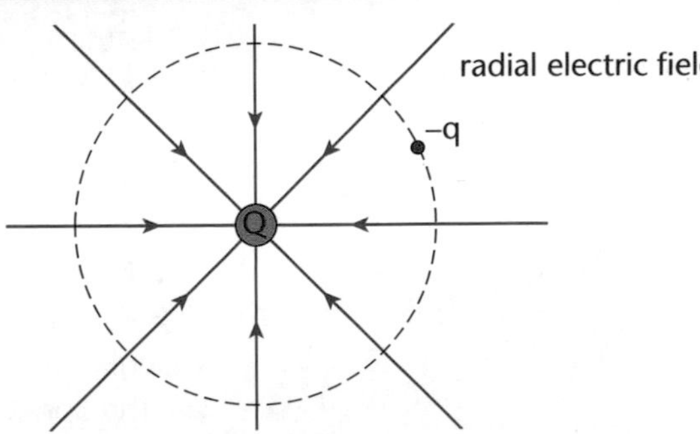

A charge moving in a uniform electric field follows a pathway that is the same as that of a mass moving without resistance in a uniform gravitational field.

A charge moving as shown between two plates, and a stone thrown horizontally, follow parabolic pathways.

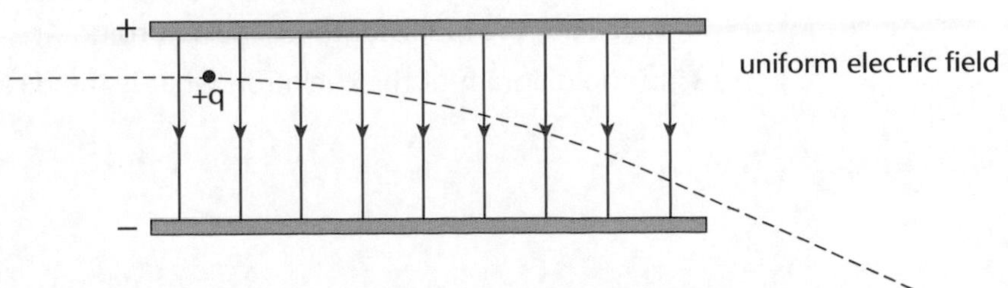

For a charged body moving in a magnetic field, the force (Bqv) is inevitably perpendicular to the velocity. So it acts as a centripetal force and the result is circular motion (i.e. the force is $\frac{mv^2}{r}$).

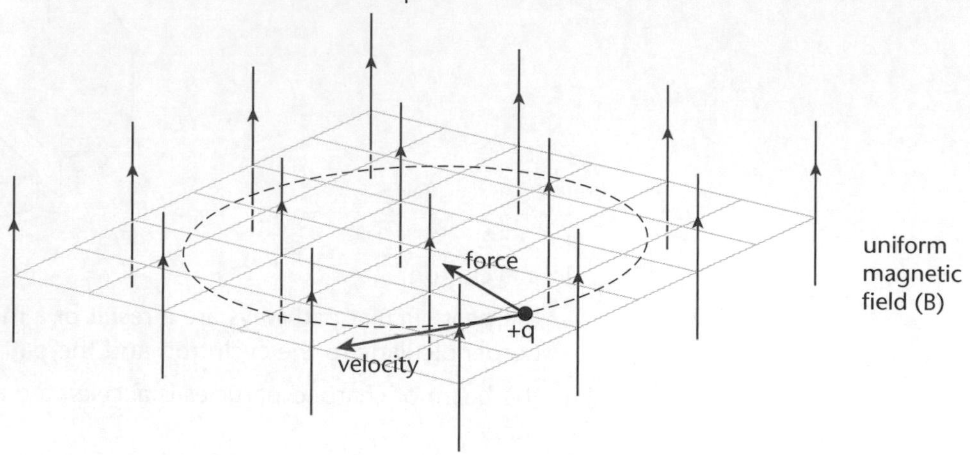

For a charged body moving in an electric field and a magnetic field that are at right angles to each other, the forces can be in opposite directions. They can balance, resulting in no deflection of the charged body's pathway. This happens when:

$Eq = Bqv$

$v = E/B$

So, by adjusting the values of the field strengths, E and B, it is possible to select charged bodies of a particular velocity to travel in a straight line.

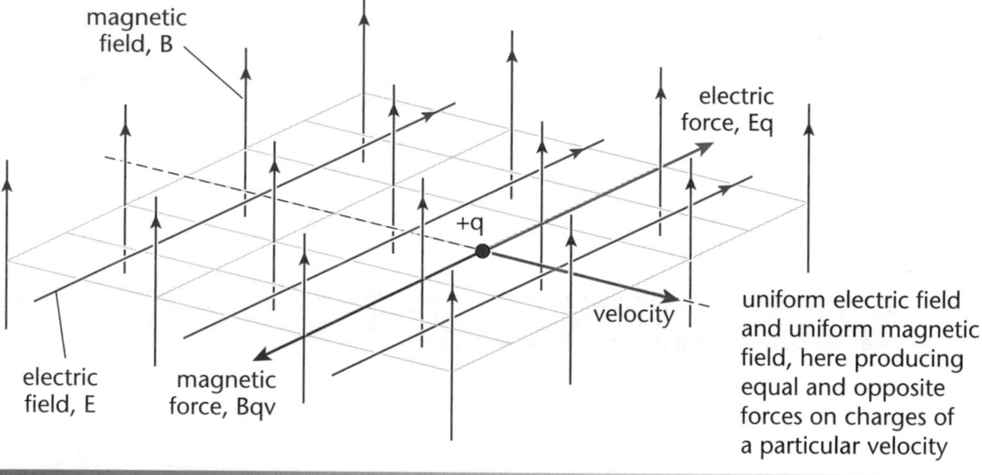

Progress check

1 The Moon orbits the Earth, mass 6.0×10^{24} kg, at a radius of 3.84×10^{8} m. $G = 6.7 \times 10^{-11}$ N m^2 kg^{-2}.
Calculate:
a the speed of the Moon in its orbit.
b the time it takes for the Moon to complete one orbit.

2 An electron of mass $m_e = 9.10 \times 10^{-31}$ kg and charge, $e = 1.60 \times 10^{-19}$ C travels at a speed of 2.10×10^{7} m s^{-1} in a circular orbit at right angles to a magnetic field.
The magnetic field strength, $B = 6.5 \times 10^{-6}$ T.
Calculate the radius of the electron orbit.

3 In the diagram of the cyclotron what is the direction of the magnetic field?

1 a 1.02×10^{3} m s^{-1}
b 2.37×10^{6} s
2 18.4 m
3 Upwards.

2.6 Electromagnetic induction

After studying this section you should be able to:

- *calculate the flux linkage through a coil of wire in a magnetic field*
- *explain how electromagnetic induction occurs due to changes in flux linkage*
- *apply Faraday's law and Lenz's law*
- *understand the transmission of electrical power at high voltages*

LEARNING SUMMARY

Flux and flux linkage

Electromagnetic induction is used to generate electricity in power stations and to transform its voltage as it passes through the distribution system.

Flux provides a useful model for explaining the effects of magnetic fields.

The effects of induction can be explained by using the concept of **flux**. Although the actual existence of flux has long been discredited, it still provides a useful modelling tool and an awareness of its meaning is useful to understand the laws of induction as set out by Faraday and Lenz.

The current view is that these forces can be attributed to 'exchange particles'.

Like gravitational and electric fields, magnetic fields act at a distance. Magnetic field patterns are used to show the forces that are exerted around a magnet or electric current. These forces are exerted without any physical contact between the magnet or current that causes the field and a magnetic material or current placed within the field. In the days of Faraday and Lenz, they were explained in terms of flux.

When drawing magnetic field patterns:

- the relative strength at different points in the field is shown by the separation of the field lines
- the closer the lines are together, the stronger the field
- these field lines represent **magnetic flux**, which is imagined as occupying the space around a magnet and being responsible for the effect of a magnet field.

To integrate the flux model with today's explanation of magnetic effects in terms of magnetic field strength, this can be thought of in terms of a flux density, being represented by the concentration of magnetic field lines. Flux density is the flux per unit area so flux is now defined in terms of the magnetic field strength and the area that the field permeates.

This definition relates the equivalence of the modern concept of magnetic field strength to that of the older 'flux density' concept.

KEY POINT

The magnetic flux, ϕ, through an area, A, is defined as the product of the magnetic field strength and the area normal to the field.

$$\phi = B \times A$$

Magnetic flux is measured in webers (Wb) where 1 Wb is the flux through an area of 1 m^2 normal to a uniform field of strength 1 T.

The diagram shows the flux through a rectangular coil in a uniform magnetic field.

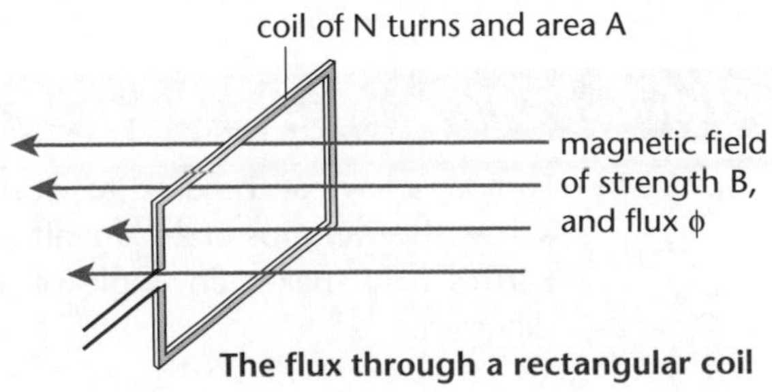

The flux through a rectangular coil

Movement of the coil parallel to the field does not induce an e.m.f., since no field lines are being 'cut'.

The induced e.m.f. has its greatest value when the movement of the coil is perpendicular to the field.

When the coil is rotated, it 'cuts' through the flux, or field lines and an e.m.f. is induced.

The size, or magnitude, of the induced e.m.f. depends on:

- the amount of flux through the coil
- the speed of rotation
- the number of turns on the coil.

Each turn on the coil has a **flux linkage** which changes as the coil rotates.

The flux linkage of a coil of N turns is $N\phi$, where ϕ is the flux through the coil.

Faraday's law

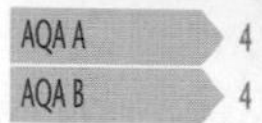

Electromagnetic induction occurs whenever the magnetic field through a conductor changes. This can be due to a conductor moving through a magnetic field or a conductor being in a fixed position within a changing magnetic field, such as that due to an alternating current. Both of these result in an e.m.f. being induced in the conductor.

In a power station, electricity is generated by an electromagnet spinning inside copper coils.

Examples of electromagnetic induction include:

- moving a magnet inside a wire coil
- generating the high voltage necessary to a) ionise the vapour in a fluorescent tube or b) cause the spark needed to ignite the explosive mixture in a petrol engine
- changing the voltage of an alternating current, using a transformer.

The diagram below shows the difference in the size of the e.m.f. when a magnet is moved at different speeds in a coil.

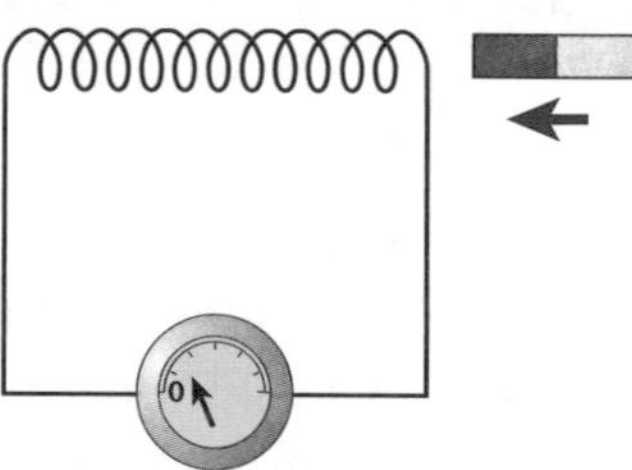

slow movement produces a small e.m.f.

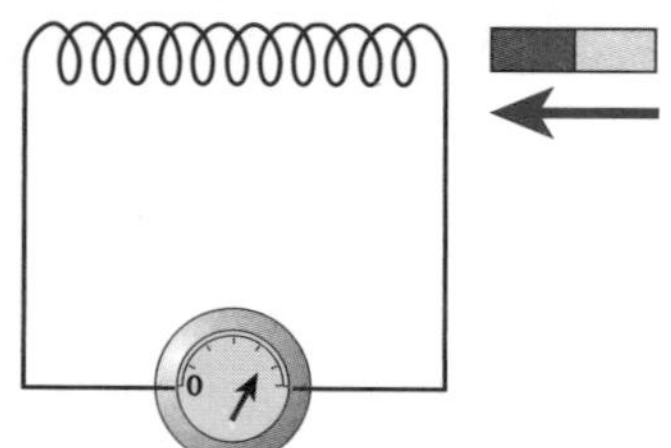

faster movement produces a bigger e.m.f.

Faraday's law relates the size of the induced e.m.f. to the change in flux linkage.

To generate the high voltage needed to cause a spark, the flux has to change rapidly. This happens when the current in an electromagnet is switched off.

KEY POINT

Faraday's law states that:

the size of the induced e.m.f. is proportional to the rate of change of flux linkage.

As the proportionality constant is equal to 1, for a uniform rate of change of flux linkage this can be written as:

$$\text{magnitude of induced e.m.f.} = N\frac{\Delta\phi}{\Delta t}$$

where $\Delta\phi$ is the change of flux in time Δt.

What direction?

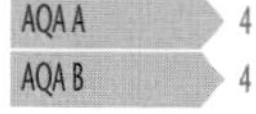

Faraday's law can be used to work out the size of an induced e.m.f. such as that across the wingtips of an aircraft flying in the Earth's magnetic field. In Britain the Earth's field makes an angle of about 20° with the vertical, see the following diagram.

Unlike that of a bar magnet, the Earth's magnetic field is from South to North. It can be considered to have two components, vertical and horizontal.

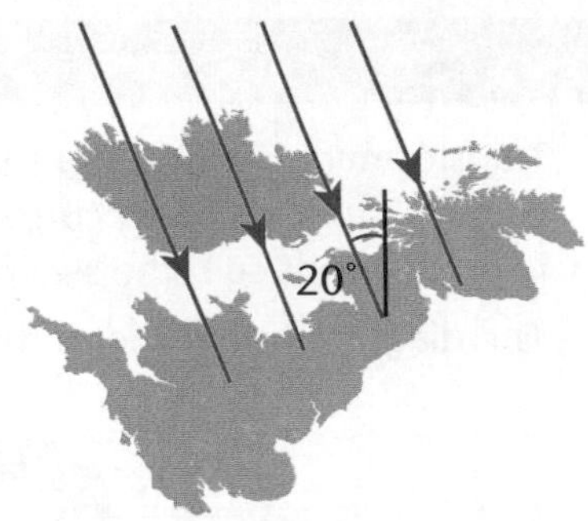

An aircraft flying in a North–South direction is cutting across the vertical component only, while flying East–West involves cutting across the horizontal component in addition.

All the charged particles experience a force due to their movement through a magnetic field, but the force is too weak to affect anything other than the free electrons.

The induced e.m.f. arises as a consequence of the force on the free electrons in the metal of the aircraft frame. Fleming's left hand rule can be used to work out the direction of the force on the electrons and hence the direction of the induced e.m.f.

In the case of an aircraft flying from North to South:

- the current is South–North (since electron travel is North–South)
- the magnetic field being 'cut' is vertically downwards
- the force on the free electrons is towards the West.

In the case of the aircraft the change producing the e.m.f. is the aircraft's motion. If the induced e.m.f. in the aircraft caused electrons to slow continuously from East to West, it would produce a force in a Northerly direction – opposite to the motion of the aircraft.

This does not happen, however, because there is no complete circuit.

This results in a charge imbalance and a voltage, or e.m.f., across the wingtips.

The direction of the e.m.f. induced in the aircraft and when a magnet moves into a coil of wire can be worked out using **Lenz's law**.

KEY POINT

Lenz's law states that:

the direction of an induced e.m.f. is always in opposition to the change that causes it.

The diagram below shows that when the North pole of a magnet is moved into one end of a coil, the induced e.m.f. causes an induced current in an anticlockwise direction. When current passes in a coil, the magnetic field is similar to that of a bar magnet, the North pole being the end where the current passes anticlockwise.

The direction of the current is always such as to repel the approaching magnet – *the induced e.m.f. opposes the change that produces it.*

The direction of the induced current is reversed by reversing the magnet or its direction of movement.

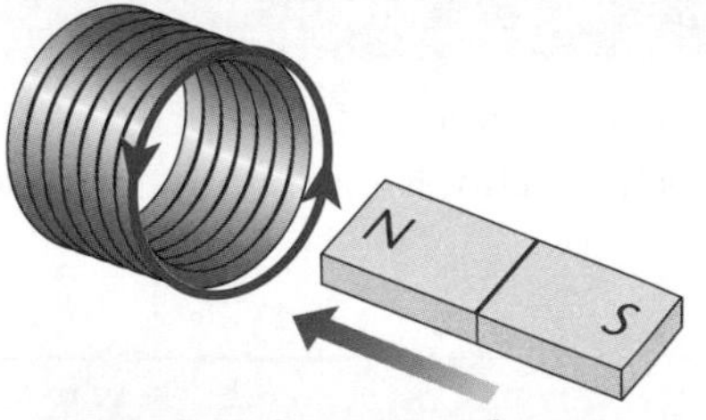

The induced current when a magnet enters a coil of wire

How to identify the poles of an electromagnet

If the induced current was in the opposite direction, it would attract the magnet into the coil and generate electricity with no energy input.

Lenz's law is a re-statement of the principle of conservation of energy; the induced current opposes the motion of the magnet so work has to be done to move the magnet against the induced magnetic field. This work is the energy transfer to the circuit needed to cause a current.

Combining Faraday's and Lenz's laws gives the equation for induced e.m.f.:

KEY POINT

$$\varepsilon = -N\frac{\Delta\phi}{\Delta t}$$

Where ε is the induced e.m.f. The negative sign shows that the induced e.m.f. is in opposition to the change of flux causing it.

Currents can be induced inside a solid conductor when it lies in a changing magnetic field (and is experiencing a change in flux). These are called eddy currents.

The transformer

AQA A 4

Transformers use changing magnetic fields to change the size of an alternating voltage. An alternating current passing in one coil (the primary) induces an e.m.f. in an adjacent coil (the secondary).

The diagram below shows the flux when the two coils are wound on an iron core.

The *e.m.f.* is induced whether or not there is a secondary circuit. If there is a complete circuit, there is also an induced *current*.

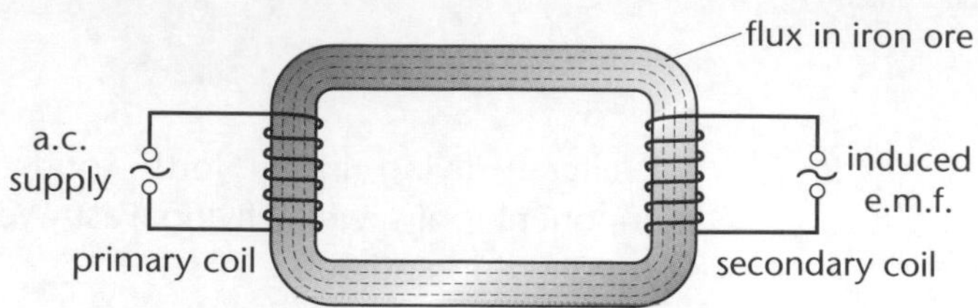

Flux linking two transformer coils

In a transformer:

Iron is easily magnetised; its magnetic domains contribute to the strength of the magnetic field.

- alternating current in the primary produces an alternating magnetic field
- this is reinforced by the high-permeability iron core
- the flux concentrates in the iron
- an e.m.f. is induced in the secondary because of the changing flux linkage.

It follows from the last bullet point that the induced e.m.f. is proportional to the number of turns on the secondary coil.

A transformer constructed from low-resistance coils on a laminated iron core is close to ideal.

KEY POINT

The relationship between the voltages and numbers of turns for an ideal transformer is:

$$\frac{V_p}{V_s} = \frac{N_p}{N_s}$$

This states that the voltages are in the same ratio as the numbers of turns. In an ideal transformer there is no energy loss in the wires or the core so the power output from the secondary is equal to the power input to the primary and the currents are in the inverse ratio to the voltages. (In reality, energy dissipation is significant.)

High voltage transmission

AQA A 4

The transmission of power from a 'supply' e.g. power station, to a 'load' e.g. consumers such as schools, hospitals, houses and industry, can be modelled by the following situation :

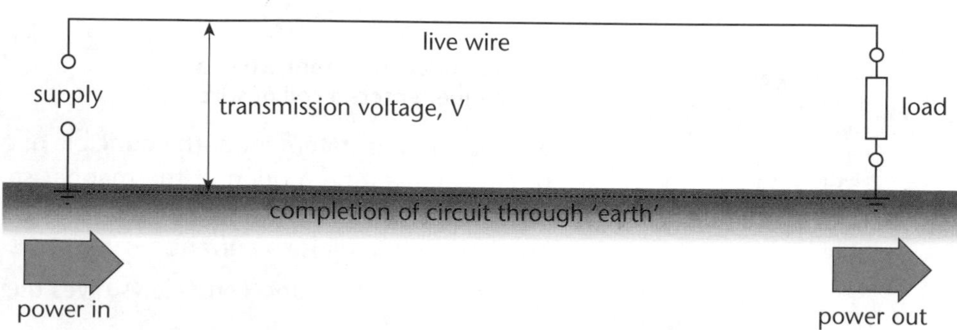

In this ideal model,

power in = power out = IV

(Note that power can be large even if the current, I or potential difference, V, are small, provided that the other is large enough to compensate).

In reality, despite the thickness of the cables used, there is still a significant amount of resistance which means that thermal energy is transferred to their surroundings resulting in heating of the air. Power, which is intended for the consumer, is lost at economic and environmental expense. The fact that the power cables are hundreds of kilometers long makes the problem worse.

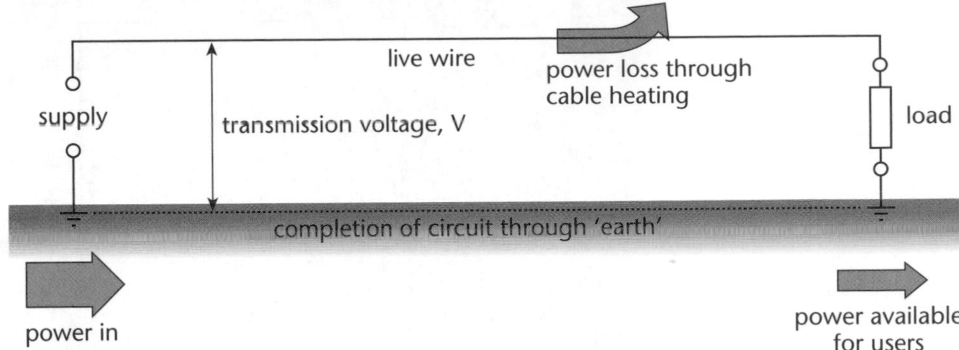

In this situation,

power in = power available for load + power loss from cables

It is therefore important to reduce the power loss from cables and this can be calculated using the equation,

power loss from cables = I^2R

where R is the resistance in the cables and I the current flowing through them.

The power loss is dependent on the square of the current, so power should be transmitted using low currents. This is achieved using transformers to 'step' the voltage up to high values (tens of thousands of volts), and then down again (to 230 volts) for transmission to the consumers.

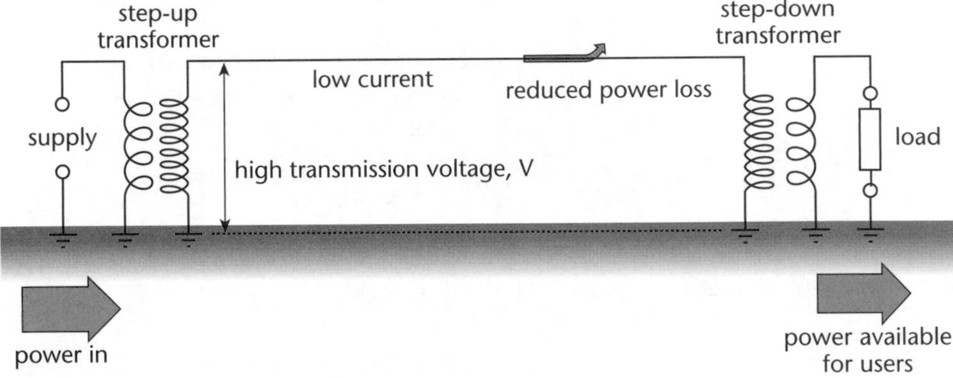

Progress check

1 A rectangular coil has 25 turns and an area of 2.5×10^{-4} m^2.
It is placed in a magnetic field of strength 6.8×10^{-6} T.
Calculate the flux linkage when the plane of the coil is

a parallel to the magnetic field

b perpendicular to the magnetic field.

2 An aircraft flies from East to West.
In what direction is the induced e.m.f. due to the 'cutting' of the Earth's horizontal magnetic field?

3 The flux linking a coil of 60 turns changes at the rate of 4.0×10^{-3} Wb s^{-1}.
Calculate the size of the induced e.m.f.

1 a 0

b 4.25×10^{-8} Wb

2 The top of the aircraft is positive and the bottom negative

3 0.24 V

Sample question and model answer

The diagram shows a capacitor connected in series with a 12.0 V power supply and a 500 Ω resistor.

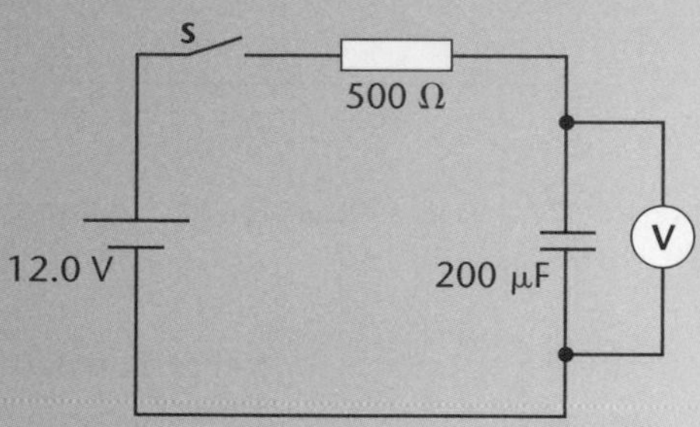

The switch is closed.

(a) When the switch is first closed, what is the initial voltage across:

(i) the capacitor? [1]

0 V — 1 mark

> At the instant that the switch is closed there is no charge on the capacitor, so there is no voltage across it. All the voltage is therefore across the resistor.

(ii) the resistor? [1]

12 V — 1 mark

(b) Calculate the value of the current in the circuit immediately after the switch is closed. [3]

> The value of the resistor determines the initial current in the circuit.

$I = V \div R$ — 1 mark

$= 12.0\ \text{V} \div 500\ \Omega$ — 1 mark

$= 0.024\ \text{A}$ — 1 mark

(c) When the capacitor is fully charged, calculate:

(i) the charge stored on the capacitor [3]

$Q = C \times V$ — 1 mark

$= 200\ \mu\text{F} \times 12.0\ \text{V}$ — 1 mark

$= 2.4 \times 10^{-3}\ \text{C}$ — 1 mark

(ii) the energy stored by the capacitor. [3]

> Alternatively, $\frac{1}{2}QV$ should give the same answer, provided that the charge, Q, has been calculated correctly.

$E = \frac{1}{2} C \times V^2$ — 1 mark

$= \frac{1}{2} \times 200\ \mu\text{F} \times 12.0^2$ — 1 mark

$= 1.44 \times 10^{-2}\ \text{J}$ — 1 mark

(d) The capacitor is discharged and a second 200 μF capacitor is connected in parallel with the capacitor shown in the diagram. The switch is then closed. Without doing any further calculations, explain how this affects:

> As the question states that further calculations are not required, give an answer based on the relationships used to calculate these quantities, with the reasons.

(i) the final charge stored [2]

This is doubled (1 mark) since the potential difference is unchanged and the capacitance is doubled (1 mark).

(ii) the final energy stored. [2]

This is also doubled (1 mark) as there is double the capacitance (and double the amount of charge stored) for the same potential difference (1 mark).

Practice examination questions

Throughout this section, use the following values of constants:
electronic charge, $e = -1.6 \times 10^{-19}$ C
universal gravitational constant, $G = 6.7 \times 10^{-11}$ Nm^2 kg^{-2}
$k = 1/4\pi\varepsilon_0 = 9.0 \times 10^9$ Nm^2 C^{-2}
gravitational field strength at the surface of the Earth, $g = 10.0$ N kg^{-1}

1

A hydrogen atom consists of a proton and an electron at an average separation of 5.2×10^{-11} m.

(a) Calculate the size of the force between them. [3]

(b) (i) Calculate the electric field strength that each experiences due to the other. [3]

(ii) What is the direction of the electric field between the proton and the electron? [1]

(c) Assuming that the electron orbits the proton, calculate the speed of its orbit.

The mass of an electron, $m_e = 9.1 \times 10^{-31}$ kg. [3]

2

The diagram shows a pair of parallel plates connected to 2500 V supply.

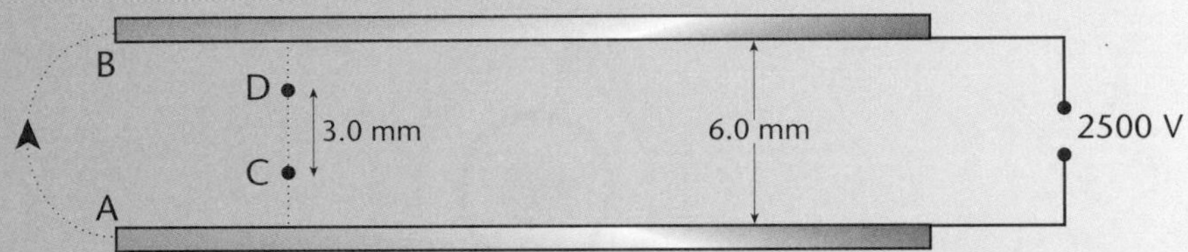

An oil drop of mass 0.050 g between the plates carries a charge of 10e.

(a) Calculate:

(i) the voltage between C and D [1]

(ii) the energy transfer when the drop moves from C to D. [3]

(b) Calculate the energy transfer when the drop moves from A to B. [1]

(c) The voltage is changed so that the drop is stationary between the plates.

(i) What must the polarity of the plates be to achieve this? [1]

(ii) Calculate the voltage required. [3]

3

Use this data to answer the questions:
The mass of the Earth = 6.0×10^{24} kg.
The mass of the Moon = 7.4×10^{21} kg.
The distance between the Earth and the Moon = 3.8×10^8 m.

(a) (i) Calculate the strength of the Earth's gravitational field at the Moon's orbit. [3]

(ii) Calculate the size of the Earth's pull on the Moon. [2]

(iii) Use the answer to (ii) to calculate the Moon's period of rotation around the Earth. [3]

Practice examination questions (continued)

(b) Tides are due to the combined effect of the Moon and the Sun.

(i) Calculate the size of the Moon's pull on 1.0 kg of water. [3]

(ii) Suggest why the pull of the Sun has much less effect than the pull of the Moon on tides. [3]

(iii) The diagram shows two positions of the Moon relative to the Earth and the Sun.

Suggest why tides are higher when the Moon is in position A than when it is in position B. [2]

4

A moving coil loudspeaker consists of a cylindrical permanent magnet and an electromagnet. The electromagnet is positioned between the poles of the fixed magnet. The diagram shows the arrangement.

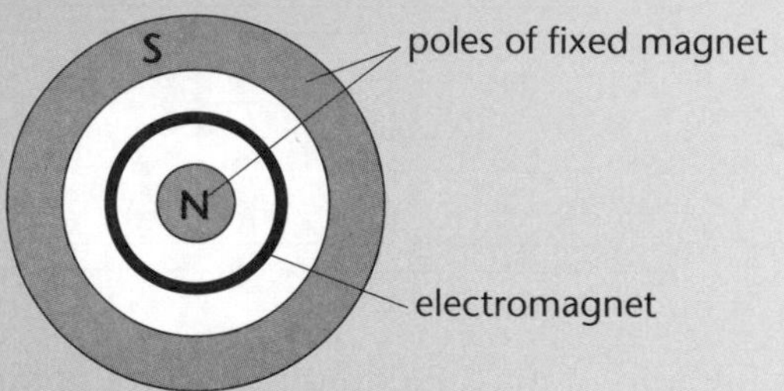

(a) What is the direction of the magnetic field between the poles of the fixed magnet? [1]

(b) The current in the electromagnet passes in a clockwise direction.
What is the direction of the force on the electromagnet? [1]

(c) Explain why the electromagnet vibrates when an alternating current passes in it. [2]

(d) The value of the magnetic field strength due to the fixed magnet is 0.85 T at the position of the electromagnet.
The electromagnet consists of a coil of wire of 150 turns and diameter 5.0 cm.
Calculate the size of the force on the coil when a current of 0.055 A passes in it. [3]

Practice examination questions (continued)

5

Electrons can be made to move in a circular path by firing them into a region where there is a magnetic field acting at right angles to their velocity.

The diagram shows a device for producing high-speed electrons.

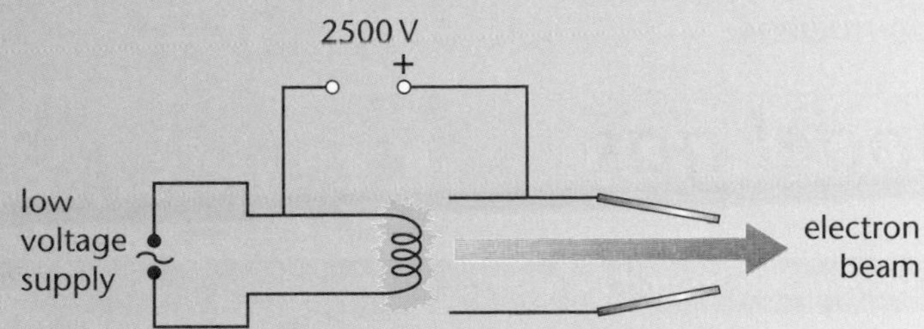

(a) Calculate:

(i) the kinetic energy of an electron in the accelerated beam [3]

(ii) the speed of an electron in the accelerated beam. The mass of an electron, $m_e = 9.1 \times 10^{-31}$ kg. [3]

(b) The electron beam passes into a magnetic field directed into the paper.

(i) What is the direction of the force on the electron beam as it enters the magnetic field? [1]

(ii) Explain why the electrons follow a circular path. [2]

(iii) The strength of the magnetic field is 2.0 mT. Calculate the radius of the circular path of the electrons. [2]

6

A rectangular coil of wire is placed so that its plane is perpendicular to a magnetic field of strength 0.15 T. This is shown in the diagram.

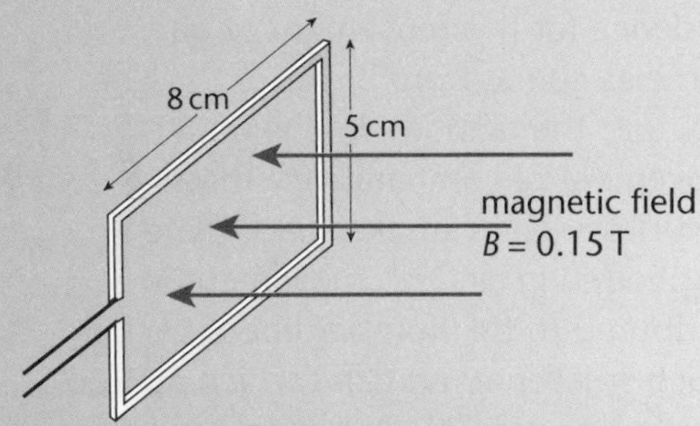

(a) Calculate the magnetic flux through the coil:

(i) when the coil is in the position shown in the diagram [3]

(ii) when the coil is turned through 90°. [1]

(b) The coil is connected to a high-resistance voltmeter.

(i) Describe and explain how the reading on the voltmeter changes as the coil is rotated through 360°. [4]

(ii) What device makes use of this effect? [1]

(c) State three ways of increasing the size of the induced voltage. [3]

Chapter 3

Thermal and particle physics

The following topics are covered in this chapter:

- A model gas
- Internal energy
- Radioactive decay
- Energy from the nucleus
- Probing matter
- Accelerators and detectors
- Quantum phenomena

3.1 A model gas

After studying this section you should be able to:

- understand the concept of absolute zero of temperature and be able to convert temperatures between the Celsius and Kelvin scales
- recall and use the ideal gas equation of state pV = nRT
- explain the relationship between the temperature and the mean kinetic energy of the particles of an ideal gas

LEARNING SUMMARY

Gas pressure

AQA A 5
AQA B 4

The kinetic model pictures a gas as being made up of large numbers of individual particles in constant motion. At normal pressures and temperatures the particles are widely-spaced compared to their size. The motion of an individual particle is:

- **rapid** – a typical average speed at room temperature is 500 m s^{-1}
- **random** in both speed and direction – these are constantly changing due to the effects of collisions.

As you will see later in this section, the average speed of the particles is different for different gases at the same temperature.

Gases exert **pressure** on the walls of their container and any other objects that they are in contact with. This pressure acts in all directions and is due to the forces as particles collide and rebound.

Brownian motion was first observed by Robert Brown while studying pollen grains suspended in water. The explanation of Brownian motion is due to Einstein.

Evidence for this movement of gas particles comes from **Brownian motion**; the random, lurching movement of comparatively massive particles such as smoke specks when suspended in air. This movement is attributed to the bombardment by much smaller air particles which are too small to be seen by an optical microscope and must therefore be moving very rapidly.

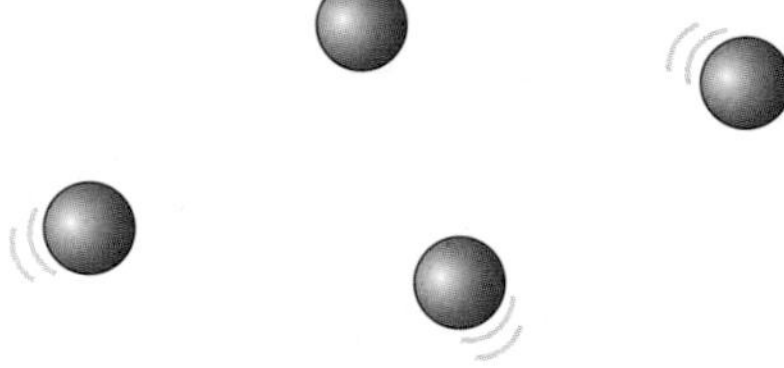

Gas particles are widely spaced. Their movement is random in both speed and direction.

Experiments show that increasing the temperature of a gas also increases the pressure that it exerts. At higher temperatures the average speed of the particles increases. This affects the pressure in two ways:

- it increases the average force of the collisions
- collisions are more frequent.

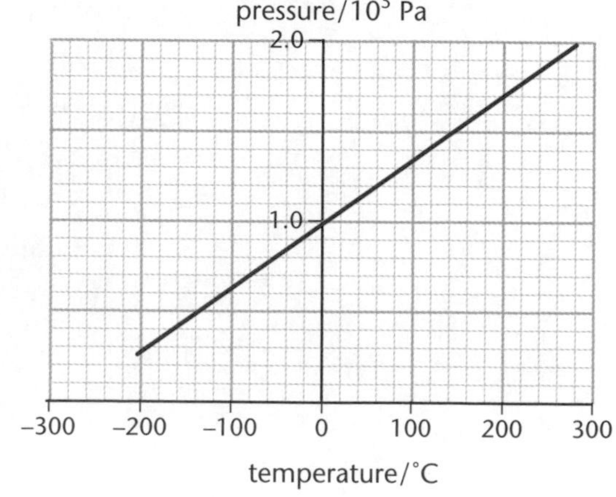

The graph shows how the pressure of a gas depends on its Celsius temperature when the volume is kept constant. Although the graph is a straight line, it does not show that pressure is directly proportional to Celsius temperature since the line does not go through the origin.

This is not surprising since 0°C does not represent 'no temperature'; it is merely a reference point that is convenient to use because it is easily reproduced.

The extrapolation assumes that the substance remains a gas as it is cooled. Real gases would reach their boiling point and liquefy before the minimum temperature was reached.

The graph does however give an indication of whereabouts 'no temperature' is on the Celsius scale. The temperature of a body determines whether it will exchange energy with its surroundings. Remember that energy transfers between bodies at different temperatures, and the net (overall) flow is from the hotter to the cooler. So the lowest possible temperature is that at which a body can only *receive* energy from the surroundings. It then has no internal energy and its particles have no motion. By extrapolating the graph back to the corresponding pressure, zero, it gives a figure for the minimum temperature at –273°C.

Knowledge of the zero of temperature allows it to be measured on an absolute scale. On an absolute scale of measurement the zero is the minimum amount of a quantity – it is not possible for a smaller measurement to exist.

Note that the kelvin is not called a degree and does not have a ° in the unit.

> **KEY POINT**
> The absolute scale of temperature is called the Kelvin scale; its unit is the kelvin (K) and its relationship to the Celsius scale, to the nearest whole number, is:
> $$T/K = \theta/°C + 273$$

The diagram shows the relationship between Celsius and kelvin temperatures.

Since there is no such thing as perfect insulation a body can never be at absolute zero. There would always be energy flow to the body.

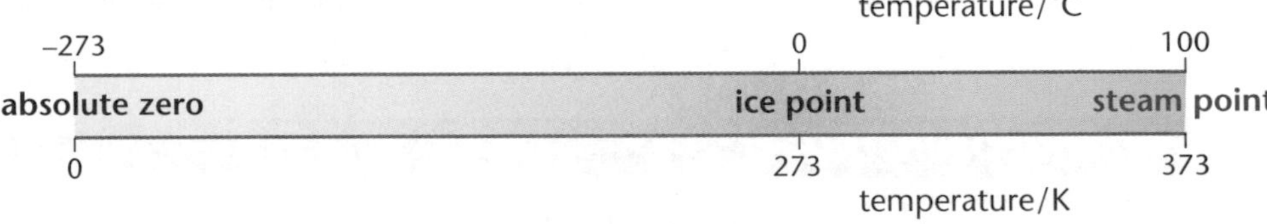

When Celsius temperatures are converted to kelvins, the experimental results show that:

The value of the constant depends on the amount of gas (measured in moles) and its volume.

Remember: the relationship is only valid when temperature is measured in kelvin.

> **KEY POINT**
> For a fixed mass of gas at constant volume:
> pressure ∝ absolute temperature
> or:
> $$p/T = \text{constant}$$

The ideal gas equation

AQA A 5
AQA B 4, 5

When experimenting with a fixed mass of gas the three variables are pressure, volume and temperature. Fixing one of these enables the relationship between the other two to be established. In addition to the relationship between pressure and temperature above:

The relationship pV = constant is known as Boyle's law.

> **KEY POINT**
> For a fixed mass of gas at constant temperature, pressure is inversely proportional to volume:
> pressure ∝ 1/volume or pV = constant
> For a fixed mass of gas at constant pressure, the volume is proportional to absolute temperature:
> volume ∝ temperature or V/T = constant

These relationships are based on experiments. They are valid for all real gases to a high level of precision provided that the gas is neither at a high pressure nor close to its boiling point. They lead to the concept of an **ideal gas** as one that obeys Boyle's law in all conditions, i.e. a gas for which pV = constant.

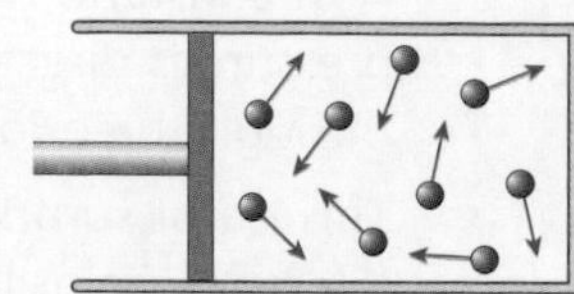

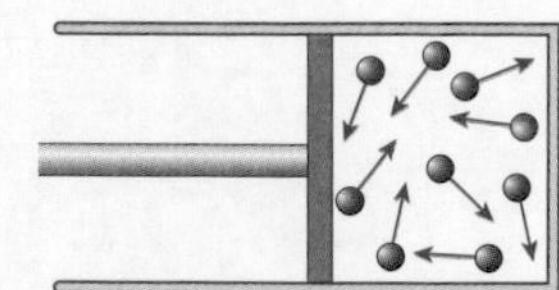

Halving the volume of a gas leads to twice the rate of collisions – doubling the pressure. This is an example of Boyle's Law.

The concept of an ideal gas is useful because the behaviour of all gases is close to ideal provided that certain conditions about temperature and pressure are met.

Real gases do not obey Boyle's law when the particles are close enough together so that they occupy a significant proportion of the gas's volume and exert appreciable forces on each other.

For an ideal gas the three gas laws can be combined into the single equation:

pV/T = constant

In this case the value of the constant depends only on the amount of gas. For 1 mole of any gas under ideal conditions it has the value of 8.3 J mol^{-1} K^{-1}. This constant is called the **molar gas constant** and has the symbol R. Doubling the number of particles by considering two moles of gas has the same effect as doubling the value of the constant, so:

1 mole of gas consists of the Avogadro constant, N_A, of particles. The value of the Avogadro constant is 6.02×10^{23} mol^{-1}.

> The equation of state for an ideal gas is:
>
> pV = nRT
>
> where n is the number of moles of gas and R is the molar gas constant.
>
> **KEY POINT**

An algebraic model

If a gas is sealed in a container, then increasing its temperature results in an increase in the pressure of the gas. This can be attributed to an increase in the average speed of the particles. The kinetic model can be extended to establish the relationship between the temperature of a gas and the speed of the particles. First of all it is necessary to establish just what is meant by 'ideal gas behaviour'.

The kinetic theory of gases involves some basic assumptions about the particles of a gas:

- a gas consists of a large number of particles in a state of rapid, random motion
- gas pressure is a result of collisions between the particles and the container walls.

In addition, for an ideal gas it is assumed that:

- the volume occupied by the particles is negligible
- intermolecular forces are negligible
- all collisions are elastic, so there is no loss in kinetic energy.

With these assumptions, it can be shown that the pressure of an enclosed sample of an ideal gas is proportional to the mean of the squares of the particle speeds:

Note that working out $<c^2>$ involves:

- listing all the individual particle speeds
- squaring them
- calculating the mean of the squares.

$$<c^2> = \frac{c_1^2 + c_2^2 + c_3^2 + \ldots c_n^2}{n}$$

> $$pV = \frac{1}{3} Nm < c^2 >$$
>
> Where N = total number of particles
>
> m = mass of each particle
>
> $< c^2 >$ = mean of the squares of the particle speeds
>
> **KEY POINT**

Alternatively, since Nm/V = mass ÷ volume = density:

$$p = \frac{1}{3}\rho <c^2>$$

Where ρ = density of the gas

KEY POINT

If the mass and volume of an ideal gas are fixed, the gas pressure is proportional to its temperature, p/T = constant. The above results show that, as pressure is proportional to $< c^2 >$ it must be proportional to the mean kinetic energy of the particles.

These factors taken together give the relationship between kinetic energy and temperature:

the mean kinetic energy of the particles of an ideal gas is proportional to the kelvin temperature

$$\frac{1}{2}m < c^2 > \propto T$$

KEY POINT

This means that doubling the kelvin temperature of a gas doubles the mean kinetic energy and the mean square speed of its particles. Remember that these relationships apply to all ideal gases, irrespective of the gas. So at a particular temperature the mean kinetic energy of the molecules in a sample of hydrogen is the same as those in a sample of oxygen. Since an oxygen molecule has sixteen times the mass of a hydrogen molecule, the mean square speed of the hydrogen molecules must be sixteen times as great as that of oxygen molecules.

The Boltzmann constant

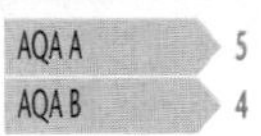

The algebraic model can be used to put a figure on the mean kinetic energy of the particles of an ideal gas at any temperature.

For one mole of gas the number of particles, N, is equal to Avogadro's number, N_A. So for one mole, combining the equations pV = RT and $pV = \frac{1}{3}N_A m < c^2 >$ gives:

$$\frac{1}{2}m < c^2 > = \frac{3}{2}RT/N_A \text{ or } N_A \times \frac{1}{2}m < c^2 > = \frac{3}{2}RT$$

This means that the total kinetic energy of one mole of an ideal gas is equal to $\frac{3}{2}RT$, since N_A is the number of particles and $\frac{1}{2}m < c^2 >$ is their mean kinetic energy.

The Boltzmann constant, k, relates the mean kinetic energy of the particles in an ideal gas directly to temperature:

$$\frac{1}{2}m < c^2 > = \frac{3}{2}RT/N_A$$
$$= \frac{3}{2}kT$$

Where $k = R/N_A$ is the Boltzmann constant and has the value $1.38 \times 10^{-23}\ J\ K^{-1}$

KEY POINT

To work out the mass of a molecule of a gas, divide the molar mass by the Avogadro constant.

So, for example, at a temperature of 20°C (293 K) the mean kinetic energy of the particles of an ideal gas is equal to $\frac{3}{2} \times 1.38 \times 10^{-23}\ J\ K^{-1} \times 293\ K = 6.07 \times 10^{-21}\ J$. This may seem a tiny amount of energy, but remember that it refers to a single gas particle which has a very small mass.

Progress check

Assume that R = 8.3 J mol^{-1} K^{-1}

1 Calculate the molar volume of an ideal gas at atmospheric pressure (1.01×10^5 Pa) and a temperature of 0°C.

2 The mean kinetic energy of the particles of an ideal gas is 6.07×10^{-21} J at a temperature of 20°C. At what temperature is their mean kinetic energy equal to 12.14×10^{-21} J?

3 2.0 mol of an ideal gas is sealed in a container of volume 5.5×10^{-2} m^3.

a Calculate the pressure of the gas at a temperature of 18°C.

b At what Celsius temperature would the pressure of the gas be doubled?

4 1.0 mol of oxygen has a mass of 32 g. Calculate the mean square speed of the molecules at 0°C and a pressure of 1.01×10^5 Pa.

5 A room measures 3.5 m × 3.2 m × 2.2 m.
It contains gas at a pressure of 1.03×10^5 Pa
and a temperature of 25°C. Calculate the
number of moles of gas in the room.

1 2.24×10^{-2} m^3
2 586 K or 313°C
3 a 8.8×10^4 Pa
b 309°C
4 2.2×10^5 m^2 s^{-2}
5 1030

3.2 Internal energy

After studying this section you should be able to:

LEARNING SUMMARY

- *explain the meaning of* specific heat capacity *and* specific latent heat *and use the appropriate relationships*
- *describe the principle of operation of a* heat engine *and a* heat pump
- *calculate the maximum efficiency of a heat engine*
- *understand the relationship between heat, work and entropy*

Energy of individual particles

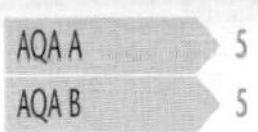

All materials are made up of particles, and particles have energy. The particles of an ideal gas have only kinetic energy since forces between the particles are negligible. However, this is not true of a solid or a liquid, and the particles in real gases exert some force on each other whenever they are close enough together and especially during collisions.

Imagine a collision between two particles in a real gas:

The speed of colliding particles is momentarily zero as their direction of travel is reversed.

- as the particles approach, the attractive forces cause an increase in the speed and kinetic energy
- the particles then get closer, the force between them becomes repulsive and the speed decreases, being momentarily zero before they start to separate
- the process is then reversed.

There is an interchange between potential (stored) energy and kinetic energy during the collision, as shown in the diagram.

As the separation between two particles decreases, the net force changes from attractive to repulsive. The equilibrium position is where these forces are equal in size, so the resultant force is zero.

particles speed up due to short-range attractive forces

particles slow down due to close-range repulsive forces

The particles of real gases, like those of solids and liquids, have both potential and kinetic energy. The distribution of energy between kinetic and potential is constantly changing and so is said to be random.

The total amount of energy of the particles in an object is known as its **internal energy**. At a constant temperature the internal energy of an object remains unchanged, but the contributions of individual particles to that energy change due to the transfer of energy during interactions between particles.

Being specific

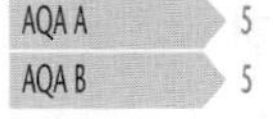

Changing the temperature of an object involves a change in its internal energy, the total potential and kinetic energy of the particles. One way of doing this is to place the object in thermal contact with a hotter or colder object, so that there is a net flow of energy from hot to cold (see diagram).

Two objects are in thermal contact if they can exchange heat. Energy flows both ways between a warm and a cold object in thermal contact, but there is more energy flow from the warm object to the cold one than the other way.

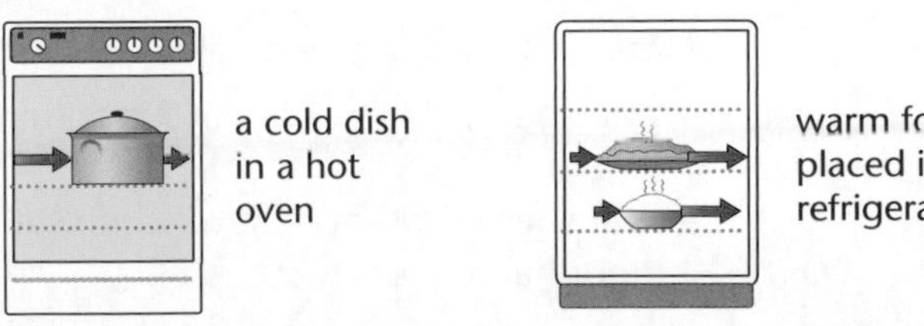

The energy flow between objects at different temperatures

The energy transfer required to change the temperature of an object depends on:

- the temperature change
- the mass of the object
- the material the object is made from.

These are all taken into consideration in the concept of **specific heat capacity**. The term 'specific' means 'for each kilogram', so any physical measurement that is described as 'specific' refers to 'per kilogram of material'.

When using this relationship the temperatures can be in either Celsius or kelvin since the temperature change is the same in each case.

KEY POINT

The specific heat capacity of a material, c, is defined as:
The energy transfer required to change the temperature of 1 kg of the material by 1°C.

$$\Delta E = mc\,\Delta\theta$$

where c, specific heat capacity, is measured in $J\ kg^{-1}\ °C^{-1}$ or $J\ kg^{-1}\ K^{-1}$

Changing phase

AQA A 5
AQA B 5

If you try to squash a solid or a liquid, the particles are pushed closer together and they repel each other. Stretching has the opposite effect; when the separation of the particles is increased the forces are attractive. At increased separations the particles have increased potential energy and this energy has to be supplied for any process that involves expansion to take place.

The term 'phase' means whether the substance is a solid, liquid or gas.

When a substance changes **phase** there is a change in the potential energy of the particles. For most substances there is a small increase in potential energy when changing from solid to liquid and a much bigger change from liquid to gas.

The increase in particle separation during a change of phase from solid to liquid is small, but that during a change of phase from liquid to gas is large.

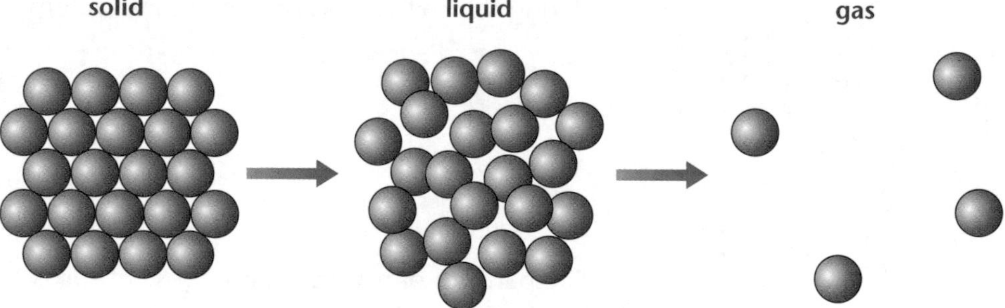

There is a small increase in potential energy of the particles when a substance changes from solid to liquid, and a much larger increase in changing from liquid to gas.

The energy absorbed or released during a change of phase is called **latent heat**. As with heat capacity, the term 'specific latent heat' refers to 1 kilogram of material.

This definition applies to a change of phase where energy is removed from the substance, as well as a change where energy is supplied.

KEY POINT

The specific latent heat of a material, l, is defined as:
The energy required to change the phase of 1 kg of the material without changing its temperature.

$$E = ml$$

where m is the mass of the material

Note that there are two values of the specific latent heat for any material:

- the specific latent heat of fusion refers to a change of phase between solid and liquid
- the specific latent heat of vaporisation refers to a change of phase between liquid and gas.

On each horizontal part of the curve the substance exists in two phases. Can you identify them?

The diagram shows a typical temperature–time graph as a solid is heated at a constant rate and passes through the three phases. Where the curve is horizontal it shows that energy is being absorbed with no change in temperature. This corresponds to a change of phase. Note that much more energy is absorbed during the change from liquid to gas than during the change from solid to liquid.

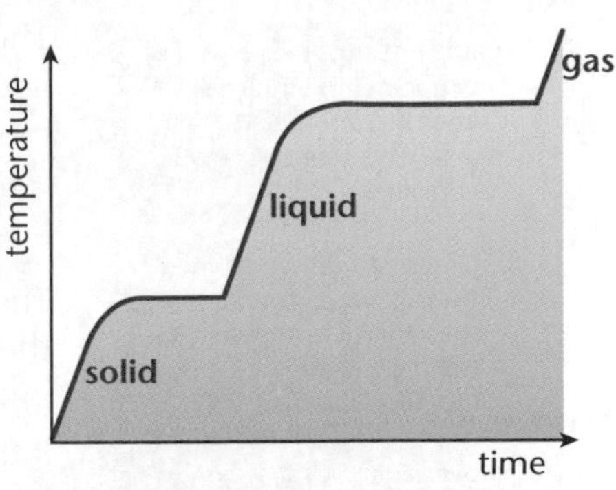

Heat and work

Placing an object in thermal contact with one that is hotter or colder, for example an immersion heater or the inside of a freezer, is one way of changing its temperature. It is a simple thermal process. Anyone who has ever pumped up a bicycle tyre is aware that compressing the air in the pump also causes it to become warmer. A person using a pump does work on the air. This could be called a mechanical process.

Compressing a gas causes heating and expanding a gas causes cooling. Whether the gas is heated or cooled depends on whether work is done **on** it or **by** it.

Other everyday examples of work causing a change in temperature include:

- striking a match, where work done against friction forces results in heating of the match head
- release of a pressurised gas, where the rapid expansion of the gas does work to push surrounding air out of the way; this requires energy transfer and causes cooling.

The **first law of thermodynamics** states the internal energy of a gas depends only on its state, i.e. the conditions of temperature, pressure and volume and the amount of gas, and not how it reached that state. Transferring energy as heat to a gas has the same effect as doing work on it – they both cause an increase in the internal energy, and therefore the temperature.

In other words, the first law states that working and heating are equivalent.

Q is the energy transferred to the gas by thermal processes.

W is the energy transferred to the gas by mechanical processes.

KEY POINT

The first law of thermodynamics can be written as:

increase in internal energy of gas = heat transferred to gas + work done on gas

$$\Delta U = Q + W$$

Note that the values of ΔU, Q and W are negative if the internal energy decreases or heat is removed from the gas or work is done by the gas.

Heat engines

In internal combustion engines:

- the burning of fuel supplies heat to a gas, increasing its internal energy
- the gas does work as it pushes the pistons, decreasing the internal energy.

Energy that is not removed from the gases by pushing the pistons is wasted in the exhaust gases and passes to the atmosphere, which is at a lower temperature.

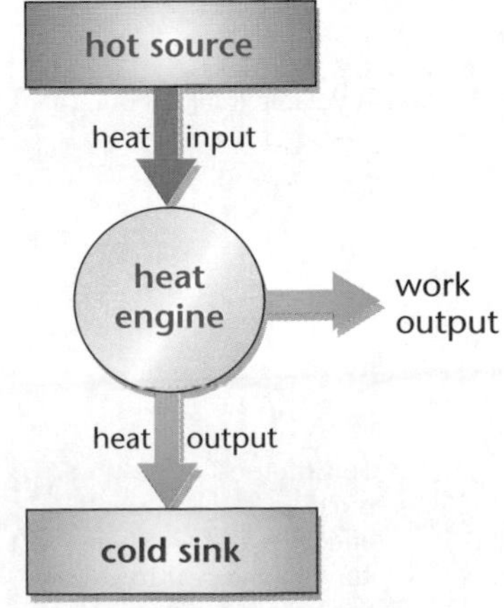

This illustrates what is meant by a **heat engine**; a device that uses the energy flow from a **hot source** to a **cold sink** to do work. The principle of operation of a heat engine is shown in the diagram.

Rockets, jet engines and steam engines are all examples of heat engines.

It is not possible to extract all the internal energy from the steam in a steam turbine. Can you explain why?

Gases can do work when they expand, provided that energy is supplied to them by heating or if they lose internal energy. For a gas to do work over and over again it must go through cycles of energy transfer. The most efficient is called the ideal Carnot cycle.

Efficiency is defined as:
$\frac{\text{useful energy output}}{\text{total energy input}}$
It is usually expressed as a fraction or decimal, which can be multiplied by 100 to give a percentage efficiency.

The steam turbines that drive the generators in power stations are also heat engines. Steam enters the turbines at high temperature and pressure and leaves at much lower temperature and pressure. Of the internal energy lost by the steam:

- some does work on the turbines
- some is lost to the surroundings.

The turbines are designed to extract as much energy from the steam as possible; the greater the proportion of the available energy extracted, the more efficient the process is.

If there are no energy losses, then the maximum efficiency of a heat engine is given by the relationship:

KEY POINT

maximum efficiency = $(T_H - T_C) / T_H$

where T_H is the temperature of the hot source and T_C is the temperature of the cold sink. Both temperatures are measured in kelvin.

For example, if steam enters a turbine at 250°C (523 K) and leaves at 120°C (393 K) then the maximum efficiency is (523 K – 393 K) ÷ 523 K = 0.25. In practice, real heat engines do not operate at their maximum efficiency since there are always other energy losses.

Heat pumps

AQA A 5
AQA B 5

Some ovens and convector heaters use fans to speed up the rate of energy transfer.

The pipes at the back of a refrigerator are painted black. How does this help to remove energy?

When two objects at different temperatures are placed in simple thermal contact, the net energy flow is from hot to cold. No other energy source is needed to drive this flow, although sometimes fans are used to speed up the natural process. In refrigeration, the direction of energy flow is from cold to hot. To achieve this, an external source of energy is needed to drive a **heat pump**.

The principle of a refrigerator is shown in the diagram.

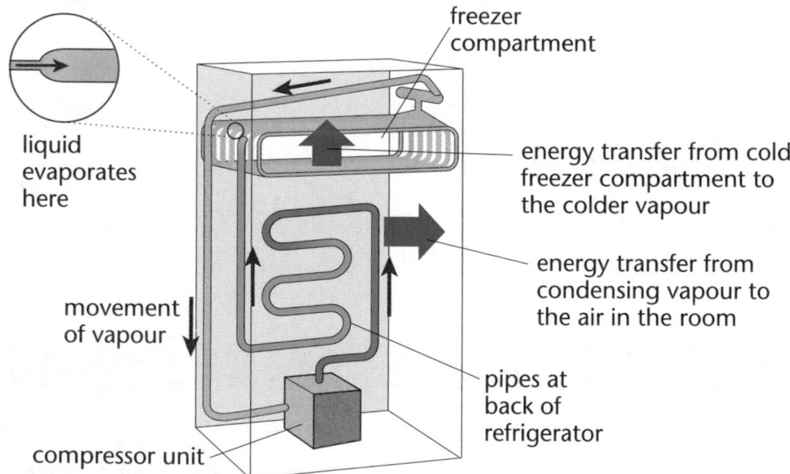

A volatile liquid is one that evaporates easily.

- Volatile liquid evaporates as it is forced through a narrow jet inside the refrigerator.
- This evaporation causes cooling and the cold vapour absorbs energy from its surroundings.
- The vapour returns to the compressor unit where it is compressed, causing it to become heated.
- Energy is dissipated to the surroundings as the hot vapour passes through pipes outside the fridge, where it condenses.

This method of heating a building can be used effectively where there is a convenient river to extract the energy from.

Heat pumps are also used to heat buildings by extracting energy from the surroundings. The energy available for heating is more than the energy needed to drive the pump, so this method is both cost-efficient and energy-efficient.

Entropy

AQA B 5

Entropy is a measure of the disorder of a system and is essentially a statement of the second law of thermodynamics.

In an isolated system, entropy increases over time. For example, particles of two gases that are neatly separated into two spaces will mix together to create a disordered mixture.

For non-isolated systems, it is possible for entropy to decrease, but this requires an input of energy.

In essence the entropy of a system increases when it absorbs heat and decreases when it rejects heat. The change in entropy is then defined as

$$\Delta S = Q/T$$

where Q is the heat absorbed (rejected) at a constant temperature T.

Progress check

1 1.5 kg water at 10°C is placed in a kettle. Calculate the energy transferred to the water in bringing it to the boil. The specific heat capacity of water, $c = 4.2 \times 10^3$ J kg^{-1} °C^{-1}.

2 Hot water enters a central heating radiator at a temperature of 52°C and leaves it at a temperature of 39°C. The specific heat capacity of water is 4200 J kg^{-1} K^{-1}.

- **a** Calculate the energy transfer from each kg of water that passes through the radiator.
- **b** If the rate of flow of water through the radiator is 3.0 kg min^{-1}, calculate the power output of the radiator.

3 The specific latent heat of vaporization of nitrogen is 2.0×10^5 J kg^{-1}. Calculate the energy required to change 0.65 kg of liquid nitrogen into vapour without changing its temperature.

4 Some air is trapped in a sealed syringe. The piston is pushed in and does 450 J of work on the air. While this is happening 300 J of heat flows out of the air.

- **a** Calculate the change in internal energy of the air.
- **b** Explain whether the air ends up hotter or colder than it was to start with.

1 5.67×10^5 J
2 a 5.46×10^4 J **b** 2.73 kW
3 1.30×10^5 J
4 a +150 J **b** hotter as the internal energy has increased.

3.3 Radioactive decay

After studying this section you should be able to:

- *describe the main types of radioactive emission*
- *explain the effects of radioactive decay on the nucleus*
- *calculate the half-life of a radioactive isotope from an activity–time graph*
- *use balanced nuclear equations*

LEARNING SUMMARY

Radiation all around us

AQA A 5
AQA B 5

Radioactive decay occurs when an atomic nucleus changes to a more stable form. This is a random event that cannot be predicted. The emissions from these nuclei are collectively called **radioactivity** or **radiation**.

The term 'background radiation' is also used to describe the microwave radiation left over from 'Big Bang'. This is a different type of radiation to nuclear radiation.

Material can absorb radiation, taking its energy from it. The 'absorbed dose' is the energy transferred per unit mass of the material. The SI unit of absorbed dose is the gray, Gy, for which 1 Gy = 1 J kg^{-1}

We are subjected to a constant stream of radiation called **background radiation.** Most of this is 'natural' in the sense that it is not caused by the activities of people. Sources of background radiation include:

- the air that we breathe; radioactive radon gas from rocks can concentrate in buildings
- the ground and buildings; all rocks contain radioactive isotopes
- the food that we eat; the food chain starts with photosynthesis. Radioactive material enters the food chain in the form of carbon-14, an unstable form of carbon that is continually being formed in the atmosphere
- radiation from space, called cosmic radiation
- medical and industrial uses of radioactive materials.

The emissions

AQA A 5
AQA B 5

Radioactive emissions are detected by their ability to cause **ionisation**, creating charged particles from neutral atoms and molecules by removing outer electrons. This results in a transfer of energy from the emitted particle, which is effectively absorbed when all its energy has been lost in this way. The four main emissions are **alpha** (α), **beta-plus** (β^+), **beta-minus** (β^-) and **gamma** (γ). Of these, alpha radiation is the most intensely ionising and has short range while gamma radiation is the least intensely ionising and has the longest range.

The range of a beta-plus particle in ordinary material is effectively zero, since it is annihilated as soon as it collides with an electron.

Alpha radiation is the most intensely ionising and can cause a lot of damage to body tissue. Although the radiation cannot penetrate the skin, alpha emitting isotopes can enter the lungs during breathing.

- In alpha emission the nucleus emits a particle consisting of two protons and two neutrons. This has the same make-up as a helium nucleus.
- Beta-minus emission occurs when a neutron decays into a proton, emitting an electron in the process.
- In beta-plus emission a proton changes to a neutron by emitting a **positron**, an anti-electron.
- Gamma emission is short wavelength electromagnetic radiation.

Radioactive emission	*Nature*	*Charge/e*	*Symbol*	*Penetration*	*Causes ionisation*	*Affected by electric and magnetic fields*
alpha	two neutrons and two protons	+2	${}^{4}_{2}He$ or ${}_{2}^{4}\alpha$	absorbed by paper or a few cm of air	intensely	yes
beta-minus	high-energy electron	−1	${}^{0}_{-1}e$ or ${}_{-1}^{0}\beta$	absorbed by 3 mm of aluminium	weakly	yes
beta-plus	positron (antielectron)	+1	${}^{0}_{+1}e$ or ${}_{+1}^{0}\beta$	annihilated by an electron		yes
gamma	short-wavelength electromagnetic radiation	none	${}^{0}_{0}\gamma$	reduced by several cm of lead	very weakly	no

Gamma emission doesn't change the composition of neutrons and protons in a nucleus, but carries energy away. A nucleus with more energy than the normal minimum is said to be in an excited state. Care is needed not to confuse this with the excited state of an atom, which again means having energy to lose, but in the atom as a whole the energy transfer involves the arrangement of the surrounding electrons and not the nucleus.

Notice that the electron emitted in beta-minus decay and the positron emitted in beta-plus decay have been allocated the **atomic numbers** –1 and +1. This is because of the effect on the nucleus when these particles are emitted.

Other types of nuclear decay include:

- **electron capture**, where a proton in the nucleus captures an orbiting electron and becomes a neutron
- **nucleon emission** when some artificially-produced isotopes decay by emitting a proton or a neutron.

Alpha and beta radiations are the most intensely ionising but they are readily absorbed and so easy to shield. Gamma radiation only reacts weakly with matter and is very penetrating. One of the safest ways of protecting against danger from gamma radiation is to maximise the distance from the source to any people. The radiation from a gamma point source is emitted equally in all directions, so the intensity decreases as an inverse square law.

In this context, intensity means the number of photons detected per square metre.

KEY POINT

The intensity, I, of gamma radiation detected from a point source is related to the intensity of the source, I_0, by the equation:

$$I = \frac{kI_0}{x^2}$$

where k = $1/4\pi$ and x is the distance from the source.

Balanced equations

AQA A 5
AQA B 5

When a nucleus decays by alpha or beta emission, the numbers of protons and neutrons are changed. Gamma emission does not change the make-up of the nucleus, but corresponds to the nucleus losing excess energy. Gamma emission often occurs alongside alpha and beta emissions, though some artificial radioactive isotopes emit gamma radiation only.

Technetium-99 is an artificial isotope that emits gamma radiation only. It is used as a tracer in medicine. The gamma radiation can be detected outside the body and there is less risk of cell damage than with an isotope that also emits alpha or beta radiation.

The changes that take place due to alpha and beta emissions are:

- **alpha**; the number of protons decreases by 2 and the number of neutrons also decreases by two
- **beta-minus**; the number of neutrons decreases by one and the number of protons increases by one
- **beta-plus**; the number of neutrons increases by one and the number of protons decreases by one.

In writing equations that describe nuclear decay, both charge (represented by the atomic number, Z) and the number of nucleons (represented by the mass number, A) are conserved. The table summarises these changes and gives examples of each type of decay.

Check that the equations given as examples are balanced in terms of charge and number of nucleons.

Particle emitted	*Effect on A*	*Effect on Z*	*Example*
alpha	–4	–2	${}^{226}_{88}\text{Ra} \rightarrow {}^{222}_{86}\text{Rn} + {}^{4}_{2}\text{He}$
beta-minus	unchanged	+1	${}^{14}_{6}\text{C} \rightarrow {}^{14}_{7}\text{N} + {}^{0}_{-1}\text{e} + \bar{\nu}$
beta-plus	unchanged	–1	${}^{11}_{6}\text{C} \rightarrow {}^{11}_{5}\text{B} + {}^{0}_{+1}\text{e} + \nu$

$\bar{\nu}$ is an anti-neutrino. ν is a neutrino.

Rate of decay and half-life

AQA A 5
AQA B 5

KEY POINT

The activity, or rate of decay, of a sample of radioactive material is measured in **becquerel** (Bq). An activity of 1 Bq represents a rate of decay of 1 s^{-1}.

Radioactive decay is a random process and the decay of an individual nucleus cannot be predicted. However, given a sample containing large numbers of undecayed nuclei, then statistically the rate of decay should be proportional to the number of undecayed nuclei present. Double the size of the sample and, on the average, the rate of decay should also double.

There are only two factors that determine the rate of decay of a sample of radioactive material. They are:

- the radioactive isotope involved
- the number of undecayed nuclei.

Unlike chemical reactions, radioactive decay is not affected by changes in temperature.

KEY POINT

The relationship between the rate of decay, or activity, of a radioactive isotope and the number of undisclosed nuclei is:

$$\text{activity, } A = \lambda N$$

Where the activity is measured in becquerel, N is the number of undecayed nuclei present and λ is the **decay constant** of the substance. λ has units of s^{-1}.

When the activity of a radioactive isotope (after deducting the average background count) is plotted against time, the result is a curve that shows the activity decreasing as the number of undecayed nuclei decreases. A decay curve is shown in the diagram.

The graph shown is a plot of activity against time. A plot of number of undecayed nuclei against time would be identical but with a different scale on the *y*-axis.

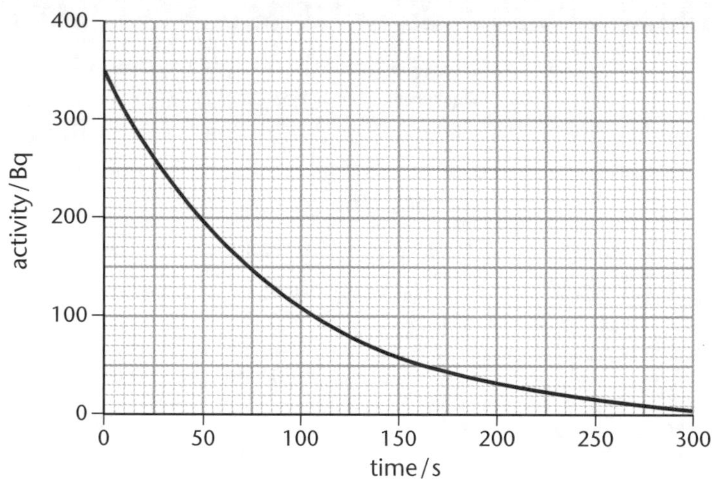

The shape of the curve is the same for all radioactive substances, but the activities and time scales depend on the size of the sample and its decay constant.

The negative sign is needed here because the *rate of change of undecayed nuclei* is always negative due to the decrease in the numbers of undecayed nuclei with increasing time.

KEY POINT

An alternative way of expressing activity is *the rate of change of the number of undecayed nuclei with time*:

$$\frac{\Delta N}{\Delta t} = -\lambda N$$

this leads to the relationship between the number of undecayed nuclei, N, and the number at time t = 0, N_0:

$$N = N_0 e^{-\lambda t}$$

The curve on the graph above shows **exponential decay**. In an exponential decay curve:

- the rate of change of y, represented by the gradient of the curve, is proportional to y
- in equal intervals of x the value of y always changes in the same ratio.

Note that although half-life is defined in terms of the number of undecayed nuclei, it is usually measured as the average time for the activity of a sample to halve.

This second point means that it always takes the same time interval for the activity to decrease to a given fraction of any particular value. The time for the activity to halve is the **half-life** of the substance.

The half-life of a radioactive isotope, $t_{1/2}$ is the average time taken for the number of undecayed nuclei of the isotope to halve.

Because radioactive decay is a random process, the results of experiments to measure activity may not fit the curve exactly and there is some variation in the change of activity when identical samples of the same material are compared. This is why the term *average* is used in the definition of half-life.

The values of half-lives range from tiny fractions of a second to many millions of years. Starting with N nuclei of a particular isotope, the number remaining after one half-life has elapsed is N/2, after two half-lives N/4 and after n half-lives it is $N/2^n$.

There is a relationship between the half-life and the decay constant of any particular radioactive isotope. The shorter the half-life, the greater the rate of decay and the decay constant.

When using this relationship, λ and $t_{1/2}$ must be measured in consistent units, e.g. λ in s^{-1} and $t_{1/2}$ in s.

The half-life of a radioactive isotope and its decay constant are related by the equation:

$$\lambda t_{1/2} = \ln 2 = 0.69$$

Stable and unstable nuclei

AQA A 5
AQA B 5

Carbon-11, carbon-12 and carbon-14 are three isotopes of carbon. Of these, only carbon-12 is stable. It has equal numbers of protons and neutrons. The graph shows the relationship between the number of neutrons (N) and the number of protons (Z) for stable nuclei.

Remember, isotopes of an element all have the same number of protons but different numbers of neutrons in the nucleus.

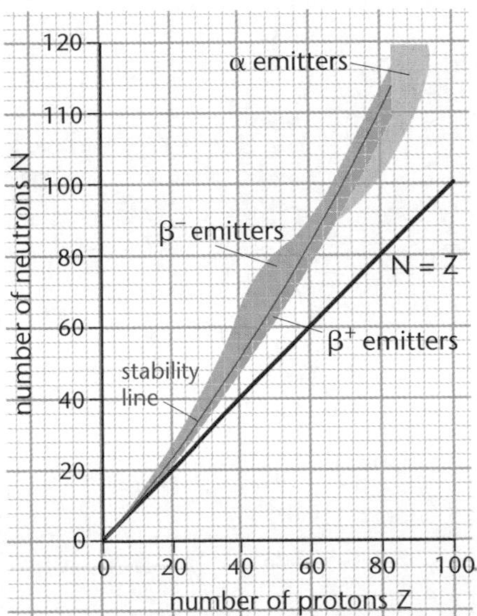

It can be seen from this graph that the condition for a nucleus to be stable depends on the number of protons:

- for values of Z up to about 20, a stable nucleus has equal numbers of protons and neutrons
- for values of Z greater than 20, a stable nucleus has more neutrons than protons.

For stable nuclei with more than about 20 protons, the neutron : proton ratio increases steadily to a value of around 1.5 for the most massive nuclei.

Unstable nuclei above the stability line in the diagram are **neutron-rich**; they can become more stable by decreasing the number of neutrons. They decay by

β^- emission; this leads to one less neutron and one extra proton and brings the neutron–proton ratio closer to, or equal to, one. An example is:

$${}^{24}_{11}Na \rightarrow {}^{24}_{12}Mg + {}^{0}_{-1}e + \bar{\nu}$$

Unstable nuclei below the stability line decay by β^+ emission; this increases the neutron number by one at the expense of the proton number. An example is:

$${}^{11}_{6}C \rightarrow {}^{11}_{5}B + {}^{0}_{+1}e + \nu$$

The N=Z line corresponds to a neutron:proton ratio of 1. As an alpha particle consists of two neutrons and two protons, its emission would hardly affect a neutron:proton ratio that is nearly 1.

The emission of an alpha particle has little effect on the neutron–proton ratio for isotopes that are close to the $N=Z$ line and is confined to the more massive nuclei. For these nuclei, emission of an alpha particle changes the balance of the proton–neutron ratio in the favour of the neutrons. In the decay of thorium-228 shown below, the neutron–proton ratio increases from 1.53 to 1.55.

$${}^{228}_{90}Th \rightarrow {}^{224}_{88}Ra + {}^{4}_{2}He$$

Progress check

1 What is the effect on atomic and mass number of decay by:
   - a an alpha particle only
   - b a beta-minus particle only
   - c a gamma photon only?

2 Technetium-99 is a gamma emitter with a half-life of 6 hours. A fresh sample is prepared with an activity A.
   - a Calculate the activity of the sample when 24 hours have elapsed since its preparation.
   - b Calculate the decay constant of technetium-99.

3 Complete the equation for the decay of phosphorus–32 by beta-minus emission.
$${}^{32}_{15}P \rightarrow S +$$

1 a The atomic number decreases by 2 and the mass number decreases by 4.
b The atomic number increases by 1 and the mass number is unchanged.
c Both the atomic number and the mass number are unchanged.
2 a $A/16$
b $3.2 \times 10^{-5}\ s^{-1}$
3 ${}^{32}_{15}P \rightarrow {}^{32}_{16}S + {}^{0}_{-1}e + \bar{\nu}$

3.4 Energy from the nucleus

After studying this section you should be able to:

LEARNING SUMMARY

- *represent nuclear reactions by nuclear equations*
- *explain the equivalence of mass and energy*
- *understand induced fission and the technology of nuclear reactors*
- *describe how energy is released in nuclear fusion*

The nucleus

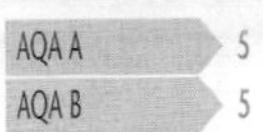

If the size of an atom is compared to that of a cathedral or a football stadium, the **nucleus** is about the size of a tennis ball. Evidence of the size of the nucleus comes from alpha particle scattering experiments (see page 97).

There are two types of particle in the nucleus, **protons** and **neutrons**.

You will not find El in the periodic table – it is fictitious.

KEY POINT

The nucleus of an element is represented as ${}^{A}_{Z}\mathrm{El}$.

Z is the atomic number, the number of protons.

A is the mass number, the number of nucleons (protons and neutrons).

The element is fixed by Z, the number of protons. A neutral atom has equal numbers of protons in the nucleus and electrons in orbit. Different atoms of the same element can have different values of *A* due to having more or fewer neutrons. As this does not affect the number of electrons in a neutral atom, the chemical properties of these atoms are the same. They are called **isotopes** of the element.

The most common form of carbon, for example, is carbon-12, ${}^{12}_{6}\mathrm{C}$, which has six protons and six neutrons in the nucleus. Carbon-14, ${}^{14}_{6}\mathrm{C}$, has the same number of protons but two extra neutrons. These two forms of carbon are isotopes of the same element.

Atomic mass and energy conservation

AQA A 5
AQA B 5

The charges and relative masses of atomic particles are shown in the table. The masses are in atomic mass units (u), where 1u = 1/12 the mass of a carbon-12 atom = 1.661×10^{-27} kg. The charges are relative to the value of the electronic charge, e = 1.602×10^{-19} C.

The phrase 'relative to the value' means compared to the actual amount of charge, ignoring the sign.

Atomic particle	*Mass*	*Charge*
proton	1.0073	+1
neutron	1.0087	0
electron	5.49×10^{-4}	–1

A carbon-12 nucleus consists of six protons and six neutrons. The mass of the atom is precisely 12u, by definition, so after taking into account the mass of the electrons, that of the nucleus is 11.9967u. The mass of the constituent neutrons and protons is:

$$6\ m_p + 6\ m_n = 6(1.0073\ \mathrm{u} + 1.0087\ \mathrm{u}) = 12.0960\ \mathrm{u}$$

The nucleus has less mass than the particles that make it up. This appears to contravene the principle of conservation of mass. Einstein established that *energy has mass.* The mass that you gain due to your increased energy when you walk upstairs is infinitesimally small, but at a nuclear level mass changes cannot be ignored. Any change in energy is accompanied by a change in mass, and *vice versa.*

Try using Einstein's equation to calculate the increase in your mass when you climb upstairs to see how small it is.

KEY POINT

Einstein's equation relates the change in energy, ΔE, to the change in mass, Δm

$$\Delta E = \Delta mc^2$$

where c is the speed of light.

It applies to **all** energy changes.

When dealing with the nucleus and nuclear particles, energy and mass are so closely linked that their equivalence, and that of their units, has been established.

KEY POINT

$$1\ u = 931.3\ MeV$$

where 1 eV (one electron volt) is the energy transfer when an electron moves through a potential difference of 1 volt.

$$1 MeV = 1 \times 10^6 eV$$

Using this relationship, the separate conservation rules regarding mass and energy can be combined into one so that (mass + energy) is always conserved in nuclear interactions.

To split a nucleus up into its constituent nucleons requires energy. It follows that a nucleus has less energy than the sum of the energies of the corresponding number of free neutrons and protons. So the fact that a nucleus has less energy than its nucleons would have in isolation means that it also has less mass.

A common misconception is that the binding energy is the energy that holds a nucleus together. It is the energy needed to split it up.

KEY POINT

The difference between the sum of the masses of the individual nucleons and the mass of the nucleus is called the **mass defect**. The matching quantity of energy is called **nuclear binding energy**. It represents the energy required to separate a nucleus into its individual nucleons.

In the case of carbon-12 the mass defect, or nuclear binding energy, is equal to 0.096u = 89.4 MeV. In calculating nuclear binding energy, the effect of the electron mass-energy is usually ignored.

As would be expected, the greater the number of nucleons, the greater the binding energy. The diagram shows how the **binding energy per nucleon** varies with nucleon number. The most stable nuclei have the greatest binding energy per nucleon.

Binding energy (or mass defect) can be thought of as the source of the energy that is made available by fission and fusion. Note that an *increase* in binding energy corresponds to a *release* of energy.

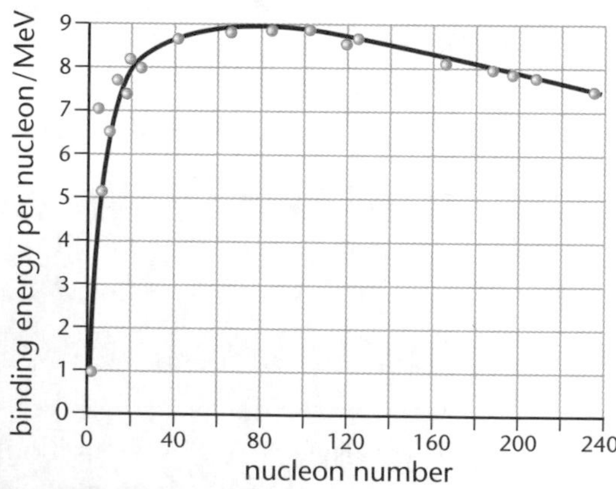

Joining and splitting

AQA A 5
AQA B 5

The nuclear reaction that releases energy in stars like our Sun is **fusion**.

In a **fusion** reaction:

- small nuclei join together to form larger, more massive ones
- the nucleus formed in fusion has less mass than those that fuse together, due to its greater binding energy per nucleon
- the mass difference is released as energy.

Despite some scientists' claims to have carried out 'cold fusion' reactions, these have not been verified.

Fusion reactions can only take place at high temperatures where the nuclei have enough thermal energy to come together despite the repulsive forces that act as they approach each other.

Large nuclei may release energy by **fission**.

In **fission**:

- a large nucleus splits into two smaller ones and two or three neutrons
- the particles formed from fission have less mass than the original nucleus, due to the greater binding energy per nucleon
- the mass difference is released as kinetic energy of the fission fragments.

Induced fission

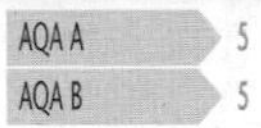

Fission can occur spontaneously in a neutron-rich nucleus or it can be caused to take place by changing the make-up of a nucleus so that it is neutron-rich. This is known as **induced fission**.

In a neutron-rich nucleus the neutron–proton ratio is greater than that required for stability.

Induced fission is triggered when a large nucleus absorbs a neutron. A slow-moving, or thermal, neutron has the right amount of energy to cause the fission of uranium-235.

A thermal neutron has energy similar to the mean kinetic energy of neutrons at room temperature.

When uranium-235 undergoes fission:

- a nucleus of $^{235}_{92}U$ absorbs a neutron to become $^{236}_{92}U$
- the resulting nucleus is unstable and splits into two approximately equal-sized smaller nuclei, together with two or three neutrons
- a large amount of energy, approximately 200 MeV, is released as kinetic energy.

The fission of uranium-235 is illustrated in the diagram.

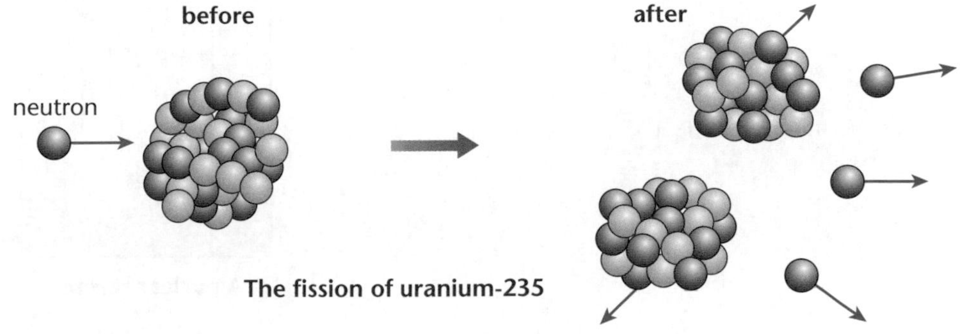

The fission of uranium-235

Check that this equation is balanced in terms of both mass and charge.

When uranium-235 fissions, the nuclei formed are not always the same. The equation represents one possible reaction, the products being barium and krypton.

$$^{235}_{92}U + ^{1}_{0}n \rightarrow ^{144}_{56}Ba + ^{90}_{36}Kr + 2^{1}_{0}n$$

Releasing the energy

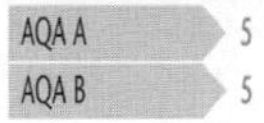

If the neutrons released from fission go on to cause further fissions, then a **chain reaction** can build up. This is illustrated in the diagram overleaf.

If every neutron produced were to cause the fission of another uranium-235 nucleus, the reaction would quickly go out of control. This does not happen with a small quantity of uranium, since the neutrons produced in the fission process are moving at high speeds. These high-speed neutrons:

- are not as likely to be absorbed by uranium-235 nuclei as thermal neutrons are
- have a good chance of leaving the material altogether due to the relatively large surface of a small mass of material.

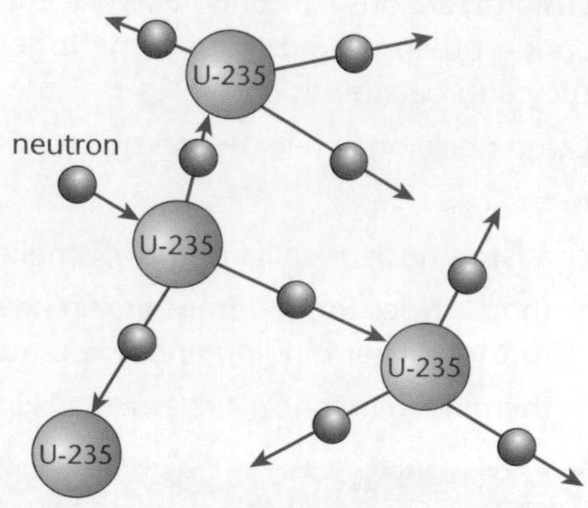

A chain reaction

The atomic bomb dropped on Hiroshima was detonated by combining three small masses of uranium into one that was above the critical mass.

For a stable chain reaction to be sustained then, on average, the neutrons released from each fission must go on to cause one further fission. The minimum mass of material needed for this to happen is called the **critical mass**. For uranium-235, the critical mass is 15 kg. This amount of uranium-235 in a spherical shape (so that it has the minimum surface area) can just sustain a reaction without it dying out.

Harnessing the energy

AQA A 5
AQA B 5

In a **nuclear reactor**, energy released from fission is removed and used to generate electricity. Three important features of a reactor are the **moderator**, **control rods**, and **coolant**. These are shown in the diagram.

This type of reactor is called a thermal reactor as it uses thermal neutrons to cause fission.

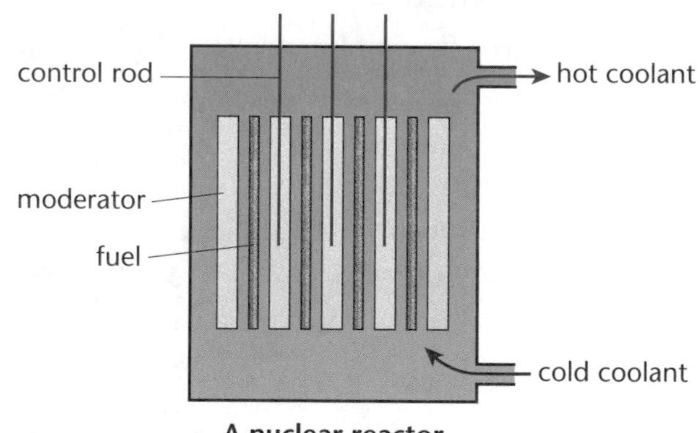

A nuclear reactor

The moderator:

- slows down the high-speed neutrons released by fission so that they are more likely to cause further fissions
- should not absorb neutrons.

Deuterium, ^2_1H, is an isotope of hydrogen that contains an extra neutron.

Graphite and heavy water, similar to normal water but using deuterium in place of hydrogen, are commonly used as moderators. Normal water is unsuitable as the hydrogen nuclei absorb neutrons.

The control rods:

- are used to control the neutron concentration by absorbing neutrons
- can be raised or lowered within the moderator material to increase or decrease the rate at which fissions take place
- must be able to absorb neutrons without becoming unstable.

The material used for control rods needs to have nuclei that are successful at capturing neutrons. Cadmium and boron are both used as they can capture neutrons over a large effective cross-section of their nuclei.

The coolant:

- should flow easily
- should not corrode the metal casings of the moderator and fuel rods
- removes energy to a heat exchanger, where steam is generated to turn turbines as in a coal-fired power station.

The coolant used depends on the temperature of the reactor core. Coolants in use include water, carbon dioxide and liquid sodium.

Nuclear fuels

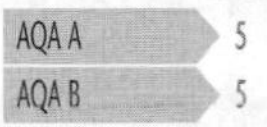

The first nuclear power stations to be built used natural **uranium** as a fuel. Natural uranium contains less than 1% uranium-235, the remainder being uranium-238 which does not fission readily. Enriched uranium fuel contains a higher proportion of uranium-235, up to 3%, so it has a longer usable lifetime than natural uranium. Fast-breeder reactors use the **plutonium**, $^{239}_{94}Pu$, formed when uranium-238, $^{238}_{92}U$, captures a neutron and then decays by β^- emission (twice). Plutonium needs fast neutrons to fission, but the spare neutrons left over from the fission process are used to create more plutonium fuel from uranium-238, so they make their own fuel from the relatively abundant isotope of uranium, uranium-238.

Fast-breeder reactors use **fast** neutrons and **breed** their own fuel.

Safety

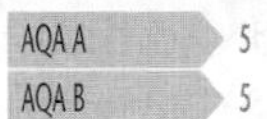

No technology is totally safe. The coolant and other materials used in building a nuclear reactor become radioactive. The products of fission, or daughter nuclei, are also radioactive. Complete removal of the control rods results in a melt-down, as happened at Chernobyl. This released vast quantities of radioactive material into the atmosphere, having disastrous effects in the immediate vicinity and affecting farmland thousands of miles away.

In normal use a thick concrete wall built around the reactor shields the outside environment by absorbing much of the radiation emitted within. In an emergency, a reactor can be shut down by lowering the control rods fully so that they absorb more neutrons, leaving insufficient neutrons to maintain the chain reaction.

The way in which active waste materials are disposed of depends on the level of radioactivity.

- Low-level waste, such as laboratory clothing and packaging, is buried underground or at sea.
- Medium-level waste, such as empty fuel casing and parts of the reactor fabrication, is kept in a concrete-lined store or underground cavern.
- High-level waste, such as the radioactive fission daughter-products and spent fuel rods, is kept in steel tanks of water to keep it cool.

Spent fuel rods contain highly radioactive materials with long half-lives. There is no safe way of disposing of these.

Artificial isotopes

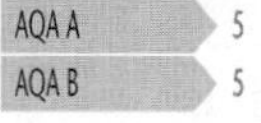

Many naturally-occurring radioactive isotopes have long half-lives and so are not suitable for use as tracers in the environment or in the body. Man-made radioactive isotopes can be produced in two ways:

- in a nuclear reactor, by irradiating a stable isotope with neutrons
- by firing charged particles at a stable isotope using a particle accelerator.

The resulting change in the nucleus is known as **artificial transmutation**. An isotope used extensively in medicine is a form of technetium-99, ^{99m}Tc, which emits gamma radiation only and has a half-life of 6 hours. It is formed when molybdenum-99, artificially created In a reactor, decays by beta-emission. This form of technetium-99 is particularly useful in medical diagnosis because:

- the gamma radiation can be detected outside the body using a gamma camera
- it does not emit the more intensely-ionising alpha and beta radiations which could cause cell damage
- the half-life is long enough for it to be detected for several hours after being injected and allowed to circulate the body
- the half-life is short enough for a minimum dose to be used
- by attaching the technetium-99 to different substances, it can be targeted towards specific body organs.

The shorter the half-life, the greater the rate of decay so the smaller the dose needed to be detectable.

More about nuclear fusion

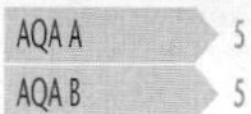

The graph on page 92 shows that there is a much larger increase in binding energy per nucleon when two small nuclei fuse than when two large ones split. Fusion can release a lot of energy from a small mass of material. Also, the raw fuel can be hydrogen. So the potential of fusion as a future source of energy is enormous.

The technical problems, however, are also considerable. Nuclei repel each other by electric force, so only at very high temperatures (such as in the Sun or in a 'hydrogen bomb' that is triggered by an initial fission explosion) can they be made to join together. Containing fusing material is very difficult, and can only be done using magnetic fields, with the material in the form of a plasma (ionised). The JET project is a European project that is working to overcome the difficulties, using deuterium (^2_1H) and tritium (^3_1H) as the fusing materials.

Progress check

1 4.0×10^5 J of energy is transferred to some water in a kettle to bring it to the boil.
Calculate the increase in mass of the water due to this energy transfer.

2 Describe the difference between nuclear *fusion* and nuclear *fission*.

3 Explain why a fast-breeder reactor does not need a moderator.

1 4.4×10^{-12} kg

2 Fusion is the joining together of two nuclei.
Fission is the splitting up of a nucleus.

3 Fission is caused by fast neutrons, so they do not need to be slowed down.

3.5 Probing matter

After studying this section you should be able to:

LEARNING SUMMARY

- *describe how the scattering of alpha particles and high-energy electrons gives evidence for the atomic model and the size of the nucleus*
- *understand that electron diffraction enables measurements of atomic spacing to be made*
- *explain that electrons, neutrinos and their antiparticles are the decay products of unstable nuclei*
- *understand the production of electron beams by thermionic emission*

Evidence for the atomic model

AQA A 5
AQA B 5

The simple atomic model pictures the atom as a tiny, positively-charged nucleus surrounded by negatively-charged particles (electrons) in orbit. Evidence for this model comes from the **scattering** of **alpha particles** as they pass through a thin material such as gold foil. The results of these experiments, first carried out under the guidance of Rutherford in 1911, can be summarised as:

- most of the alpha particles travel straight through the foil with little or no deflection
- a small number are deflected by a large amount
- a tiny number are scattered backwards.

An alpha particle is a positively-charged particle consisting of two protons and two neutrons.

The experiments were carried out in a vacuum so that the alpha particles were not scattered by air particles.

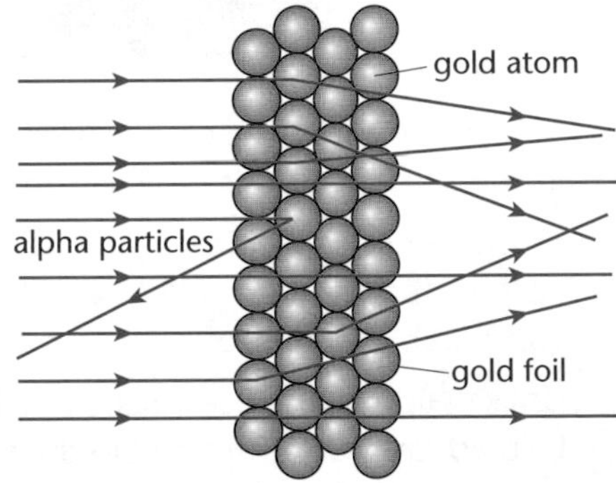

The distance of closest approach of an alpha particle to a nucleus depends on its initial kinetic energy. The kinetic energy lost as the alpha particle approaches the nucleus and is repelled is equal to the potential energy it gains.

$E_k = E_p = Qq/4\pi\varepsilon_0 r_c$

r_c is the distance of closest approach. Bombarding material in this way provides a value for the largest possible radius of its nuclei.

Rutherford concluded that the atoms of gold are mainly empty space, with tiny regions of concentrated charge. This charge must be the same sign as that of alpha particles (positive) to explain the **back-scattering** as being due to the repulsion between similar-charged objects.

The size of the nucleus

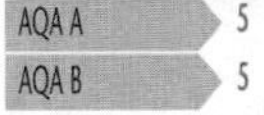

Rutherford's experiments established the existence of a nucleus that is very small compared to an atom. They also enabled the first estimates of the size of a nucleus to be made. When an alpha particle is scattered back along its original path:

- as it approaches the nucleus its kinetic energy is transferred to potential energy due to its position in the electric field
- at the closest distance of approach all its energy is potential
- the energy transfer is reversed as it moves away from the nucleus.

It is a maximum value as the nucleus cannot be larger than this but is probably smaller, as the repulsive force reverses the direction of motion of the alpha particles before they reach the nucleus.

The closest distance of approach to the nucleus is therefore calculated from the kinetic energy of the alpha particle, assuming that Coulomb's law applies to the nucleus. This gives a maximum value for the size of the gold nucleus of about 10^{-14} m.

Electron diffraction

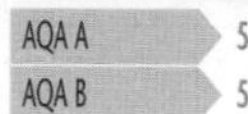

AQA A 5
AQA B 5

More precise evidence comes from the scattering of electrons, known as **electron diffraction**. Moving electrons show wave-like behaviour (see section 3.6) with a wavelength that depends on their speed.

Electrons accelerated through a potential difference of a few hundred volts have a wavelength similar to that of X-rays and gamma rays.

This wavelength is also similar to the spacing of the atoms in crystalline materials, so these materials provide suitable sized 'gaps' to cause diffraction.

Diffraction patterns formed by a beam of electrons after passing through thin foil or graphite show a set of 'bright' and 'dark' rings on photographic film, similar to those formed by **X-ray diffraction** (see below).

Analysis of these patterns is used to calculate the spacing between rows of atoms in the sample. This is typically around 10^{-10} m.

Higher energy electrons are used to estimate the size of a nucleus. Electrons accelerated through a voltage of a few hundred million volts have a wavelength comparable to that of a nuclear diameter.

When probing matter, the choice of particle used depends on its interaction with the matter being probed.

Unlike alpha particles, electrons are not repelled as they approach a nucleus. Instead, they are diffracted in the same way that light is diffracted around a circular or spherical obstacle and a diffraction pattern similar to that shown in the diagram above emerges, with the rings having fuzzy edges.

Analysis of the diffraction pattern gives a nuclear size of about 10^{-15} m.

Unlike their nuclei, the atoms of different elements can have different densities, and differences in inter-atomic spacing result in a huge range of material densities.

The results of electron diffraction experiments show that:

- all nuclei have approximately the same density, about 1×10^{17} kg m^{-3}
- the greater the number of nucleons, the larger the radius of the nucleus
- the nucleon number, A, is proportional to the cube of the **nuclear radius**, R.

An alternative way of writing this last point is:

the nuclear radius, R, is proportional to the cube root of the nucleon number, A.

KEY POINT

The relationship between nucleon number, A, and nuclear radius, R, is:

$$R = r_0 A^{1/3}$$

where r_0 has the value 1.2×10^{-15} m.

This relationship means that if the nuclei of ^{64}Cu and ^{32}S are compared, the radius of the copper nucleus is $2^{1/3} = 1.26$ times that of the sulphur, as the nucleon number of copper is twice that of sulphur. Similarly, the copper nucleus has a radius which is $4^{1/3} = 1.59$ times that of ^{16}O.

Electron beams

Electron diffraction investigations require beams of energetic electrons. Cathode ray oscilloscopes and older (vacuum tube) televisions also use electron beams, as do electron microscopes. 'Cathode ray' is an old name for an electron beam.

Beams can be produced by **electron guns**. An electron gun has a heated metal, and the heating provides energy for the escape of some electrons from the metal surface. This is called **thermionic emission**. The metal electron source is given a negative electric charge, and is called a **cathode**.

A metal **anode**, which is not heated, has a positive electric potential. The potential difference between the anode and cathode accelerates the electrons, and if the anode is symmetrical with a fine hole then some electrons pass through this to create a narrow beam. The electron gun is enclosed in a vacuum and the beam can be deflected by electric or magnetic fields.

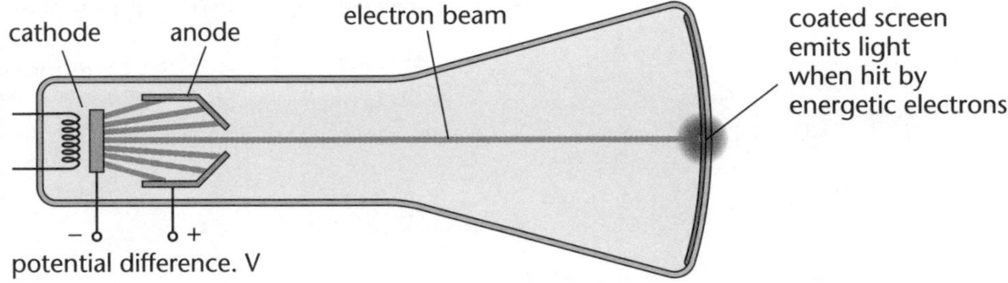

A coated screen will glow when hit by the electron beam. The kinetic energy gained by an electron is $\frac{1}{2}m_e v^2 = eV$, where m_e is electron mass, e is electron charge, v is the velocity acquired and V is the potential difference.

Particles and their antis

AQA A 5
AQA B 5

For every type of particle there is an antiparticle. An antiparticle:

- has the same mass as the particle, but has the opposite charge if it is the antiparticle of a charged particle
- annihilates its particle when they collide, the collision resulting in energy in the form of gamma radiation or the production of other particles. This process is called **pair annihilation**
- can be created, along with its particle, when a gamma ray passes close to a nucleus. This process is called **pair production**.

Where a particle has zero charge, for example the neutron, its antiparticle also has zero charge.

Anti-electrons, or **positrons**, are emitted in some types of radioactive decay and can be created when cosmic rays interact with the nuclei of atoms. Like all antiparticles, they cannot exist for very long on Earth before being annihilated by the corresponding particles.

The symbols β^- and β^+ are normally used to refer to electrons and positrons given off as a result of radioactive decay.

KEY POINT

Using symbols for particles

An electron is represented by the symbol e^- whilst an electron originating from the nucleus is symbolic by β^-. Both particles are identical.
The positron can be represented by either of the symbols e^+ and β^+.
The proton and antiproton are written as p^+ and p^-.
More generally, a line drawn above the symbol for a particle refers to an antiparticle.

The greek letter ν is pronounced as 'new'.

The **antineutrino**, $\bar{\nu}$, is a particle of antimatter emitted along with an electron when a nucleus undergoes β^- decay. It is a particle that has almost no mass and no charge. Its corresponding particle the **neutrino**, ν, is emitted with a positron in β^+ decay.

Progress check

1 In alpha particle scattering by a gold nucleus, what feature of the results of the experiment shows that:
   a most of the atom is empty space
   b the nucleus has the same charge as the alpha particles?

2 What is the speed of an alpha particle at its closest distance of approach to a nucleus?

3 Explain why it is more difficult to accelerate neutrons than it is to accelerate electrons.

4 State one piece of evidence for:
   a the particle nature of matter
   b the existence of nuclei.

1 **a** Most of the particles pass through undeviated.
  b Some alpha particles are back-scattered.

2 Zero.

3 Electrons are charged so they can be accelerated by an electric field. Neutrons are uncharged.

4 Possible answers include: **a** Brownian motion **b** alpha scattering

3.6 Quantum phenomena

After studying this section you should be able to:

- *explain how the photoelectric effect gives evidence for the particulate nature of electromagnetic radiation*
- *understand wave-particle duality and the de Broglie hypothesis*
- *describe the use of Stokes' law in determining the quantum of electronic charge*

LEARNING SUMMARY

Waves or particles

AQA A 5

The **photoelectric effect** provides evidence that electromagnetic waves have a particle-like behaviour which is more pronounced at the short-wavelength end of the spectrum. In the photoelectric effect electrons are emitted from a metal surface when it absorbs electromagnetic radiation.

The **threshold wavelength**, λ_0, is the wavelength of the waves that have the threshold frequency.
$\lambda_0 = c \div f_0$

The results of photoelectricity experiments show that:

- there is no emission of electrons below a certain frequency, called the **threshold frequency**, f_0, which is different for different metals
- above this frequency, electrons are emitted with a range of kinetic energies up to a maximum, $(\frac{1}{2}mv^2)_{max}$
- increasing the frequency of the radiation causes an increase in the maximum kinetic energy of the emitted electrons, but has no effect on the photoelectric current, i.e. the rate of emission of electrons
- increasing the intensity of the radiation has no effect if the frequency is below the threshold frequency; for frequencies above the threshold it causes an increase in the photoelectric current, so the electrons are emitted at a greater rate.

The photoelectric effect can be demonstrated using a zinc plate connected to a gold leaf electroscope. An ultraviolet lamp discharges a negatively-charged plate but has no effect on a positively-charged plate.

The wave model cannot explain this behaviour; if electromagnetic radiation is a continuous stream of energy then radiation of all frequencies should cause photoelectric emission. It should only be a matter of time for an electron to absorb enough energy to be able to escape from the attractive forces of the positive ions in the metal.

The explanation for the photoelectric effect relies on the concept of a **photon**, a quantum or packet of energy. We picture electromagnetic radiation as short bursts of energy, the energy of a photon depending on its frequency.

A lamp emits random bursts of energy. Each burst is a photon, a quantum of radiation.

The word quantum refers to the smallest amount of a quantity that can exist. A quantum of electromagnetic radiation is the smallest amount of energy of that frequency.

KEY POINT

The relationship between the energy E of a photon, or quantum of electromagnetic radiation, and its frequency, f, is:

$$E = hf$$

where h is Planck's constant and has the value 6.63×10^{-34} J s.

The energy of a photon can be measured in either joules or **electronvolts**. The electronvolt is a much smaller unit than the joule.

The conversion factor for changing energies in eV to energies in joules is
1.60×10^{-19} J eV^{-1}

KEY POINT

One electron volt (1 eV) is the energy transfer when an electron moves through a potential difference of 1 volt.

$$1 \text{ eV} = 1.60 \times 10^{-19} \text{ J}$$

The work function is the **minimum** energy needed to liberate an electron from a metal. Some electrons need more than this amount of energy.

Radiation below the threshold frequency, f_0, no matter how intense, does not cause any emission of electrons.

Einstein's explanation of photoelectric emission is:

- Light arrives at the metal surface in separate 'packages' or quanta of energy, each with energy hf.
- An electron needs to absorb a minimum amount of energy to escape from a metal. This minimum amount of energy is a property of the metal and is called the **work function**, ϕ.
- If the photons of the incident radiation have energy (hf) less than ϕ then there is no emission of electrons.
- Emission becomes just possible when $hf = \phi$.
- For photons with energy greater than ϕ, the electrons emitted have a range of kinetic energies, those with the maximum energy being the ones that needed the minimum energy to escape.
- Increasing the intensity of the radiation increases the number of photons incident each second. This causes a greater rate of emission of electrons, but does not affect their maximum kinetic energy.

KEY POINT

Einstein's photoelectric equation relates the maximum kinetic energy of the emitted electrons to the work function and the energy of each photon:

$$hf = \phi + (\tfrac{1}{2}mv^2)_{max}$$

At the threshold frequency, the minimum frequency that can cause emission from a given metal, $(\frac{1}{2}mv^2)_{max}$ is zero and so the equation becomes $hf_0 = \phi$.

Stopping the emission

A stopping potential is positive compared to earth.

The emitting surface can be given positive potential, so that it acts as a positive electrode and attracts the electrons that are emitted from it by photoelectric emission. A second plate above this can act as a negative electrode, repelling the emitted electrons. The potential difference between the plates can be varied. The **stopping potential**, V_s, is the potential difference that is just large enough to prevent even the fastest electron from reaching the negative electrode.

KEY POINT

The **stopping potential** V_s, is the potential difference that just stops the fastest electrons from passing between the emitting surface (which acts as a positive electrode) and a negative electrode.

$$eV_s = (\tfrac{1}{2}mv^2)_{max}$$

Particles or waves

AQA A 5

If waves can show particle-like behaviour in photoelectric emission, can particles also behave as waves? Snooker balls bounce off cushions in the same way that light bounces off a mirror, so reflection is not a test for wave-like or particle-like behaviour. Diffraction and interference are properties unique to waves, so particles can be said to have a wave-like behaviour if they show these properties.

Electrons can also be made to interfere when two coherent beams overlap. They produce an interference pattern similar to that of light, but on a much smaller scale.

Other 'particles' such as protons and neutrons also show wave-like behaviour.

Try calculating the de Broglie (pronounced de Broy) wavelength of a moving snooker ball. Is it possible for the ball to show wave-like behaviour?

All particles have an associated wavelength called the **de Broglie** wavelength:

> **KEY POINT**
>
> The wavelength, λ, of a particle is related to its momentum, p, by the de Broglie equation:
>
> $$\lambda = h/p = h/mv$$
>
> where h is the Planck constant.

The de Broglie wavelength of such an electron is of the order of 1×10^{-10} m.

An electron that has been accelerated through a potential difference of a few hundred volts has a wavelength similar to that of X-rays and gamma rays.

This wavelength is also similar to the spacing of the atoms in crystalline materials, so these materials provide suitable sized 'gaps' to cause diffraction.

Particles and waves are the models that we use to describe and explain physical phenomena. It is not surprising that the real world does not fit neatly into our models.

There are two separate models of how matter behaves. The particle model explains such phenomena as ionisation and photoelectricity, while the wave model explains interference and diffraction. It is not appropriate to classify matter absolutely as 'waves' or 'particles' as photons and electrons can fit either model, depending on the circumstances.

Quantum of charge

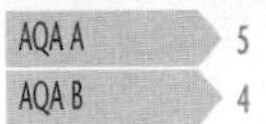

The fundamental unit of charge, 'e' was experimentally determined by Millikan in his 'oil-drop' experiment. A mist of oil drops fell through a hole in an upper plate and viewed to determine their precise velocity. The only forces on the drop are its weight and the viscous force from the air and these will balance when the drop experiences terminal velocity.

> **KEY POINT**
>
> The viscous force (F) on a spherical drop is given by Stokes' law
>
> $$F = 6\pi\eta r v$$
>
> where η is the viscosity of the fluid, r is the radius of the drop and v its velocity.

From this the radius of an oil drop can be determined. An applied voltage was used to set-up a uniform electric field between the upper and lower plates to provide a force that held the oil drop stationary. The viscous force had disappeared and the two remaining forces were in balance.

The balance equation is:

$$mg = QV/d$$

where d is the distance between the plates and m is calculated from the drop's radius and density.

From this the charge Q could be determined. It was found that Q = ne for n = ±1, ±2, … in which 'e' turned out to be the fundamental unit or quantum of charge.

Progress check

1 The work function of potassium is 2.84 eV. Calculate:

a the minimum frequency of radiation that causes photoelectric emission.

b the maximum energy of the emitted electrons when ultraviolet radiation of frequency 8.40×10^{14} Hz is absorbed by potassium.

2 Calculate the de Broglie wavelength of a neutron, mass = 1.67×10^{-27} kg, travelling at a speed of 200 m s^{-1}.
$h = 6.63 \times 10^{-34}$ J s

1 a 6.85×10^{14} Hz
b 1.03×10^{-19} J
2 1.99×10^{-9} m

Sample question and model answer

Note that the two marks awarded here are for describing the structure of an alpha particle and describing the emission.

The nuclear equation should be balanced in terms of both mass (top number) and charge (lower number).

Radon-220 decays by *alpha-emission* to polonium-216.

(a) Explain what is meant by *alpha-emission*. [2]

The nucleus emits a particle (1 mark) which consists of two protons and two neutrons (1 mark).

(b) Complete the decay equation, [2]

$$^{220}_{86}\text{Rn} \rightarrow \text{Po} + \text{He}$$

$^{220}_{86}\text{Rn} \rightarrow {}^{216}_{84}\text{Po} + {}^{4}_{2}\text{He}$ (1 mark for each correct symbol)

(c) The graph shows how the activity of a sample of radon-220 changes with time.

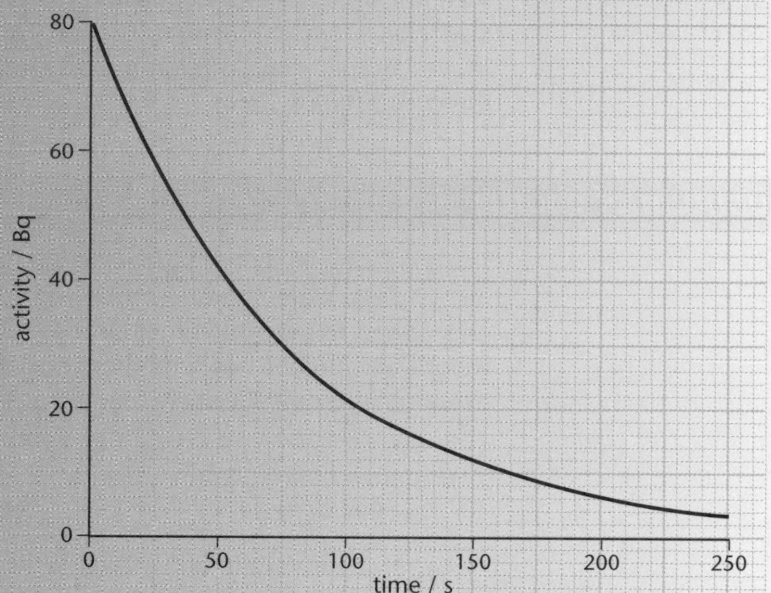

It is important to take at least two readings here to determine the *average* time for the rate of decay to halve. Had the two readings been different, a third one would be necessary.

If these relationships are not given in the question, they will be provided on a data sheet. Always check a data sheet, it may contain the formula that you cannot remember!

(i) Use the graph to determine the half-life of radon-220. [2]

time taken for activity to halve from 80 Bq to 40 Bq = 52.5 s — 1 mark

time taken for activity to halve from 40 Bq to 20 Bq = 52.5 s

half-life = average of these readings = 52.5 s — 1 mark

(ii) The relationship between half-life, $t_{\frac{1}{2}}$ and decay constant, λ, is:

$$\lambda t_{\frac{1}{2}} = \ln 2$$

Calculate the decay constant of radon-220. [2]

$\lambda = \ln 2 \div t_{\frac{1}{2}}$ (1 mark) $= \ln 2 \div 52.5\ \text{s} = 1.32 \times 10^{-2}\ \text{s}^{-1}$ — 1 mark

(iii) The activity, A, of a radioactive sample is related to the number of undecayed nuclei, N, by the formula:

$$A = \lambda N$$

Calculate the number of undecayed nuclei in the sample of radon-220 at time = 0 on the graph. [2]

activity at time t = 0 is 80 Bq — 1 mark

$N = A \div \lambda = 80\ \text{Bq} \div 1.32 \times 10^{-2}\ \text{s}^{-1} = 6060$ — 1 mark

Practice examination questions

1

The diagram shows how an electric immersion heater can be used to measure the specific heat capacity of a metal block.

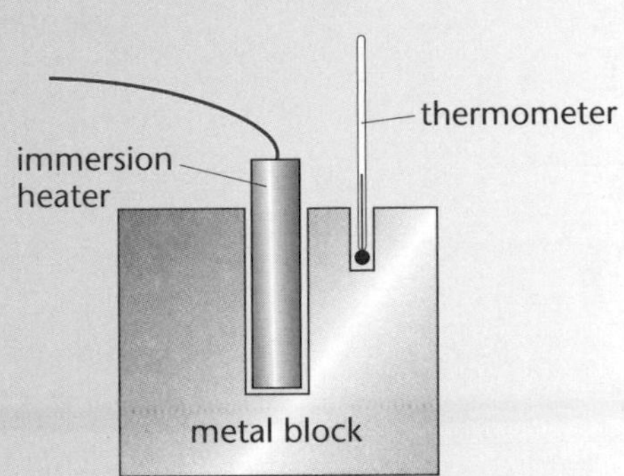

(a) (i) What is the advantage of heating the block through a temperature difference of 10 K rather than 1 K? [1]

(ii) What is the advantage of heating the block through a temperature difference of 10 K rather than 100 K? [1]

(iii) Explain whether you would expect the experiment to give a higher value or a lower value than that given in a data book. [2]

(iv) Suggest one way of improving the reliability of the result. [1]

(b) Use the data to calculate a value for the specific heat capacity of the metal. [3]

heater power = 24 W
time heater switched on = 350 s
mass of block = 0.80 kg
initial temperature = 22°C
final temperature = 36°C

2

The first law of thermodynamics can be written as:

$$\Delta U = Q + W$$

(a) State the meaning of each symbol in the equation. [3]

(b) As a gas is being compressed, 4500 J of work are done on it and 4800 J of heat are removed from it. Explain what happens to the temperature of the gas as a result of this. [2]

3

The diagram shows a heat engine.

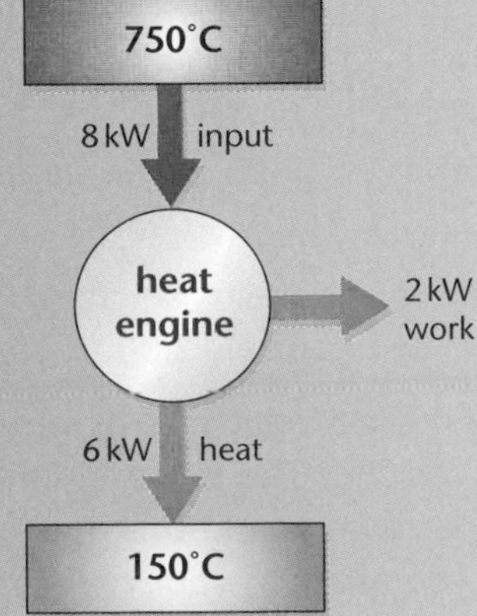

Practice examination questions (continued)

(a) Calculate the efficiency of the heat engine. [2]

(b) What is the maximum efficiency of the engine? [2]

(c) Suggest one reason why the efficiency of the heat engine is less than its theoretical maximum. [1]

4

A sample of an ideal gas has a volume of 2.20×10^{-6} m^3 at a temperature of 25°C and a pressure of 2.02×10^5 Pa.

(a) Calculate the temperature of the gas in kelvin. [1]

(b) Calculate the number of moles of gas in the sample. [3]
$R = 8.3$ J mol^{-1} K^{-1}

(c) The absolute temperature of the gas is doubled, while the pressure remains constant.

(i) What is meant by the *internal energy* of the gas? [1]
(ii) In what form is this energy in an ideal gas? [1]
(iii) What happens to the internal energy of the gas when its absolute temperature is doubled? Explain your answer. [2]
(iv) Calculate the new volume occupied by the gas after its absolute temperature has doubled and constant pressure. [2]

5

In a nuclear reactor uranium-238, $^{238}_{92}U$, undergoes fission when it captures a neutron.

(a) (i) Write down the symbol for the isotope formed when uranium-238 captures a neutron. [1]
(ii) Explain what happens when a nucleus of this isotope undergoes fission. [3]
(iii) Explain how fission can lead to a chain reaction. [2]

(b) (i) Explain why nuclear fission results in the release of energy. [2]
(ii) In what form is the energy released? [2]
(iii) Outline how the energy released in fission is used to generate electricity. [3]

(c) State the purpose of each of the following parts of a nuclear reactor:
(i) control rods [1]
(ii) moderator. [1]

6

A sample of an ideal gas has a volume of 1.50×10^{-4} m^3 at a temperature of 22°C and a pressure of 1.20×10^5 Pa.

(a) Calculate the temperature of the gas in kelvin. [1]

(b) Calculate the number of moles in the sample.
$R = 8.3$ J mol^{-1}. [3]

(c) If the volume of the gas remains constant, calculate the temperature at which the pressure of the gas is 2.00×10^5 Pa. [3]

(d) (i) Calculate the mean kinetic energy of the gas particles at a temperature of 22°C. The Boltzmann constant, $k = 1.38 \times 10^{-23}$ J K^{-1}. [3]
(ii) At what temperature would this mean kinetic energy be doubled? Explain how you arrive at your answer. [2]

Practice examination questions (continued)

7

Calculate the energy released when 1.8 kg of steam at 100°C is changed into water at 85°C. [3]

specific latent heat of vaporisation of water = 2.25×10^6 J kg^{-1}

specific heat capacity of water = 4.20×10^3 J kg^{-1} K^{-1}

8

A kettle is filled with 1.50 kg of water at 10°C.

The kettle is fitted with a 2.40 kW heater.

The specific heat capacity of water is 4.2×10^3 J kg^{-1} K^{-1}.

(a) Calculate the time it takes for the water to boil after the kettle is switched on. [3]

(b) Suggest three reasons why the time taken is longer than the answer to (a). [3]

9

Iodine-123 and iodine-131 are two radioactive isotopes of iodine.

Iodine-123 emits gamma radiation only with a half-life of 13 hours.

Iodine-131 emits beta and gamma radiation and has a half-life of 8 days.

Both can be used to monitor the activity of the thyroid gland by measuring the count rate using a detector outside the body.

(a) Which of the radioactive emissions from iodine-131 can be detected outside the body? Give the reason for your answer. [2]

(b) In terms of the radiation emitted, explain why iodine-123 is often preferred to iodine-131. [2]

(c) Explain why a smaller dose is needed when iodine-123 is used than for iodine-131. [2]

(d) A patient is given a dose of iodine-123. After what time interval can medical staff be sure that the activity of the patient's thyroid gland is less than 2% of its greatest value? [2]

10

Information about the size of a nucleus can be obtained by firing high-energy electrons at the nucleus and examining the subsequent diffraction pattern.

(a) (i) Why are electrons suitable for this purpose? [2]

(ii) Explain why the electrons need to be high-energy. [3]

(b) The relationship between the radius, R, of a nucleus and its atomic number, A is: $R = r_0 A^{1/3}$

where r_0 has the value 1.20×10^{-15} m.

Calculate the radius of the nuclei of:

(i) ${}^{4}_{2}He$ [2]

(ii) ${}^{32}_{16}S$ [1]

(c) Explain what the answers to (b) show about the densities of the helium and sulphur nuclei. [3]

Practice examination questions (continued)

11

The diagram shows some of the results of firing alpha particles at gold foil.

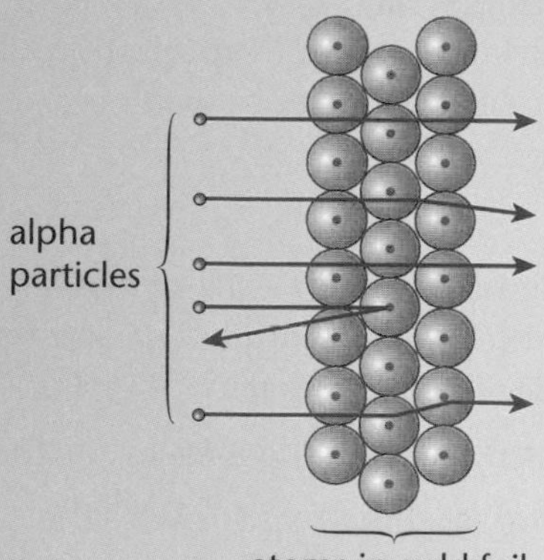

Explain how each of the following provides evidence for the atomic model.

(a) Most alpha particles pass through the foil undeviated. [2]

(b) A very small number of alpha particles are scattered backwards. [3]

12

Einstein's equation for photoelectric emission is:

$$hf = \Phi + (\tfrac{1}{2}mv^2)_{max}$$

(a) (i) State the meaning of each term. [3]

(ii) What is the significance of the subscript 'max'? [1]

(b) Electromagnetic radiation of frequency 2.02×10^{15} Hz is directed at a metal surface. The maximum kinetic energy of the electrons emitted is 4.05×10^{-19} J. The intensity of the radiation is 4.5 μW m^{-2}.

Calculate:

(i) the energy of each photon [2]

(ii) the number of photons reaching the surface each second if the area is 2.5×10^{-6} m^2 [2]

(iii) The work function of the metal. Give your answer in eV. [3]

Chapter 4

Further physics

The following topics are covered in this chapter:

- *Physics of the senses*
- *Medical and geophysical imaging*
- *Starlight*
- *The expanding Universe*

4.1 Physics of the senses

LEARNING SUMMARY

After studying this section you should be able to:

- *use ray diagrams and the lens formula to describe image formation by lenses*
- *explain how the quality of vision depends on the intensity of the ambient light*
- *explain that the ear is sensitive to a range of frequencies and intensities of sound*
- *compare sound intensities in W m^{-2} and in decibels*

Lenses and the eye

AQA A 5

When light leaves a lamp or other source of light it spreads out or diverges, reducing in intensity with increasing distance from the source. A **concave** or **diverging lens** exaggerates this effect, making the light more divergent. A **convex** or **converging lens** acts against this effect, so that the light either converges or becomes less divergent.

Light from a source converges after passing through a convex lens if the source is at a greater distance from the lens than the focal length.

The dotted lines in the diagram of a concave lens show where light **appears** to have come from.

The diagrams show the effect of these lenses on a parallel beam of light.

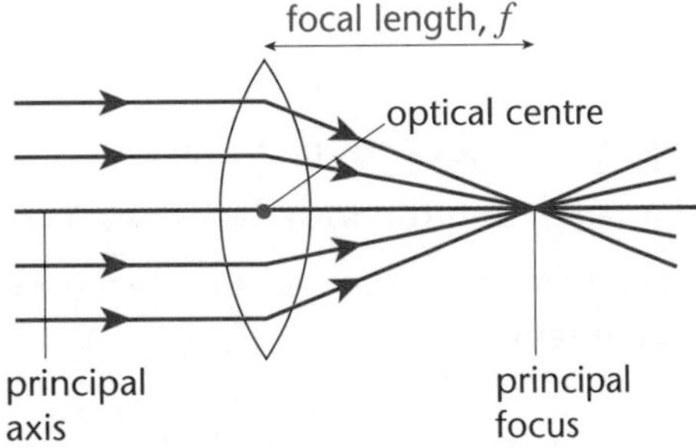

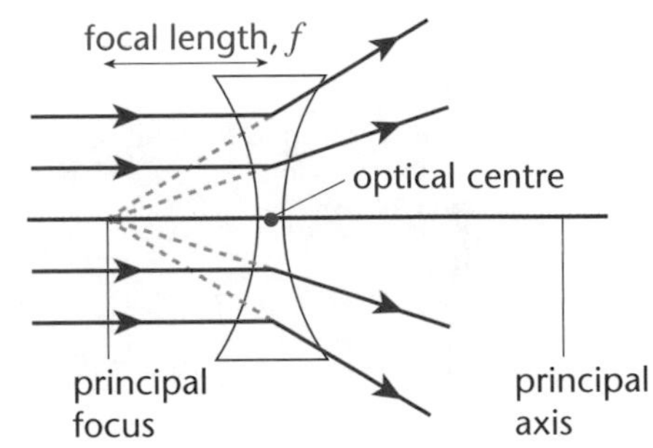

The diagrams also show some important features of the lenses:

- the **optical centre** is at the centre of the lens
- the **principal axis** is a line drawn through the optical centre, perpendicular to the plane of the lens
- the **principal focus** is the image position for light that is initially parallel to the principal axis
- the **focal length** is the distance between the optical centre and the principal focus.

The image is virtual in the case of a concave lens, and real for a convex lens.

The shorter the focal length of a lens, the greater the effect on converging or diverging light that passes through it. This effect is measured by the power of the lens.

Lens power is calculated using the relationship:

power = 1/focal length

The unit of power is the dioptre (D) when the focal length is measured in m.

KEY POINT

A virtual image (such as in a plane mirror) can be seen by the eye but cannot be projected onto a screen.

The diagram below shows ray diagrams for an object placed at a distance of 8 cm from converging and diverging lenses each of focal length, f = 5 cm.

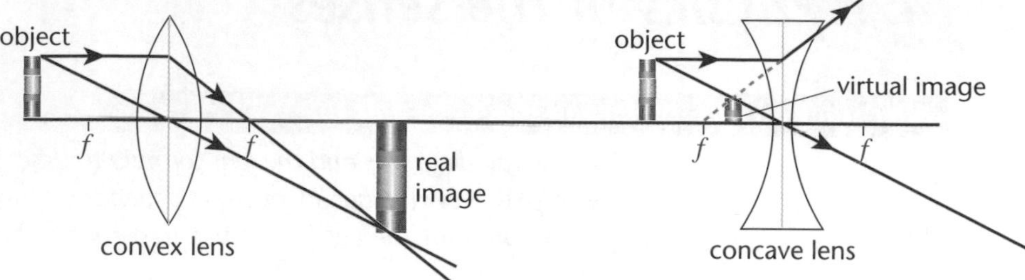

A real image is formed if the light converges after passing through the lens. If the light is still diverging, then the image is virtual.

The position, nature and size of the images can be determined from the diagrams. They can also be calculated using the **lens formula**:

The magnification, i.e. the size of the image compared to the object, is equal to v/u.

The lens formula applies to both convex and concave lenses. It relates the positions of the object and image to the focal length of the lens.

$$\frac{1}{u} + \frac{1}{v} = \frac{1}{f}$$

Where u is the distance between the object and the lens, v is the distance between the image and the lens, and f is the focal length of the lens.

KEY POINT

This is known as the 'real is positive' sign convention.

When using the lens formula:

- real objects, images and the focal length of a convex lens have positive values
- virtual images, and the focal length of a concave lens, have negative values.

The eye uses two converging lenses, the cornea and the eye lens, to produce a real image on the retina.

Adjustment of the eye to focus on objects at different distances is called **accommodation**; this is achieved by changing the shape of the lens.

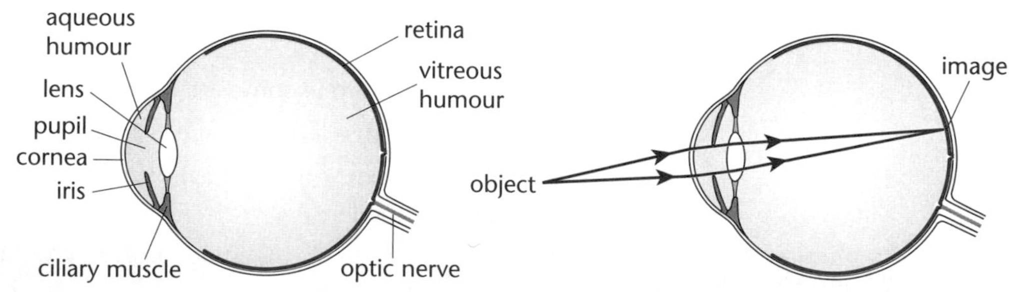

Defective vision

The common defects in vision are:

- **short sight** or **myopia** when an eye cannot focus on distant objects but can focus on near objects. It is caused by the eyeball being too long or the lens, at its weakest, being too strong.
- **long sight** or **hypermetropia** is the opposite of short sight; distant objects can be seen clearly but near ones cannot. It is caused by the eyeball being too short or the lens, at its strongest, being too weak.

An eye lens is elastic tissue which becomes more difficult to stretch with increased use.

- **presbyopia** is the condition of being both long-sighted and short-sighted. It often occurs in older people due to the lens becoming stiffer with age.
- **astigmatism** is caused by the surface of the cornea not being spherical. This results in objects in one direction being sharp and in focus while those in a different direction are blurred.

The diagrams show the conditions of short sight and long sight and their correction with converging and diverging lenses respectively.

These diagrams show light from a distant object arriving at a short-sighted eye.

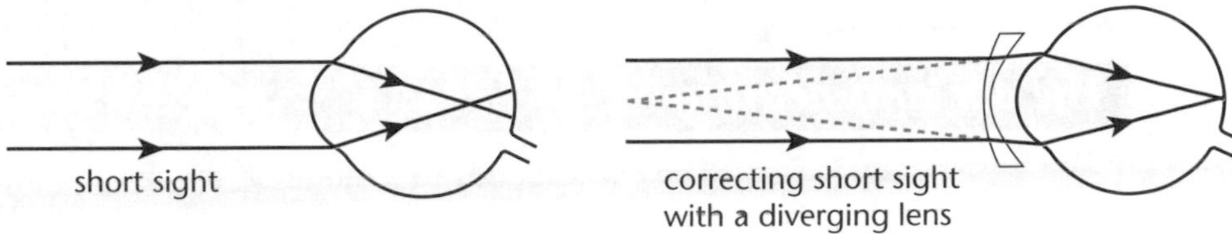

These diagrams show light from a near object arriving at a long-sighted eye.

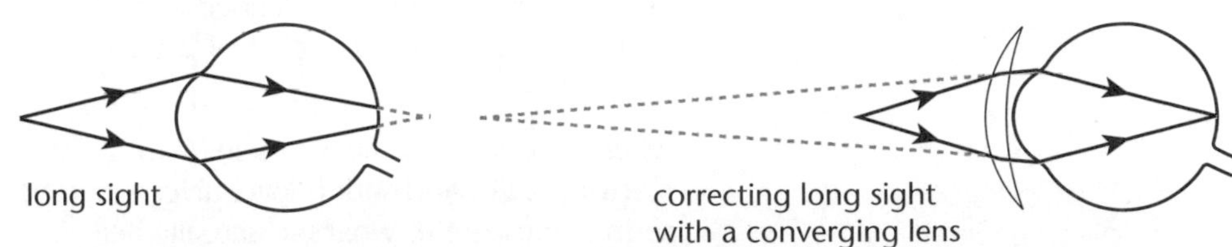

The lens formula can be used to calculate the focal length, and hence power, of the correcting lens needed. The object distance, u, is the distance of the actual object and the image distance, v, is the desired distance for the eye to focus on it.

People who suffer from presbyopia often wear bifocal spectacles. The lower half of each lens is convex for reading and the upper half is concave for distant viewing.

To correct astigmatism cylindrical lenses are used. These are less curved in one direction to compensate for the unsymmetrical curvature of the cornea.

Colour vision

AQA A 5

The light-sensitive cells that make up the retina have different sensitivities:

- **rods** are sensitive to light of different intensities but not of different wavelengths
- there are three types of **cones**, called red, green and blue. Each of these responds to a different range of wavelengths and frequencies.

This explains why, in low-intensity light such as moonlight, we see in monochrome.

Cones enable **colour vision** but they do not respond to low-intensity light. The graph shows the relative sensitivities of the three types of cone.

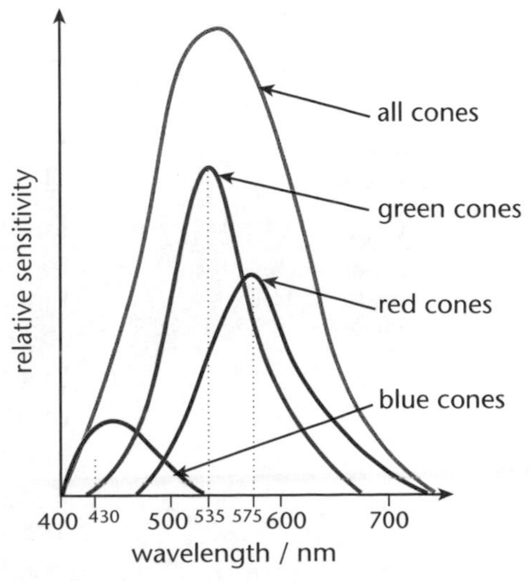

The diagram shows why blue-coloured lamps always appear dim and yellow-coloured lamps appear bright.

The low sensitivity of the blue cones explains why we never perceive this colour as bright as red or yellow. An advertising hoarding designed to catch the eye would not use blue to do so. However, it is a more acceptable colour for 'low-key' signs and buildings.

Vision in low-intensity light is **scotopic** – we cannot detect colour but only differences in brightness. In scotopic vision the cones are not responsive so the rods are the only detectors operating.

Vision is most acute at the fovea, or yellow spot. This is the point on the retina where the light-sensitive cells are most closely-packed.

In more intense light the cones take over and vision is **photoplc**. Photopic vision allows differentiation of colour in addition to intensity. The ability of the eye to distinguish between two objects that are close together, called its **spatial resolution**, is also greater in photopic vision. Spatial resolution requires at least one unstimulated cell between those that are stimulated. Resolution is greatest at the centre of the retina where the cells are most concentrated. It is worst at the edge of the retina where the cells are more widespread and several rods share a single nerve fibre, so the brain cannot distinguish between individual rods.

The ear and hearing

AQA A 5

The ear responds to a range of frequencies and a range of intensities.

> **KEY POINT**
>
> The intensity of a wave is defined as the power incident per unit area perpendicular to the direction of wave travel.
>
> $$I = P/A$$
>
> where intensity, I is measured in W m^{-2}. Sound level is measured in **decibels** (dB), and sound level gives a better, if still rather crude and simple, measure of what we actually hear.

The threshold of hearing is the minimum intensity required to be able to detect a sound.

intensity in W m^{-2}	10^{-12}	10^{-11}	10^{-10}	10^{-9}	10^{-8}	10^{-7}	10^{-6}	10^{-5}	10^{-4}	10^{-3}	10^{-2}	10^{-1}	10^{0}	10^{1}	10^{2}
sound level in decibels	0	10	20	30	40	50	60	70	80	90	100	110	120	130	140

The correspondence of sound intensity to the decibel scale at a frequency of 1 kHz

On a logarithmic scale, equal distances on the scale are used to represent multiplication by 10.

A logarithmic scale is used on the diagram because of the way in which the ear responds to relative changes in intensity. It has a **logarithmic response**. This means that equal factor changes in intensity (e.g. ×100) produce equal simple addition changes in perceived sound level (e.g. +20). This leads to the decibel scale for measuring sound level.

> **KEY POINT**
>
> sound level in decibels = 10 log (I/I_0)
>
> where I_0, the threshold of hearing, is taken to be 1.0×10^{-12} W m^{-2}.

Normal conversation has an intensity of about 1×10^{-6} W m^{-2} or 60 dB.

On the decibel scale, an increase of 10 dB corresponds to the intensity becoming ten times as great.

A sound of intensity 120 dB or 1 W m^{-2} causes discomfort; one of 140 dB or 100 W m^{-2} causes extreme pain in the ear.

By definition, loudness corresponds exactly with the decibel scale of sound level at 1 kHz.

Because our ears are most sensitive to frequencies around 3000 Hz, higher and lower frequencies need a greater intensity to sound equally loud. A more sophisticated method of measuring actual perceived loudness is in phons, where 1 phon is equivalent to a sound level of 1dB at a frequency of 1kHz. The graph shows the variation in intensity required to cause equal loudness at four loudness levels.

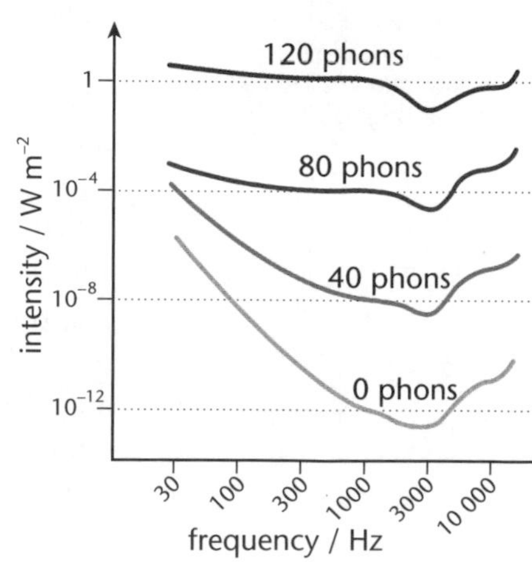

This graph shows that:

- the frequency range of a human ear is approximately 20 Hz to 20 000 Hz, but the upper limit falls with increasing age and exposure to noise due to wear of the ossicles (bones that transmit sound through the middle ear) and damage to the sensing hair cells in the inner ear
- the minimum intensity required to detect a sound, I_0, depends on frequency
- the ear is at its most sensitive at frequencies around 3000 Hz.

Progress check

1 An object of height 2.0 cm is placed on the principal axis at a distance of 12.0 cm from a convex lens of focal length 5.0 cm.
By drawing or calculation, work out the position, size and nature of the image.

2 Which **two** parts of the eye converge light as it passes through them?

3 Explain why the brain does not perceive colour in low-intensity ambient light.

4 State **two** factors that reduce the upper limit of the hearing range.

5 A sound has an intensity of 5.0×10^{-7} W m^{-2}. Calculate the sound level in dB.

1 The image is real, 8.6 cm on the other side of the lens and 1.4 cm high.
2 The cornea and the eye lens.
3 The cones do not function in low-intensity light, and the rods only detect brightness.
4 Age and exposure to noise.
5 57 dB.

4.2 Medical and geophysical imaging

After studying this section you should be able to:

LEARNING SUMMARY

- *describe and compare the techniques used for imaging the body, including ultrasound*
- *understand the physical principles and diagnostic use of X-rays*
- *describe processes of imaging the Earth*

The heart and ECG

AQA A 5

The valves are necessary to ensure that blood does not flow from the ventricle to the atrium when the ventricle contracts.

The **heart** is a pump controlled by electrical signals. One side takes in de-oxygenated blood from the body and pumps it to the lungs; the other side takes in oxygenated blood from the lungs and pumps it to the body. The diagram shows the heart and the one-way valves that only allow blood to flow in the directions shown by the arrows, from the **atrium** to the **ventricle**.

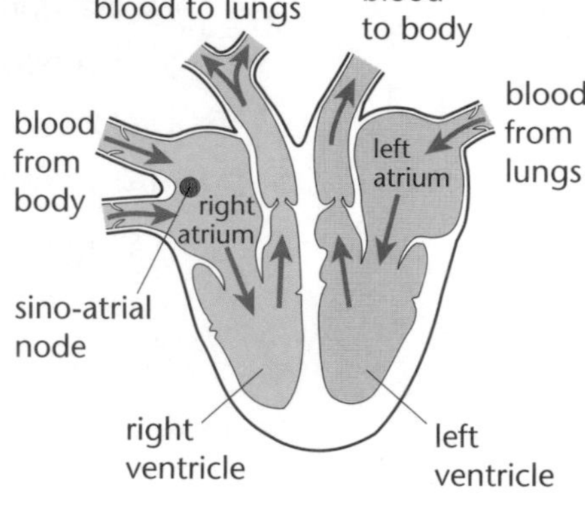

Electrical signals that control the action of the heart:

- are produced in the **sino-atrial node**
- are carried by nerve fibres
- cause contraction of the atria, followed by contraction of the ventricles to pump blood through the heart and round the body.

Body fluids also conduct the electrical signals, enabling them to be detected at the surface of the skin by electrodes. A display, such as that produced on a cathode ray oscilloscope (CRO) of the change in voltage with time, is called an **electrocardiogram** (**ECG**).

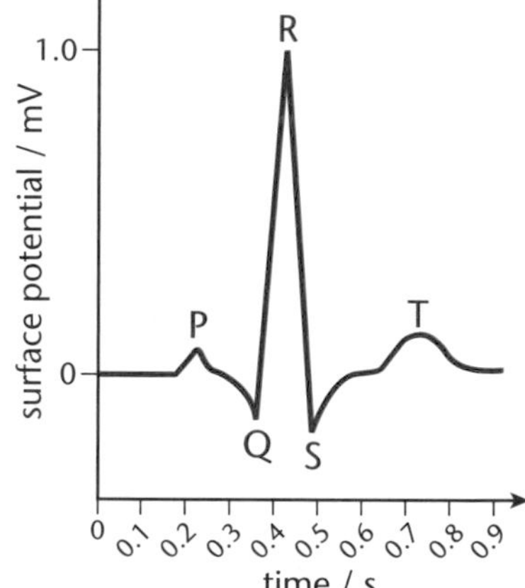

In the ECG shown in the diagram:

The pulse that causes relaxation of the atria occurs at the same time as, and is masked by, the QRS pulse.

- the pulse at P causes the atria to contract
- the QRS pulses causes the ventricles to contract
- the T pulse causes the ventricles to relax.

Using ultrasound

AQA A 5
AQA B 4

Ultrasound consists of compression waves with a frequency above that of human hearing. It is useful for examining the internal structure of the body because it passes through flesh, but some reflection takes place at tissue boundaries. These reflections are used to generate images.

Pulses of ultrasound are produced and detected by a **piezoelectric crystal**:

The alternating voltages due to the reflected waves are interpreted by a computer to produce an image.

- an alternating voltage applied to the crystal causes it to vibrate
- the voltage is removed to allow the crystal to detect any reflections
- when the crystal is made to vibrate by the reflected pulses it generates an alternating voltage.

The scanning of a fetus is now a routine procedure carried out on all pregnant women to check for abnormalities and determine the precise date when birth is due to take place.

Types of ultrasound images include:

- an **A-scan** (amplitude modulated) which uses the echoes to show how the tissue boundaries are related to depth below the surface of the skin, A-scans are used to make precise determinations of the position of an object such as a brain tumour
- a **B-scan** (brightness modulated) which gives a two-dimensional picture, such as that of a fetus.

Ultrasound scans have the advantage in that ultrasound is non-invasive and does not cause ionisation and damage to body tissue.

Acoustic impedance and reflections

AQA A 5

The decrease in intensity of a wave as it travels through a material is known as **attenuation**. Attenuation is due to a number of factors, including:

- the energy of the wave becoming spread over a wider area as it travels from its source
- interactions with the material including absorption, diffraction and scattering.

Sound travels faster in solids than in liquids and faster in liquids than in gases.

The opposition of a material to sound passing through it is called **acoustic impedance**. The acoustic impedance depends on the density of the material and the speed at which sound travels in it. Maximising the transfer of sound from one material to another depends on matching the impedances as closely as possible.

The acoustic impedance, Z, of a material depends on its density, ρ, and the speed of sound within it, c.

$$Z = \rho c$$

It is an important quantity in ultrasound technology because the proportion of sound energy reflected from a boundary depends on the acoustic impedances of the two materials.

The greater the difference in acoustic impedance, the more intense the reflection. For this reason a jelly-like substance is spread over the patient's skin during scanning, so that there is no air between probe and skin. The difference in acoustic impedance of skin and air is large, but that between the jelly and skin is small so it improves the impedance matching and maximises the amount of energy transmitted into the body.

High frequencies of ultra sound needed for good resolution are attenuated more than lower frequencies, so they can only produce high-resolution scans close to the surface of the body.

$$\frac{\text{reflected intensity}}{\text{incident intensity}} = \frac{(Z_2 - Z_1)^2}{(Z_2 + Z_1)^2}$$

where Z_2 is the acoustic impedance of the second material and Z_1 is that of the first.

Note that when $Z_1 = Z_2$ the formula predicts zero reflection.

Producing X-rays

AQA A 5
AQA B 4

X-rays are produced when high-speed electrons hit a target. In the X-ray tube shown in the diagram:

- electrons are emitted from the cathode by thermionic emission
- they are accelerated through the vacuum by the high-voltage anode

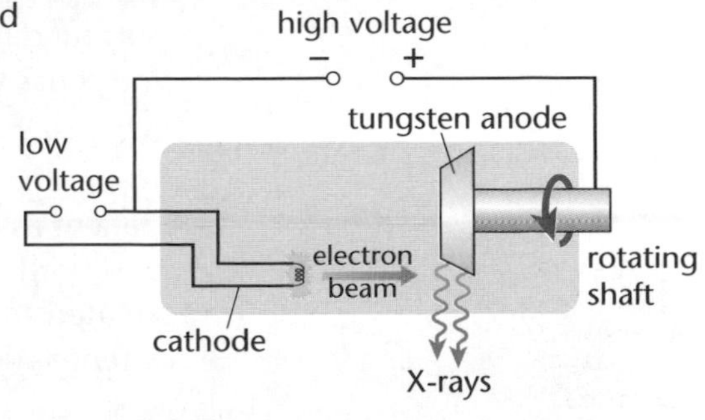

A rotating anode X-ray tube

X-rays are emitted when electrons strike the screen of a television or cathode ray tube.

- when they strike the anode about 1% of the energy is transferred to X-rays, the rest heats the anode
- the anode rotates to avoid overheating of one small area.

The **intensity** of the beam is determined by the filament current which causes thermionic emission. The energy of the X-ray photons is determined by the anode voltage; increasing this voltage produces higher-energy, shorter-wavelength X-rays which are more penetrative.

Using X-rays in diagnosis

Whereas ultrasound images are produced from reflections, X-ray images rely on some absorption and some transmission by the body. As an X-ray beam passes through tissue it becomes attenuated. The intensity of a parallel beam varies exponentially with thickness.

In exponential attenuation, the intensity of the beam is reduced by the same proportion in passing through equal thicknesses of material. This means that the half-value thickness is independent of the initial intensity of the beam.

KEY POINT

The intensity of an X-ray beam, I, after passing through a distance x of absorbing material, is given by:

$$I = I_0 e^{-\mu x}$$

Where I_0 is the initial intensity and μ is a constant, called the **linear attenuation coefficient**. The value of μ depends on the material and the energy of the X-rays.

The **half-value thickness** of a material, $x_{\frac{1}{2}}$, is the thickness that halves the original intensity of the beam. This relationship between μ and $x_{\frac{1}{2}}$ is:

$$\mu x_{\frac{1}{2}} = \ln 2$$

The linear attenuation coefficient changes with changing density, ρ, of the material. The mass attenuation coefficient, μ_m, defined as $\frac{\mu}{\rho}$, is independent of density.

An X-ray photograph is a shadow picture. A detector placed behind the part of the body being X-rayed shows areas of varying greyness, with the dark areas representing the least attenuation of the beam. When bones and teeth are X-rayed, they show up as white. Suitable detectors are:

- black and white photographic film, but this requires a beam of high intensity
- an **intensifying screen** which contains a material that absorbs energy from the X-rays and re-emits it as light, a process known as **fluorescence**. The light then produces a photographic image. Film is more sensitive to light than to X-rays, so less exposure is required.
- a fluoroscopic image intensifier; the X-rays cause electrons to be ejected from a photocathode. These electrons are then accelerated in the same way as in a cathode ray tube, where they produce a bright image on a fluorescent screen. This technique is used when a 'moving' picture is required.

Studying a moving picture enables medical staff to diagnose faults in the way that organs act or bones move.

When intestines are X-rayed, there is very little absorption by the soft tissue and the result is a low-contrast image. The contrast between the intestines and surrounding tissue is increased by giving the patient a 'barium meal'. This lines the intestines with a material that is opaque to X-rays, so they show up as white.

A CT or CAT (**computerised axial tomography**) scan uses a rotating X-ray beam to produce an image of a slice through the body. The beam is rotated around the patient as the patient moves through the beam. The scattered X-rays are detected and used by a computer to build up a picture. Three-dimensional images can be created from images of the individual slices. CAT scans detect slight changes in attenuation and are often used for detecting tumours and cancers.

Radioactive tracers

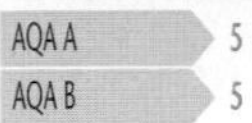

5
5

Technetium-99m is a particularly useful isotope for medical purposes. It is produced as a fission product in nuclear reactors, as excited state, or metastable, nuclei, and it changes to technetium-99 by gamma emission, with a half-life of about 6 hours. It can be separated from the other materials in spent fuel rods. (The procedure is a specialist one and only a few nuclear power stations in the world are equipped to carry it out.)

It's usefulness is due to its emission of gamma radiation only, without alpha or beta emission, and due to its suitable half-life.

Alpha and beta emissions produce strong ionization in any material, including human tissue, and this is potentially harmful. Alpha radiation is particularly strongly ionising and therefore particularly dangerous. Gamma radiation, as emitted by technetium-99m, is less ionizing and more penetrating, and so can be detected from outside the body.

When any sample of Technetium-99m is isolated its activity halves roughly every 6 hours, but that is enough to allow for it to be delivered to the medical centres where it is used. It also means that material that is not excreted by the body's natural processes depletes acceptably quickly, halving in quantity every 6 hours.

The emitted gamma radiation can be detected by gamma cameras. These can detect individual photons when they cause excitation of atoms in crystals of iodine. The return of an atom to its ground state produces a flash of light, and the effect of this is amplified by a photomultiplier system.

Iodine-131 decays by beta- emission, and this is usually accompanied by gamma emission. The gamma-ray photons are detectable from outside the body by a gamma camera. It is especially useful for studying the thyroid gland, since iodine accumulates naturally there, so that the rest of the body receives a modest radiation dose.

The half-life of the nuclide is just over 8 days, and this adds to its usefulness since high concentrations do not linger in the body. As a beta-emitter it delivers a higher radiation dose than an emitter of gamma-radiation only, and so is useful for killing cells in radiotherapy, especially in and around the thyroid gland.

A gamma camera

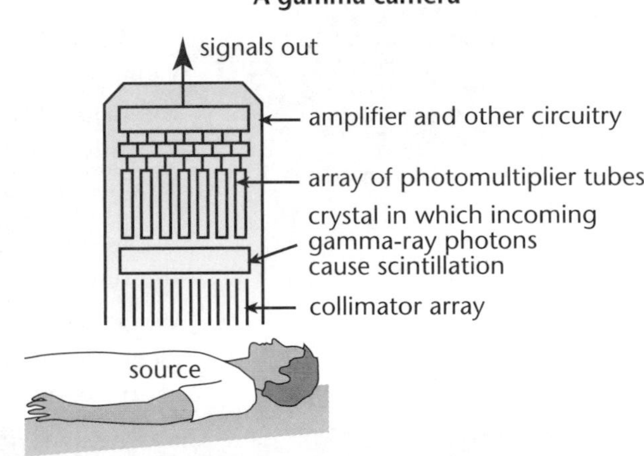

Fibre optics

5
4

Fibres are normally clad with a glass of lower refractive index than the core in order to minimise the value of the critical angle.

An **optical fibre** can transmit light round corners using **total internal reflection**. When light strikes the interface between the fibre core and its cladding it is reflected back inside the fibre provided that the angle of incidence is greater than the critical angle. This is shown in the diagram.

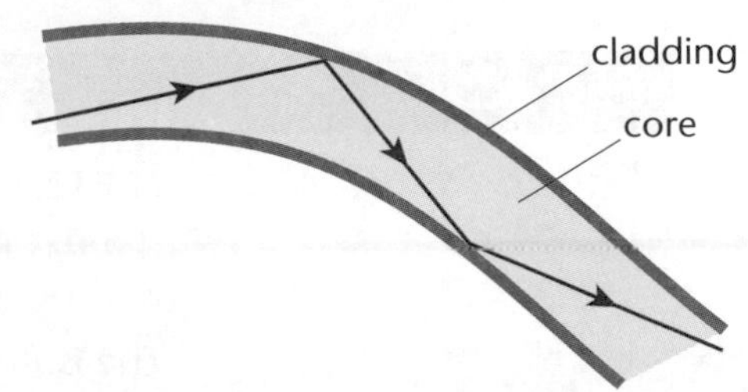

An **endoscope** is a bundle of fibres that can be used for viewing inside the body or performing surgery.

For viewing inside the body two sets of fibres are used:

- the fibres that take light into the body are **non-coherent**, they do not need to stay in the same position relative to other fibres
- the fibres that transmit the image out of the body are **coherent**, they are in the same positions relative to each other at each end of the bundle. The narrower the fibres used, the greater the resolution of the image.

Tissue becomes heated by absorption of the energy from laser radiation.

Lasers are sources of high-intensity electromagnetic radiation that can be passed into the body along an optical fibre. This allows heating to be applied to small areas of tissue by absorption of the laser beam. Used in this way optical fibres can:

- act as a scalpel by cutting through tissue but without causing bleeding
- repair tears and stop bleeding by coagulating body fluids.

CCDs

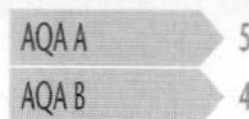

5
4

Digital cameras as well as astronomical, medical and other imaging systems use CCDs, or charge-coupled devices. A CCD is made up of small sandwiches of semiconductor in arrays, each member of the array providing the information for the creation of one pixel. Incoming photons can each liberate an electron from a bound state (bound to an atom) so that it is free to move within the semiconductor pixel. The resulting changes in voltage can be detected.

Magnetic resonance imaging

AQA A 5
AQA B 4

Magnetic resonance imaging (MRI) detects the presence of hydrogen nuclei. Since body tissue has a high water content, it can be used to produce an image of the body. It works by applying pulses of a high-intensity and high-frequency magnetic field which results in the emission of electromagnetic radiation from the nuclei by the following process.

Very powerful electromagnets are needed to produce the magnetic fields required by an MRI scanner.

- Hydrogen nuclei spin, giving them a magnetic field.
- When an external magnetic field is applied, they rotate around the direction of the field. This rotation is called **precession**, it is like that of a spinning top when it 'wobbles'.
- The frequency of precession is called the **Larmor frequency.**
- Application of a pulse of radiation at the Larmor frequency causes **resonance**, the nuclei absorb the energy into the precession.
- When the pulse is removed the nuclei lose the energy, emitting it as electromagnetic radiation. This takes place over a short period of time known as the **relaxation time**.

The emissions from the hydrogen nuclei are detected by a radio aerial and processed by a computer to give three-dimensional body imaging. MRI scanning is non-invasive and does not cause ionisation, but it is very expensive in terms of the capital equipment and running costs.

PET scans

AQA A 5

PET scans, like MRI and CAT scans, produce images of the inside of the body. PET stands for positron emission tomography. By injection or drink, a β^+ emitting isotope is introduced to the body.

The isotopes emit positrons (β^+) which meet electrons in the surrounding material, resulting in mutual annihilation and emission of gamma radiation which can be deflected from outside the body to allow an image to be produced.

The radioisotopes can be chemically bonded to substances such as glucose, so that parts of the body where glucose is dense show up brightly on the image.

Imaging the Earth's surface

AQA B 4

Satellites can use different parts of the electromagnetic spectrum to image the surface of the Earth. They can use 'radar' processes, in which the satellite emits radiation and detects the reflections. They can use reflected sunlight. They can use emitted infrared radiation. A satellite might scan the Earth and measure different intensities of infrared radiation, for example. The data can be presented as a map, using different colours to represent different intensities. Since these colours are not the natural colours of the surface, such an image is called a **false-colour image**.

False colours can also be used to represent gravitational anomalies, which are patterns of varying gravitational field strength due, for example, to different densities of rock. A satellite's height varies very slightly as is passes over different areas, and the variation is enough to be measured to produce gravitational anomaly maps of the whole of the Earth.

Seismic surveys

AQA B 4

Seismic surveys use reflected waves created by earthquakes or deliberate explosions to build up images of the Earth's crust, often vertical cross-sections of the ground. The principles are similar to those of sonar in water or ultrasound in the human body.

Surveying the ground

AQA B 4

Resistivity of a material, as for other properties of materials, such as density, can be found by measurements on a sample. Length l, parallel to the current, cross-sectional area A, perpendicular to the current, and the resistance of the sample are the required quantities. Resistivity is then given by:

$\rho = RA / l$

Resistivity surveys are useful in fields such as pollution monitoring and archaeology. Resistivity, ρ, is a property of material, related to the resistance of a given cross sectional area normal to the current and the length of material.

Two probes stuck in to the ground, with a potential difference applied to them, can be used to measure the resistance of the ground, and the distance between them (l) and the effective cross sectional area through which the current flows can be used to determine the resistivity of the ground material.

The presence of former human habitation, evidenced by regular patterns of varying resistivity due to the presence of old timber or charcoal in the ground, or to different water-holding properties of different material, can be detected. So also can more recent polluting materials.

There are subtle local variations in magnetic field for similar reasons. A proton magnetometer is a sensitive instrument that measures magnetic field by detecting the resonant frequency of oscillating (or 'precessing') the hydrogen nuclei (protons), and this frequency is highly dependent on the local magnetic field. (The principle is related to that of MRI imaging.)

Metal detectors transmit radio waves to detect buried (or other) metal objects. The radio waves induce e.m.f.s in metal objects, which then also act as transmitters. The metal detector has a receiver to detect these responses to its own signals.

Progress check

1 In an X-ray tube:
  a what determines the intensity of the X-ray beam
  b what determines the wavelength of the X-rays produced?

2 Why is ultrasound preferred to X-rays for producing an image of a fetus?

1 a The current in the filament.
  b The anode voltage and the anode material.
2 Ultrasound does not cause ionisation or damage body tissue.

4.3 Starlight

After studying this section you should be able to:

- describe the structure of reflecting, refracting and radio telescopes
- explain how the resolution of a telescope depends on its aperture
- understand Stefan's law and Wien's displacement law for black bodies
- explain how the evolution of a star depends on its mass
- compare the apparent magnitudes and calculate absolute magnitudes of stars
- describe how stars are classified in terms of their brightness, temperature and composition
- explain how an absorption spectrum gives information about the elements present in a star

LEARNING SUMMARY

Optical telescopes

AQA A 5

A **refracting telescope** uses two converging lenses. The objective lens produces a real image at its principal focus. The weaker this lens, the greater the size of the image produced and the longer the telescope. The eyepiece lens acts as a magnifying glass; by placing it so that its principal focus is at the position of the real image being magnified, the final image is formed 'at infinity' and the telescope is said to be in **normal adjustment**. This is shown in the diagram.

Refracting telescopes are limited in the diameter of the objective lens that can be used, since the weight of a large lens can cause the glass to flow and change its shape.

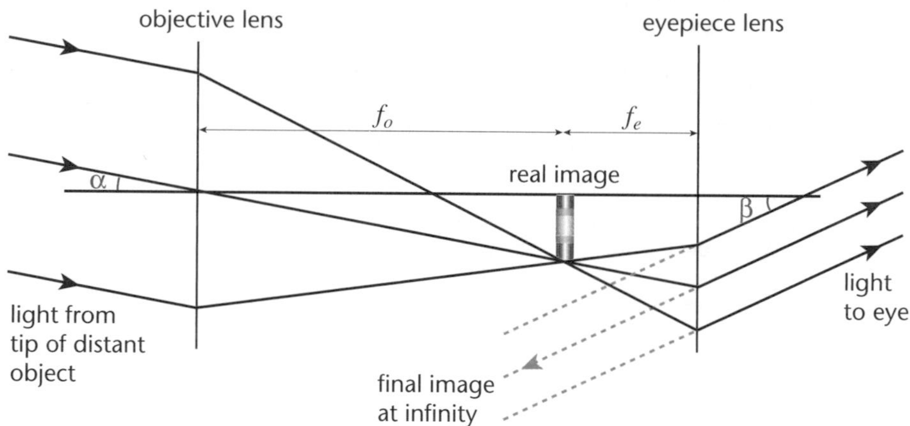

Since the size of the image on the retina depends on the angle subtended by the light reaching the eye, **angular magnification** is used to measure the factor by which the telescope increases this angle.

KEY POINT

$$\text{Angular magnification, M,} = \frac{\textit{angle subtended at eye by image}}{\textit{angle subtended at eye by object}} = \frac{\beta}{\alpha}$$

For a telescope in normal adjustment, $M = \frac{f_o}{f_e}$

As with a converging lens, the principal focus is the point where light parallel to the principal axis converges.

In the Cassegrain **reflecting telescope** a concave mirror is used to converge parallel light from a distant object. This would produce a real image at its principal focus, but the light is intercepted by a small convex mirror which produces an enlarged image for magnification by the eyepiece lens. This is shown in the diagram.

A front-silvered mirror is used to prevent loss and possible distortion of light as it passes through the glass of a rear-silvered mirror.

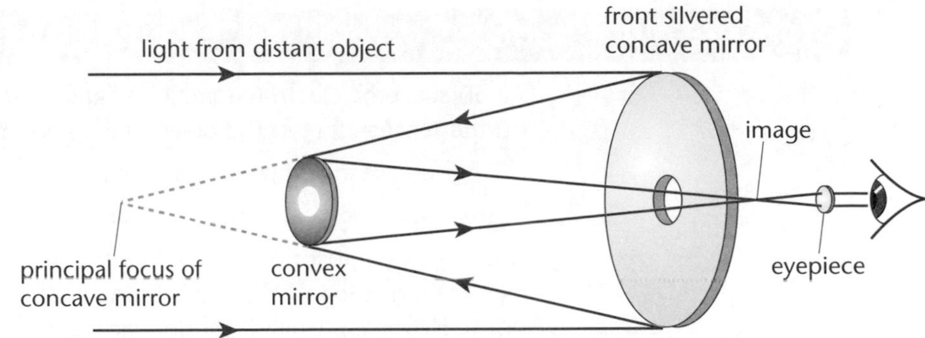

Both types of telescope can produce blurring of the image called **aberration**.

- **Chromatic aberration** is due to light of different wavelengths being focused at different points; it is reduced by using combination lenses made with two different types of glass.
- **Spherical aberration** is caused by light that strikes the outer edges of a lens or mirror being brought to a focus slightly closer than those from the centre; it is reduced by using plano-convex lenses (only one face is curved) in refracting telescopes and parabolic mirrors in reflecting telescopes.

Radio Telescopes

AQA A 5

The operation of a radio telescope is basically the same as that of an optical reflector; the radio waves are collected by a large parabolic dish and brought to a focus. The dish is steerable and scans back and forth across the source of radio emission. The received radio signals are amplified and processed by a computer to produce a visual image. The collecting power of a radio telescope is proportional to the square of the area of the dish.

Resolution

AQA A 5

The **resolving power** of a telescope describes its ability to distinguish two stars that are close together, rather than identify them as a single star. When light passes through a circular aperture it is diffracted. The bright central ring (known as the **Airy disc**) is surrounded by a number of low-intensity rings. The diagram shows the variation of intensity with angle from the centre of the Airy disc.

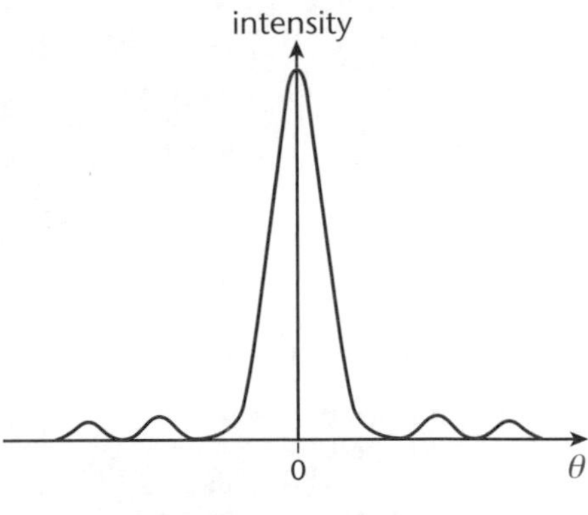

This is the separation required for both Airey discs to be discernible.

The symbol ≈ means approximately rather than precisely equal to.

To be able to distinguish between two stars:

- the minimum separation of the images is when the central maximum of each is at the same place as the first minimum of the other
- the **Rayleigh criterion** for this to happen is that the angle between the stars $\theta \approx \frac{\lambda}{D}$, where D is the diameter of the telescope and θ is in radians
- θ is known as the **resolving power** of the telescope.

Radio telescopes have much poorer resolving power than their optical counterparts because of the wavelengths involved.

Stars are black bodies

AQA A 5

All objects emit electromagnetic radiation over a range of wavelengths that depends on the temperature of the object. Increasing the temperature of an object results in:

- an increase in the power radiated
- the wavelength at which the maximum power is radiated becoming shorter.

A **black body** absorbs all wavelengths of electromagnetic radiation and reflects none. However, it need not appear black, as solid objects at a temperature greater than about 600°C emit visible radiation. The rate of emission of radiation from a black body at different temperatures is shown in the graph opposite.

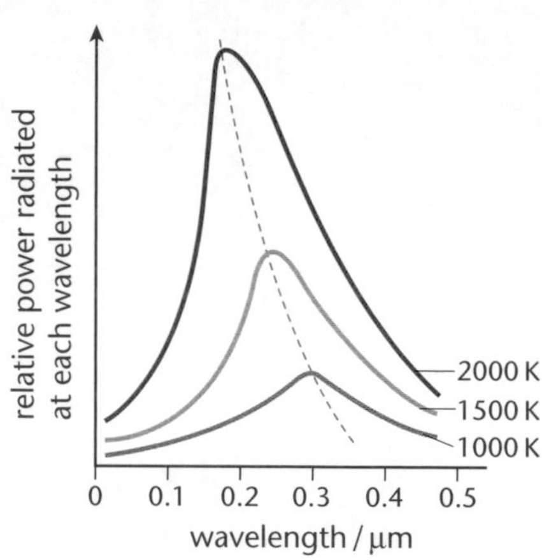

The area between the graph line and the wavelength axis represents the total power radiated. The graph shows that this increases with increasing temperature.

The dotted line shows how the wavelength at which the maximum power is emitted depends on the temperature of the black body. Knowledge of this wavelength enables the surface temperature of a star to be estimated using **Wien's law**.

Note that the unit 'm K' is 'metre × Kelvin', not milliKelvin, which would be written as mK.

KEY POINT

Wien's law relates the wavelength at which the maximum power is emitted by a black body, λ_{max}, to its temperature:

$$\lambda_{max}T = 2.90 \times 10^{-3} \text{ m K}$$

Where T is the temperature of the object in K.

In practice it may be difficult to determine λ precisely because:

- the instrument used to detect the radiation may not have the same sensitivity for all wavelengths
- the atmosphere absorbs some wavelengths of light more than others, so the intensity of these wavelengths may be reduced by more than that of others as the light travels through the atmosphere.

The power radiated by a black body depends on both its surface area and its temperature.

When applying Stefan's law to stars, it is assumed that stars are black bodies in that they are unselective in the wavelengths of radiation that they emit and absorb.

KEY POINT

Stefan's law states that:

the power radiated per m^2 of a black body is proportional to its (absolute temperature)^4

$$\frac{P}{A} = \sigma T^4$$

where σ is Stefan's constant and has the value 5.67×10^{-8} W m^{-2} K^{-4}.

When applying Stefan's law:

- the power radiated is calculated from measurements of the intensity of the light received (*I*) and the distance of the star from the Earth (d), using the **inverse square law** $I \propto \frac{1}{d^2}$

  If the temperature of a star is calculated from Wien's law, then Stefan's law can be used to estimate the area needed for the star to have the same power output as the Sun (3.8×10^{26} W). The surface area, and hence the radius of the star, can be calculated by comparing the absolute magnitudes (see page 125).

Spectral classes

AQA A 5

Analysis of the emission and absorption spectra of a star enables it to be classified according to its temperature and composition. The table shows the characteristics of the seven classes of star in order of decreasing surface temperature.

Type	*Colour*	*Surface temp/K × 10^3*	*Absorption of hydrogen Balmer lines*	*Elements detected in absorption spectrum*
O	blue	40	weak	helium ions
B	pale blue	20	fairly strong	helium atoms
A	white	10	strong	hydrogen
F	pale yellow	7	medium	ionised metals such as calcium and iron
G	yellow	6	weak	calcium ions and iron atoms
K	orange	5	weak	calcium and iron atoms, together with some molecules
M	red	3	weak	molecules of metal oxides

Note that observed absorption spectra are produced as light passes through the outer layers of stars. This table shows that:

- in the surface layers, ionised helium is only present in the hottest stars where the temperature is high enough to ionise helium atoms
- stars in groups B and A show strong absorption by hydrogen in the Balmer series; this is the part of the hydrogen emission spectrum due to exciting electrons from the first excited energy level above the ground state. This energy level is given the value n = 2, where n = 1 corresponds to the ground state. Absorption of these spectral lines shows that the temperature is cool enough for hydrogen atoms to exist but hot enough for them to be in an excited state
- the relatively cool stars in classes F and G show absorption lines due to ionised metals
- stars in classes K and M are cool enough for molecules such as those of metal oxides to exist without being separated into atoms or ionised.

Analysing an absorption spectrum obtained at the Earth's surface can be misleading, as the following diagram shows.

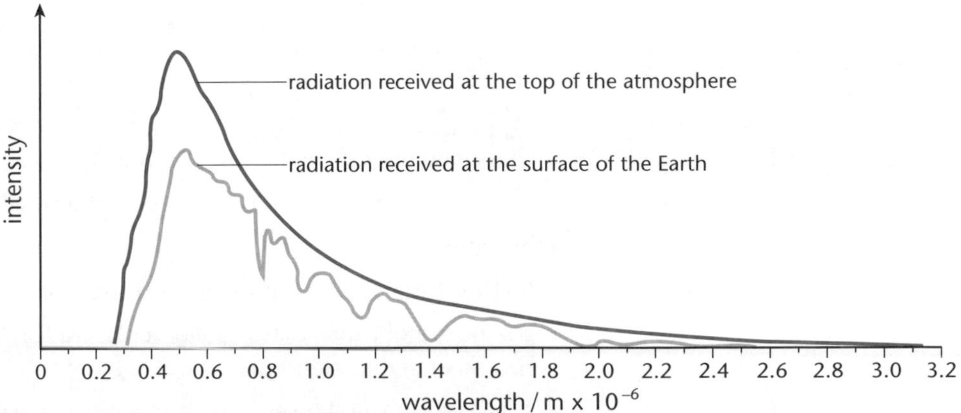

This graph shows the reduction in intensity of all wavelengths as the radiation passes through the atmosphere, with particular wavelengths being absorbed more than others. Information from spectra obtained from satellites gives a fuller picture than that from spectra obtained at ground level.

Quasars

AQA A 5

Quasars are the most distant objects that can be detected. They are very bright objects which were first discovered by detection of the radio signals that some of them emit. Analysis of the light from a quasar reveals that:

- the spectrum is an emission spectrum, with very bright lines superimposed on a continuous spectrum
- these lines show enormous red-shift, indicating that quasars are moving with very high speeds.

Analysis of the red-shift indicates that some quasars are moving at speeds greater than 90% of the speed of light at distances of more than ten billion (1×10^{10}) light years. They are relatively small and yet they appear to emit thousands of times as much energy as a typical galaxy.

The life of a star

AQA A 5

When protons fuse together the resulting nucleus has less mass than its constituent particles. Einstein's equation $E = mc^2$ is used to calculate the energy released in fusion.

Fusion to form nuclei more massive than iron requires energy to be supplied, so energy cannot be released by further fusion reactions when the core consists of iron.

When a star is formed:

- clouds containing dust, hydrogen and helium collapse due to gravitational forces (called a protostar)
- the contraction causes heating
- the temperature in the core becomes hot enough for fusion reactions to occur
- energy is released as hydrogen nuclei fuse, eventually forming the nuclei of helium.

The star is now in its **main sequence**. Its position and how long it remains on the main sequence depends on its mass; the more massive the star, the shorter the main sequence.

When there is no longer sufficient hydrogen in the core of a star to maintain its energy production and brightness the star cools:

- the core contracts but the outer layers expand, forming a **red giant** or **red supergiant**, depending on its size
- increased heating of the core creates temperatures hot enough for the helium nuclei to fuse, forming carbon and oxygen.

In a small star like our Sun, once the helium has fused the core contracts to form a **white dwarf**, and the outer layers are ejected. The star then cools to an invisible **black dwarf**.

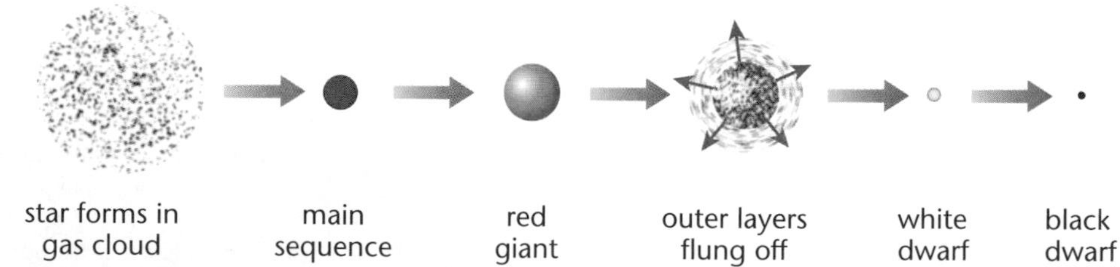

The life cycle of a small star

In a larger star:

- further fusion reactions occur and the star becomes a **blue supergiant**
- the nuclei of more massive elements are formed until the core is mainly iron, the most stable element
- the core now collapses and the heating causes the star to increase in brightness, it is a **supernova**, an exploding star

- in the explosion the outer layers are flung off to form a new dust cloud and the dense core remains as a **neutron star** or, in the case of a very massive star, a **black hole**, whose gravitational field is so strong that light cannot escape from it.

The boundary around a black hole, beyond which nothing can be seen, is called the **event horizon**. Its size is related to the mass of the black hole.

The Schwarzschild radius

AQA A 5

> KEY POINT
>
> The **Schwarzschild radius,** R_s is the radius of the event horizon around a black hole:
>
> $$R_s = \frac{2GM}{c^2}$$
>
> where M is the mass of the black hole, G is the universal gravitational constant and c is the speed of light.

Star magnitudes

AQA A 5

The brightness of the light received from a star, known as its **apparent magnitude**, depends on both its actual brightness and its distance from the Earth. A star of magnitude 1 is defined as appearing 100 times as bright as a star of magnitude 6, so one unit on the magnitude scale corresponds to a change in brightness by a factor of 2.5.

The magnitude scale is a logarithmic scale, equal changes on this scale correspond to the brightness changing by the same factor.

> KEY POINT
>
> The apparent magnitude, m, of a star is related to observed brightness, b, (sometimes called intensity) by the expression:
>
> $$m = -2.5 \log_{10} b$$

When comparing two stars the difference in their apparent magnitudes is given by:

$$m_2 - m_1 = -2.5 \log_{10}\left(\frac{b_2}{b_1}\right)$$

where m_1, m_2 are the apparent magnitudes with corresponding brightness b_1 and b_2

Apparent magnitudes are compared by the intensities of the images on photographic film or other detector exposed for a fixed time.

To compare the actual brightness between stars the effect of distance on the observed brightness needs to be taken out. The method adopted is to compare their brightness based on a fixed distance from the Earth. This standard distance is taken as 10 parsecs or 32.6 light years.

> KEY POINT
>
> The **absolute magnitude**, M, of a star is the apparent magnitude it would have if it were a distance of 10 pc from the Earth.
>
> It is calculated from the apparent magnitude using the expression
>
> $$m - M = 5 \log(d) - 5$$
>
> where d is the distance of the star from the Earth in pc.

When the absolute magnitudes of stars are plotted against their surface temperatures, the result is a **Hertzsprung–Russell** diagram (H-R diagram).

The H-R diagram shows that:

- different types of star occupy specific regions on the diagram
- most stars are in the main sequence band, with the more massive stars in the upper left corner
- the white dwarfs are hot but dim because of their small size.

Note that, on the diagram, movement towards the right corresponds to decreasing temperature.

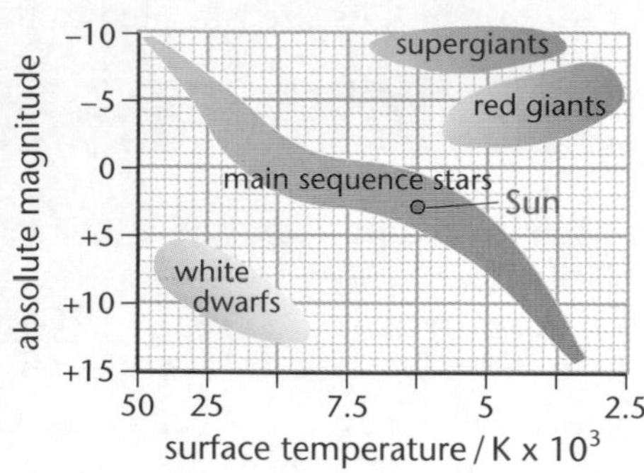

Progress check

1 What is the advantage of a reflecting optical telescope over a refracting one?

2 A radio telescope has a dish diameter of 75 m. What is its resolving power when detecting 0.50 m wavelength radio waves?

3 What causes the release of energy from a star in its main sequence?

4 What is meant by an *event horizon*?

5 The apparent magnitude of Venus is –4, and that of the Sun is –27. By what factor does the Sun appear brighter than Venus?

6 Suggest why the magnitudes of red giants are greater than some main sequence stars that have a higher surface temperature.

1 It can be made of much larger diameter, allowing increased resolution and brighter images.

2 6.7×10^{-3} radians

3 The fusion of protons into more massive nuclei, eventually forming helium.

4 The boundary around a black hole, beyond which nothing can be seen.

5 1.4×10^{9}

6 The red giants have a much greater surface area.

4.4 The expanding Universe

After studying this section you should be able to:

- *describe how Doppler shift can be used to measure the speed of a star and galaxy relative to the Earth*
- *explain how Hubble's law can be used to estimate the age of the Universe*
- *appreciate the evidence that supports the Big Bang theory*

LEARNING SUMMARY

Stars and galaxies

AQA A 5

The Sun is a small star situated on one arm close to the outer edge of a spiral galaxy called the Milky Way. The photos show the three shapes of galaxy: spiral, elliptical and irregular.

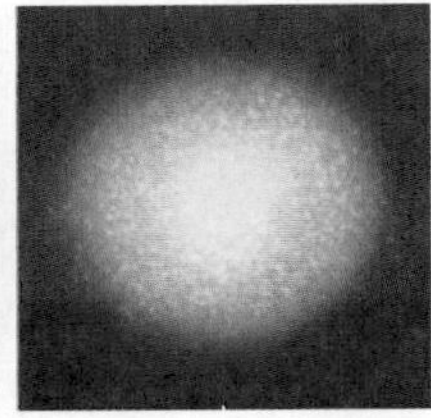

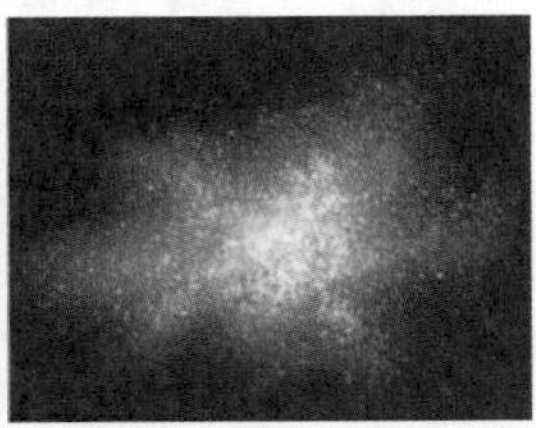

The Milky Way Galaxy whirls around at enormous speeds but, like the planets in the Solar System, the orbital time of stars increases with increasing distance from the centre. When Newton's law of gravitation is applied to the movement of the Solar System within the Galaxy, the mass of the planets is ignored as this has only a small effect. The mass of all the other stars in a galaxy cannot be ignored when applying the law to an individual star, but only those within the orbit of a star are taken into account as it is the gravitational attraction of these stars that provides the centripetal force.

KEY POINT

Newton's law applied to the orbit of a star gives the relationship between the orbital time, T, and the orbital radius, r:

$$T^2 = \frac{4\pi^2 r^3}{GM}$$

where M is the total mass contained within the orbit of the star.

Astronomical distances

AQA A 5

Different units are used to measure distances in our Solar System, our galaxy and beyond our galaxy. The table shows these units and gives their equivalent in m.

Unit	*Definition*	*Equivalent distance in m*
astronomical unit (AU)	mean distance between Earth and Sun	1.5×10^{11}
light year	distance travelled by light in 1 year	9.5×10^{15}
parsec (pc)	distance between observations in AU / angle subtended (in arc seconds, where 1 arc second = 1/3600 degrees)	3.1×10^{16}

The AU is used for measuring distances within the Solar System, the light year for close stars and the parsec for more distant stars.

The next table shows typical sizes and masses of the Universe and its contents.

Object	*Distance in light years*	*Mass in kg*
nearest star	4.3	10^{30}
Milky Way galaxy	10^5 measured across from opposite extremes	10^{40}
Universe	10^{10} measured across from opposite extremes	unknown, but thought to be around 10^{55}

Emission and absorption spectra

AQA A 5

Line spectra are emitted when an electric current passes in an ionised gas. Light is emitted when an electron loses energy either by being captured by a gas ion or by a transition to a lower energy level.

A similar effect causes a gas that is hot (typically in excess of 1000°C) to emit light. When this happens:

- electrons gain energy in the collisions due to thermal motion of the particles
- electrons can only absorb the amounts of energy associated with movement to a higher energy level
- electromagnetic radiation is emitted when an electron loses energy and moves to a lower energy level.

As each element has its own characteristic **emission spectrum**, the composition of a gaseous object that emits its own light can be determined by analysing this spectrum.

This is how helium was discovered in solar flares, but the spectrum obtained from the body of the Sun and other stars is not a line emission spectrum but a line **absorption spectrum**. It contains the full range of visible wavelengths with a large number of 'black' lines, where wavelengths are missing.

Inside a star the particles are very close together, and their energy levels overlap. So unlike a gas at normal temperatures and pressures on Earth, there are no limits to the energy level transitions that are possible. This explains why a star emits the full frequency range of the visible spectrum. The explanation of the black lines is:

- a gas can only absorb the wavelengths of electromagnetic radiation that it emits, since these correspond to the same energy level transitions in the opposite directions
- when white light from the body of the Sun passes through the cooler, gaseous surface layers, the elements present absorb their own characteristic wavelengths
- this energy is re-emitted in all directions leaving a dark line in the spectrum that reaches the Earth.

The composition of a star is determined from the absorption spectrum and the wavelengths missing from that spectrum, rather than the wavelengths it contains.

Doppler shift

AQA A 5

You can tell whether an unseen aircraft is moving towards or away from you by the pitch of the engines. You receive higher frequency sounds when it is moving towards you than when it is moving away. This is due to the **Doppler effect**, the difference between the frequency of a wave emitted and that of the wave received when there is relative movement of the wave source and an observer.

Radar systems, used to measure the speed of moving cars, also use the Doppler effect.

The spectral lines in the absorption spectrum of a star observed from the Earth are subject to Doppler shift:

- a shift towards the blue end of the spectrum corresponds to a reduction in wavelength and an increase in frequency; this happens when a star is moving towards the Earth
- a shift towards the red end of the spectrum corresponds to an increase in wavelength and a reduction in frequency.

KEY POINT

Doppler red shift, z, is the ratio of speed of recession (or relative speed of moving apart), and is related to the change in wavelength due to the relative motion:

$$z = \frac{v}{c} = \frac{\Delta\lambda}{\lambda}$$

where v is the relative speed of recession, c is the speed of light, $\Delta\lambda$ is the observed wavelength shift, and λ is wavelength expected when there is no relative motion.

Hubble and red-shift

AQA A 5

We live in an expanding Universe. Evidence for this comes from the shift in frequency and wavelength of the light received from other galaxies. Light from the relatively nearby Andromeda galaxy is shifted towards the blue end of the spectrum, but for every other galaxy the shift is towards the red. The amount of shift depends on the speed of the galaxy relative to the Earth, known as the speed of recession.

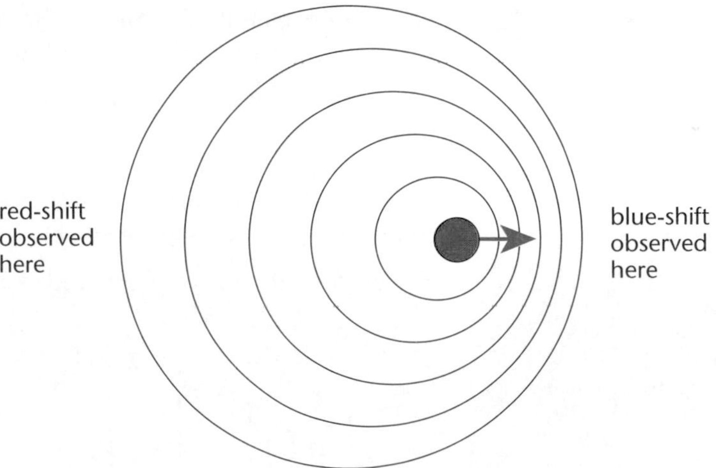

How movement of a star or galaxy causes a difference between the wavelengths emitted and those received

Hubble discovered that:

the speed of recession of a galaxy is proportional to its distance from the Earth.

This is known as **Hubble's law**, and it gives rise to Hubble's constant.

Hubble's law seems to state that the Earth is the centre of the expansion. This is not the case. Measurements taken from any point in an expanding Universe would give the same pattern.

> Hubble's law can be written as
>
> $$v \propto d \text{ or } v = Hd$$
>
> where v is the speed of a galaxy relative to the Earth, d is its distance from the Earth and H is Hubble's constant.
>
> H has units s^{-1} but its value is not known precisely.
>
> An alternative unit for H is km s^{-1} Mpc^{-1}.
>
> KEY POINT

The expansion of the Universe gives evidence to support the **Big Bang** theory. According to this theory:

- the Universe started with an enormous 'explosion' more than 13 thousand million years ago
- initially the Universe was very small and very hot
- the Universe has been cooling as the galaxies have been moving away from each other ever since.

The energy of the background radiation is equivalent to that radiated from an object at a temperature of about 3 K.

The theory is also supported by the existence of microwave background radiation that fills the whole Universe. This is thought to be electromagnetic radiation left over from the initial 'explosion'. Due to the expansion of the Universe its wavelength has increased and its energy has decreased.

The age of the Universe

AQA A 5

Measurements of the value of the Hubble constant enable the age of the Universe to be estimated. For individual stars, this involves measuring both the relative speed and the distance of a star from the Earth. To estimate the distance to a star:

The age of the Universe is estimated from $\frac{1}{H}$.

- for close stars (up to 300 light years) the apparent movement, called parallax movement, of the star across the fixed background of more distant stars is used. This involves measuring the position of the star when the Earth is at opposite points in its orbit.
- for more distant stars and galaxies observations of their spectra, brightness and periodic variation in brightness are needed.

Both methods only produce an estimate. In the case of close stars, movement of the star itself and the Sun also affect the apparent change in position due to the Earth's movement. For more distant stars, scattering of the light as it passes through the Earth's atmosphere leads to uncertainty in the measurements.

Since the distance of a star or galaxy, d, cannot be measured very precisely it follows that there is uncertainty about the value of the Hubble constant, H. Probably the only certain thing about it is that it is not constant! Its value is probably decreasing due to galaxies slowing down because of the effect of their gravitational pulls on each other.

The diagram shows how the speed of a galaxy relative to the Earth varies with distance.

gradient of graph = Hubble's constant

Each galaxy contributes a point on the graph. The galaxies do not all lie close to the line so there is significant uncertainty in the value of the gradient.

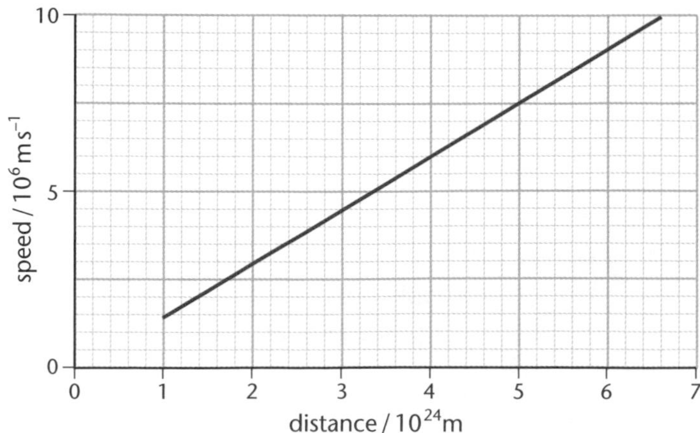

The Hubble constant can be calculated from any pair of values obtained from the graph. An estimate of the age of the Universe is then obtained by calculating the reciprocal of this constant.

In the beginning

AQA A 5

There is little direct evidence to support any theory about what happened in the first fraction of a second after the 'Big Bang'. The energies of the particles that existed would be so vast that it is impossible at the present time to accelerate particles to these energies in order to study their behaviour.

An anti-particle has the same mass as its corresponding particle, but the opposite charge (if any). When they collide they cease to exist, forming gamma rays or other particles.

Before the first 0.01 s:

- electrons and positrons (anti-electrons) and quarks existed
- quarks combined to form nucleons and their anti-particles.

After the first 0.01 s.

- annihilation of nucleons and anti-nucleons occurred, leaving an excess of nucleons
- after about 15 s, electron–positron annihilation left an excess of electrons
- after about 100 s, fusion of protons formed light nuclei such as helium
- after about 5×10^5 years, atoms formed from nuclei and electrons
- stars and galaxies began to form as there was less electrostatic repulsion between particles to oppose the gravitational attraction
- after about 5×10^6 years the formation of clusters of galaxies continued to increase the local variations in the density of the Universe.

The formation of neutral atoms resulted in less electrostatic repulsion.

Progress check

1 Explain why dark lines in an absorption spectrum correspond to bright lines in an emission spectrum of the same element.

2 A star is moving away from the Earth at a speed of 1.0×10^6 m s^{-1}.

Calculate the fractional difference in wavelength, $\Delta\lambda/\lambda$ between the waves received and those emitted.

3 Use data from the graph on page 130 to calculate a value for the Hubble constant.

1 A gaseous element can only emit and absorb certain frequencies or wavelengths. These frequencies or wavelengths are the same for emission and absorption. When white light passes through the gas it absorbs the frequencies in its emission spectrum, leaving dark lines.

2 0.0033

3 1.5×10^{-18} s^{-1}

Sample question and model answer

Ultrasound is used for examining delicate organs and tissue. The diagram shows how an ultrasound probe is used to examine the retina.

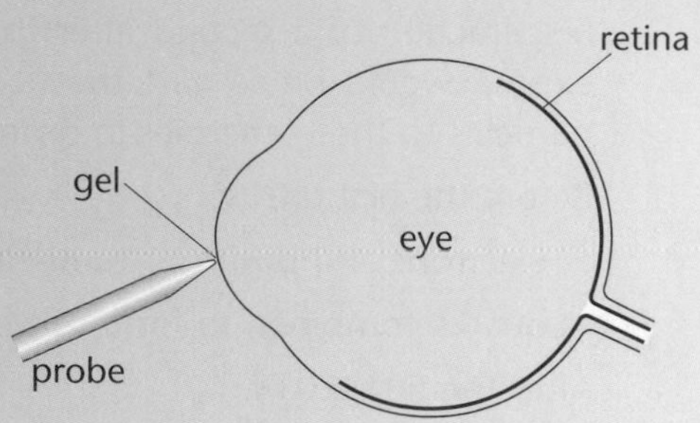

The depth of the eye is 2.5 cm.
The speed of sound in the eye tissue is 1.50×10^3 m s^{-1}.

(a) For the best results, the wavelength used for an ultrasound scan should be 1/200 of the depth of the tissue being scanned.
Calculate the optimum frequency for scanning a retina. [3]

> This is an example of where a question on an extension topic requires recall of material from the core, in this case the wave equations.

optimum wavelength = $1/200 \times 2.5 \times 10^{-2}$ m = 1.25×10^{-4} m — 1 mark
optimum frequency, $f = v/\lambda = 1.50 \times 10^3$ m s^{-1} ÷ 1.25×10^{-4} m — 1 mark
= 1.20×10^7 Hz — 1 mark

(b) Explain why the gel is used between the probe and the eyeball. [2]

> This is an example of impedance matching to maximise the energy transferred between two materials.

This maximises the energy transmitted into the eye (1 mark) by reducing the amount reflected (1 mark).

(c) In an ultrasound scan a piezoelectric crystal acts as both transmitter and detector of the ultrasound. It emits short pulses of waves, with a time gap between each pulse.

(i) Explain why there is a time gap between each pulse of waves. [2]

To allow the reflected waves to return to the crystal and be detected (1 mark) without interfering with the waves emitted by the crystal. 1 mark

(ii) Estimate the minimum time gap that should be left between pulses in this example. [3]

The minimum time should be that taken for the pulse to travel to the retina and for the reflection to be detected. — 1 mark

> The factor of two is needed here because the pulse has to travel to the back of the eye and return.

time = distance ÷ speed = $(2 \times 2.50 \times 10^{-2}$ m$)$ ÷ $(1.50 \times 10^3$ m s$^{-1})$ — 1 mark
= 3.33×10^{-5} s — 1 mark

Practice examination questions

Throughout this section use the following values of the electronic charge and Planck's constant:

$e = -1.60 \times 10^{-19}$ C
$h = 6.63 \times 10^{-34}$ J s

1

Sound is attenuated as it travels in the air.

(a) Explain the meaning of *attenuation*. [1]

(b) Sound from a loudspeaker has an intensity of 6.0×10^{-5} W m^{-2}, measured at a distance of 1 m from the loudspeaker.
The threshold of hearing is 1.0×10^{-12} W m^{-2}.

(i) Calculate the intensity of the sound at a distance of 1 m from the loudspeaker in decibels. [3]

(ii) Assuming that the sound is radiated in all directions, calculate the distance from the loudspeaker where the sound becomes inaudible. [3]

(iii) What is the intensity, in dB, of a sound at the threshold of hearing? [1]

(c) The ear is most sensitive to sounds with a frequency around 3000 Hz.
Explain how the perceived loudness of sounds of equal intensity depends on their frequency. [3]

2

For ultra-sound scanning, a piezoelectric crystal produces a series of pulses of ultrasound and then detects the reflections.

An ultrasound scan is used to detect a tumour at a depth of 6.0 cm below the surface of the skin. The speed of sound in tissue is 1500 m s^{-1}.

(a) Explain whether an A-scan or B-scan is appropriate for this application. [2]

(b) Calculate the time that elapses between the pulse being sent and the echo being detected. [2]

(c) Explain why it is necessary to use a short pulse of ultrasound. [2]

(d) Explain why it is necessary to use pulses of ultrasound rather than a continuous wave. [2]

(e) The optimum frequency is one that produces a wave with a wavelength that is 1/200 the depth of the organ.
Calculate the optimum frequency for this application. [3]

(f) Before a scan is carried out, the skin of the patient is coated with a gel.
Explain the purpose of this gel. [2]

(g) CAT scans are also used to detect tumours.

(i) Outline the principle of operation of a CAT-scanner. [3]

(ii) Describe the advantages and disadvantages of using a CAT scan rather than an ultrasound scan to detect a tumour. [3]

Practice examination questions (continued)

3

(a) (i) Explain the meaning of the terms *apparent magnitude* and *absolute magnitude.* [2]

(ii) Explain why the absolute magnitude is preferred to the apparent magnitude when comparing the magnitudes of stars. [2]

(b) Sirius is at a distance of 2.7 pc from the Earth. Its apparent magnitude is –1.5. Calculate the absolute magnitude of Sirius. [2]

(c) The absolute magnitude of the Pole star is –4.9. Its apparent magnitude is 2.0. Calculate the distance of the Pole Star from the Earth. [2]

(d) White dwarfs are hotter than red giants, but they are less bright. Suggest a reason for this. [2]

4

The hydrogen absorption spectrum contains a dark line corresponding to a wavelength of 6.563×10^{-7} m.

(a) Explain why the lines in an absorption spectrum are dark. [3]

(b) In the spectrum detected from a star, this line has a wavelength of 6.840×10^{-7} m.

(i) Explain this apparent increase in wavelength. [2]

(ii) Calculate the speed of the star relative to the Earth. $c = 3.00 \times 10^{8}$ m s^{-1}. [3]

(iii) The value of the Hubble constant, $H_0 = (2.12 \times 10^{-18} \pm 0.07)$ s^{-1}. Calculate the range of distances from the Earth within which the star lies. [3]

5

The diagram shows three possible ways in which the size of the Universe could change in the future.

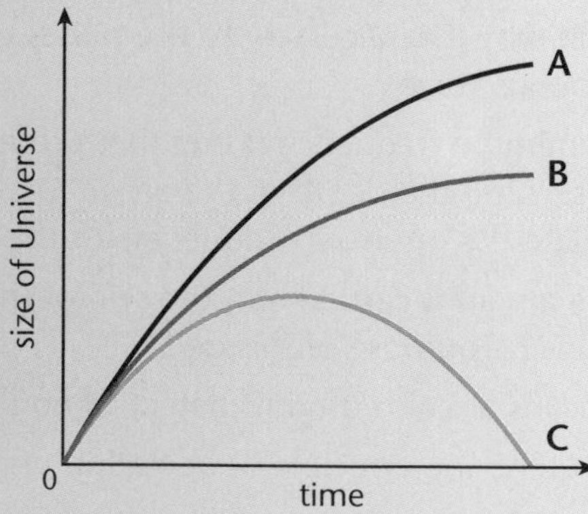

With reference to the graph, explain how the future of the Universe depends on the amount of mass that it contains. [6]

Chapter 5

Synthesis and How Science Works

5.1 How Science Works

You will also be assessed on your understanding of How Science Works. This includes awareness of the following points, shown here with examples or additional explanation.

HOW SCIENCE WORKS	EXAMPLE OR EXPANSION
Observation is at the centre of how science works.	Science studies the observable world and does not study what cannot be directly or indirectly observed.
New ideas and new observations are debated amongst scientists.	The discovery of a global layer of dust made 65 million years ago and a large crater of the same age provides evidence for a major event that could have made the dinosaurs extinct. Many scientists, however, say that more evidence is needed before becoming confident that the event killed the dinosaurs.
Theories that are supported by replicated observations may become accepted as the best available sources of explanations.	There have been very many observations that have supported the idea that atoms exist.
Scientific knowledge can be supported but not proved by observations.	No human observation is ever ABSOLUTELY reliable. Many observations, however, add together to become convincing.
The best available sources of explanations become 'established' scientific ideas, but they are still subject to tests by more observations. 'Established' scientific ideas have been subjected to many tests, and never proved wrong.	Any one observation that showed that atoms cannot exist would be enough to destroy atomic theory (but no such observation has yet been made).
Scientists who succeed in proving established ideas to be wrong make scientific breakthroughs.	Scientists who carry out an experiment that shows atoms couldn't exist would become famous.
Scientists communicate with each other by publishing accounts of their work in journals and by attending conferences.	There are many scientific journals.
Journals are available to the general public, though often only scientists and specialist journalists can understand them.	Journalists look for 'stories' that non-scientists will find particularly interesting.
In order for work to be published it must be reviewed by other scientists to see if it meets required standards of methods and conclusions.	This is called peer review.

HOW SCIENCE WORKS	EXAMPLE OR EXPANSION
Scientific knowledge has a direct impact on how people in general see the world.	The idea that the Earth is not the centre of the Universe (the Copernican revolution) changed the way we see ourselves. Images of the Earth taken from the Moon reinforced this change. We tend to be less conscious of such profound cultural changes than we are of shorter-term changes in fashion.
Scientific knowledge can make new technologies possible.	Experiments on electromagnetism made generation of electricity possible. Experiments on electric circuits and on materials made transistors possible, leading to computer technology.
New technologies can change the way people live and think.	Such technologies include: • Electromagnetism and electricity generation – allowing our nights to be lit up and allowing many labour-saving devices, helping to reduce poverty but also helping to produce pollution in general and climate change in particular • Electromagnetic radiation and communication technology – TV and radio • Mobile telephones – these have made big changes in business and social life • New technologies provide new models and metaphors. People sometimes compare the human brain to a computer, for example.
New technologies often have ethical implications.	There is uncertainty about the safety of mobile phones, especially to children. New technologies make new weapons possible. New technologies raise questions about spending money, especially on healthcare equipment such as MRI scanners.
There may be ethical considerations involved in deciding whether particular scientific research should be done.	Animal experimentation, work on human cells (especially fertilised embryos), work that could lead to new weapons of mass destruction.

5.2 Synoptic assessment

All examinations for AQA A and AQA B at A2 are synoptic. These are questions that require you to bring ideas together from different physics topics. It favours people who really know their physics and who can make links and spot patterns. This chapter provides several themes of physics that will help you to do this.

Themes of physics – energy

Many kinds of system can store energy. A stretched spring and a charged capacitor may not look very similar, but they are both energy storage systems. The formulae for the amount of energy stored are not that different.

For the stretched spring:

energy stored = average applied force × extension

$= \frac{1}{2} Fx$

For the charged capacitor:

energy stored = average applied potential difference × charge stored

$= \frac{1}{2} VQ$

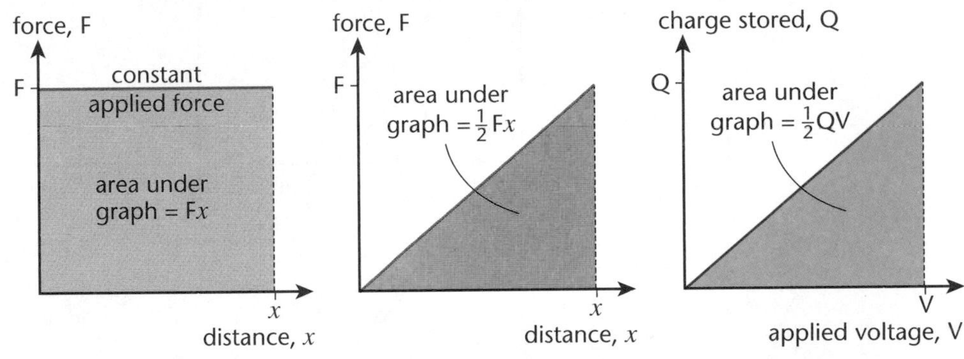

For a body moving under the action of a constant force,

work done = Fx

For a spring being stretched, force is not constant.

work done to stretch spring
= average force × distance
$= \frac{1}{2} Fx$

For a capacitor, energy transferred to charge Q
= energy stored
= average voltage × charge
$= \frac{1}{2} QV$

Write down the names of as many energy storage systems as you can think of. Wherever possible, write down a formula that can be used to predict the amount of energy stored by the system.

Try to think of processes of energy transfer from system to system that do not involve either heating or (even if at a particle level) doing work.

Energy transfers from one kind of source to another, by processes of working and/or heating. Some energy may become thinly spread in the wider environment, or dissipated, but total energy before the process is always the same as total energy after.

This equation is useful for solving problems and making predictions in many areas of physics:

total energy before = total energy after

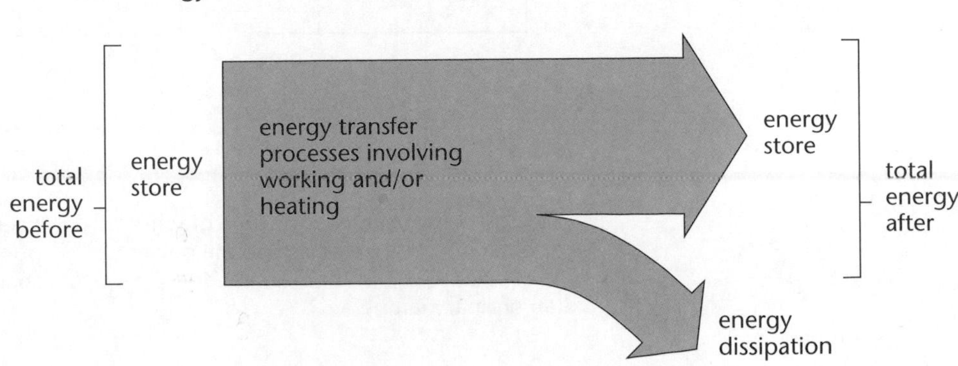

Themes of physics – rates of change with time

The world is not static and if we want to study its changes, we quickly find that it is useful to know about how quickly the changes take place. We study rates of change in many systems.

Displacement, velocity and acceleration

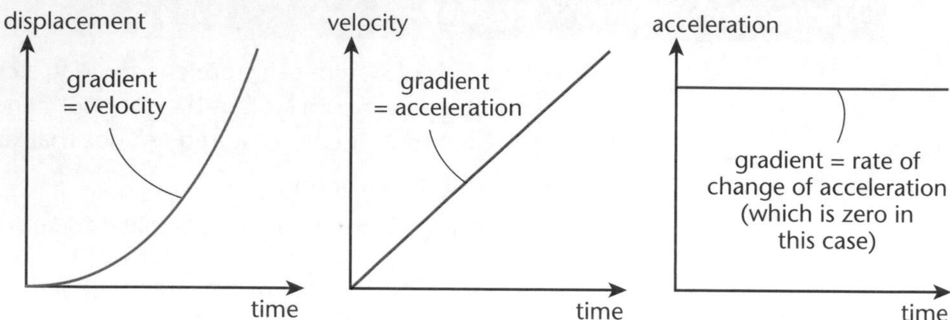

Displacement, velocity and acceleration for a motion that starts from rest and has uniform acceleration

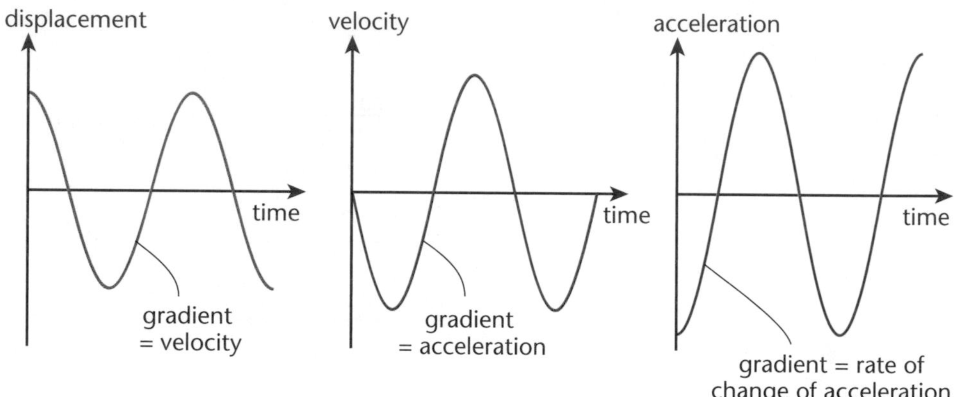

Displacement, velocity and acceleration for a simple harmonic motion with initial displacement equal to the amplitude of the motion.

Energy and power

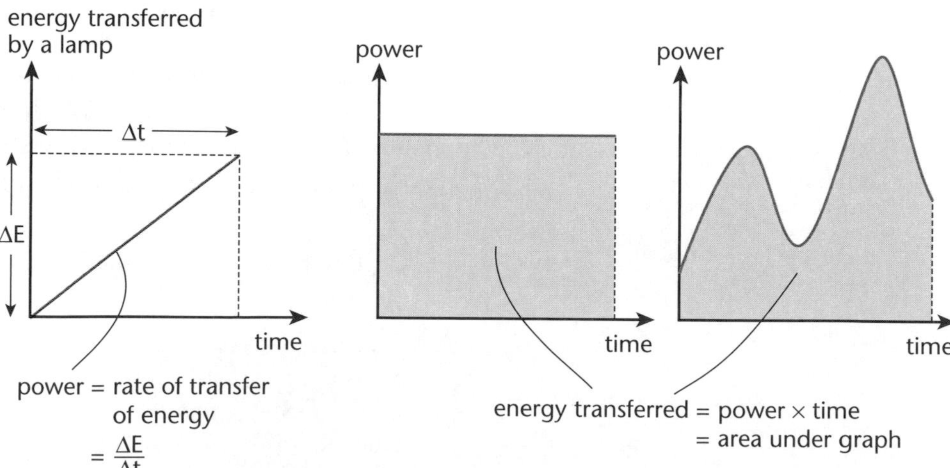

The gradient of an energy–time graph is equal to power. In this case the power is constant (as it is, for example, for an ordinary lamp).

Energy is equal to the area under a power–time graph, whatever the shape of the graph.

Proportionality and exponential rates of change

There are many possible relationships between variables. A simple one is proportionality.

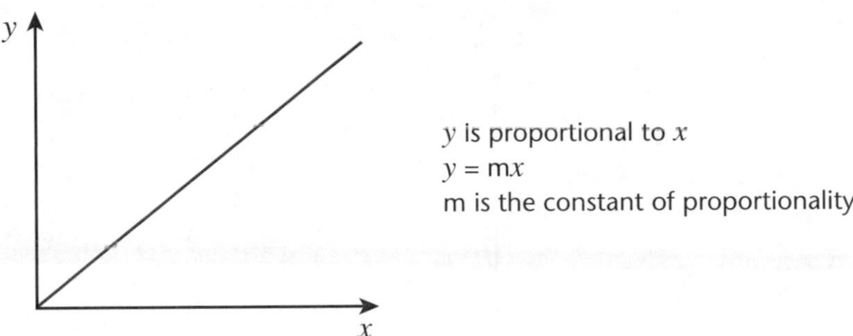

Quite often the rate of change of a quantity depends on the size of the quantity itself. That is, there is a relationship between the quantity, x, and its rate of change, $\Delta x/\Delta t$. Sometimes the relationship is simple – the rate of change of x is proportional to x.

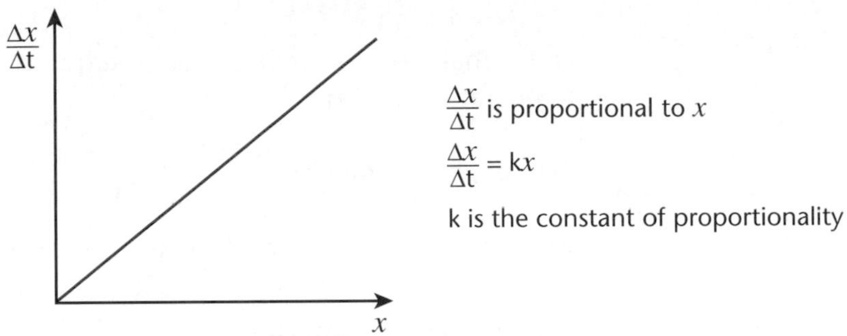

This simple relationship always leads to an exponential relationship between the quantity, x, and the time, t.

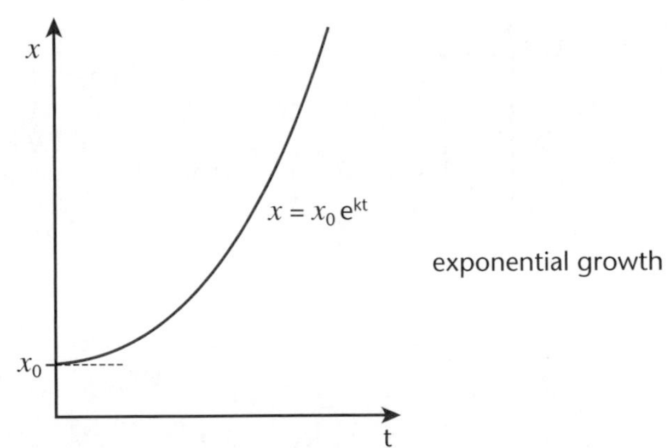

If the constant of proportionality is negative:

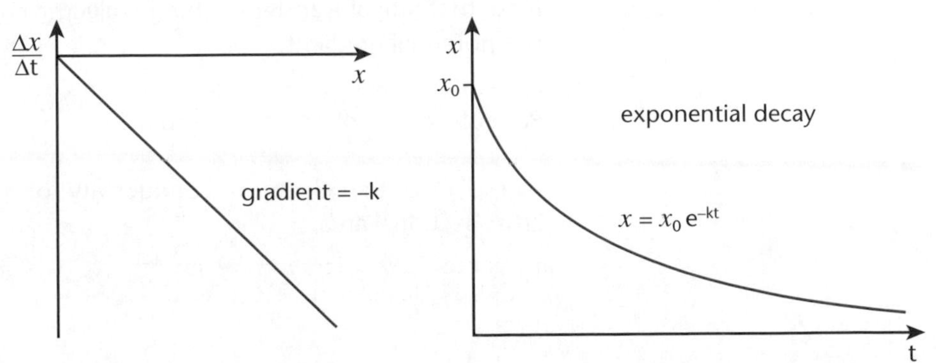

Themes of physics – rates of change with distance

Temperature gradient

temperature, T

ΔT

Δx

distance, x

Δx is the thickness of material

ΔT is the temperature difference across the material

Temperature gradient = $\dfrac{\Delta T}{\Delta x}$

Note that rate of transfer of energy through a material depends on the temperature gradient:

Rate of energy transfer, $\dfrac{\Delta Q}{\Delta t} = k\,A\,\dfrac{\Delta T}{\Delta x}$

where k is the thermal conductivity of the material and A is its cross-sectional area.

Potential gradient

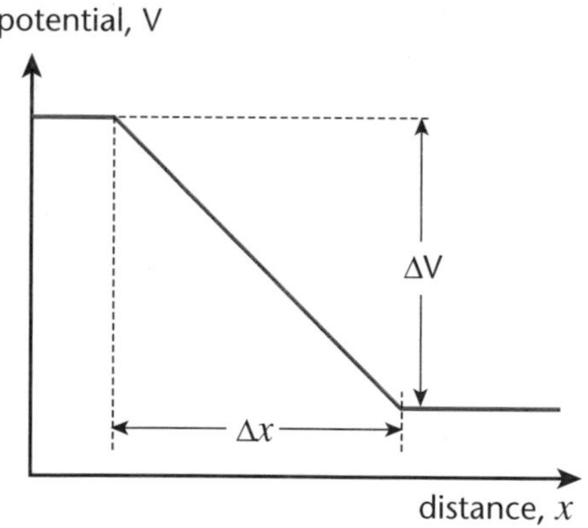

Δx is the length of the resistor

ΔV is the potential difference across the resistor

Potential gradient = $\dfrac{\Delta V}{\Delta x}$

Note that rate of transfer of charge (electric current) through a resistor depends on the potential gradient:

Rate of transfer of charge = current, I = $\dfrac{\Delta Q}{\Delta t} = \sigma\,A\,\dfrac{\Delta V}{\Delta x}$

where σ is the electrical conductivity of the resisting material and A is its cross-sectional area.

Pressure gradient

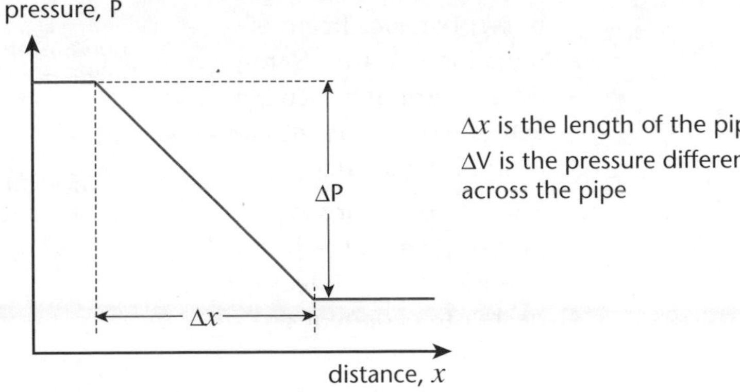

$$\text{Pressure gradient} = \frac{\Delta P}{\Delta x}$$

Note that rate of transfer of liquid through a pipe depends on the pressure gradient:

$$\text{Rate of transfer of liquid} = \frac{\Delta m}{\Delta t} = c\,A\,\frac{\Delta P}{\Delta x}$$

where m is the mass of liquid, c is a constant that depends on the properties of the liquid, and A is the cross-sectional area of the pipe.

The inverse square law

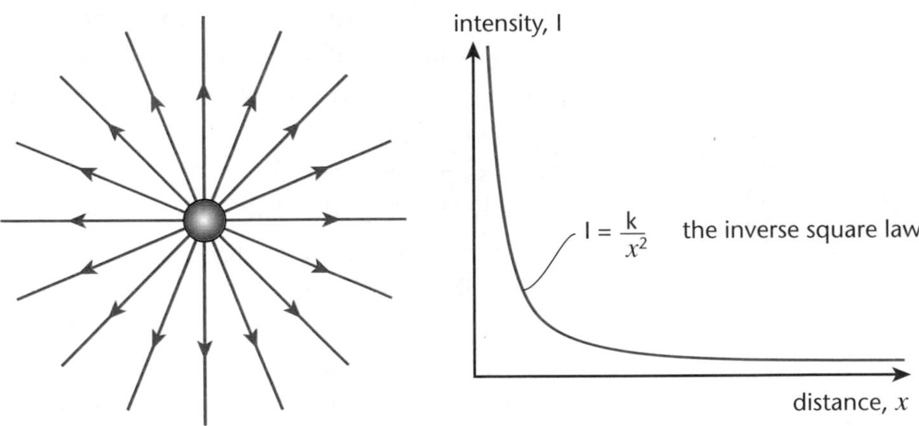

Any situation that involves intensity that radiates from a point into three dimensions also involves an inverse square law of intensity.

The 'intensity' can be intensity of field lines in an electric or gravitational field, or it can be intensity of electromagnetic radiation emitted by a point source.

Exponential decrease with distance

Consider a parallel (that is, non-spreading) beam of radiation that travels from a non-absorbing medium across a boundary into an absorbing material.

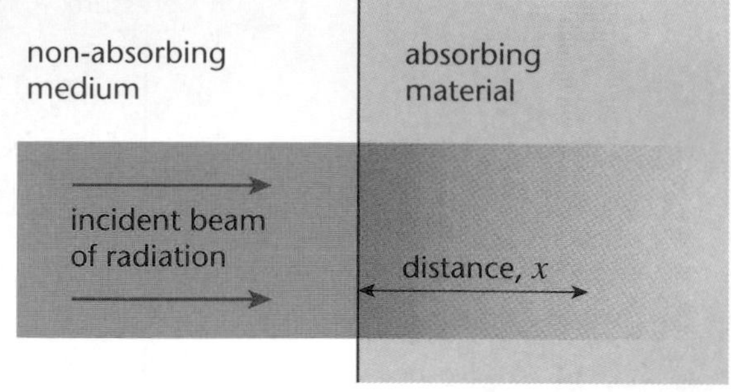

The rate of absorption is proportional to the beam intensity. So the intensity decreases exponentially with distance:

$$I = I_0 e^{-cx}$$

The value of x that corresponds with beam intensity that is half of the initial value is called half-value thickness, $x_{\frac{1}{2}}$.

The mathematics of half-value thickness is similar to that of half-life of radioactive decay.

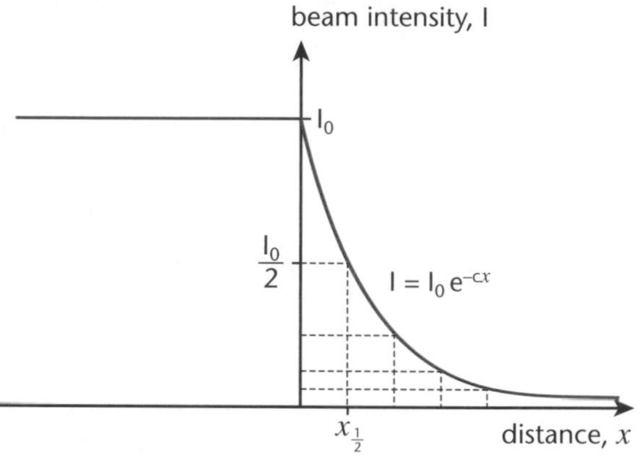

Themes of physics - Using logarithms for wide ranges of values

Our normal counting system uses equal increases in value, such as from 1 to 2 or from 27 to 28. From one step to the next, equal values are added.

A logarithmic scale can be thought of as using equal factors, not equal additions, from one step to the next. To the base 10, a 'step' can be thought of as a multiplication by ten.

The 'steps', then, are:
0.1, 1, 10, 100, 1000 and so on.

We need to be able to write these down in a neat and concise format, so we have:
10^{-1}, 10^0, 10^1, 10^2, 10^3, and so on.

This is very useful for dealing with quantities that vary over very wide ranges, such as distances in space, wavelengths and frequencies of the electromagnetic spectrum, intensities of sound and resistivity/conductivity values.

A logarithmic scale

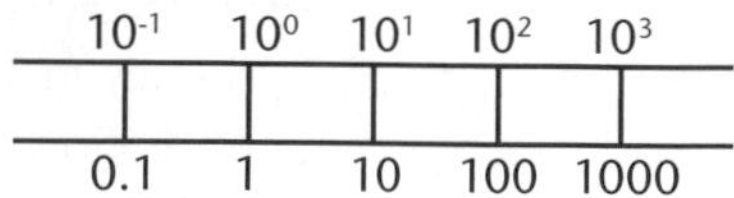

Equal distances on a logarithmic scale represent equal factors, in this case x10

Note: An ordinary counting scale, with equal distances representing equal increases of increments, is called a linear scale.

Themes of physics - predictions, models and simplifications

Physics attempts to model 'real' situations allowing it to predict future outcomes with known uncertainties. Formulae provide predictive answers. To take a simple example, knowing the length of a journey and an expected average speed, it is possible to calculate the expected journey time. Or, knowing the weight to be supported by a cable it is possible to calculate the thickness of cable needed. Meteorology is a branch of physics, and weather forecasters must use complex formulae to make predictions of the complex behaviour of weather systems. They combine their formulae into a computer model – a simulation of the weather.

A model is a simplified version of a physical system. It is intended to mimic the essential aspects of the 'real' system in a sufficiently simple form to allow calculations of specific outcomes. A simple formula is a kind of simple model – it copies the behaviour that is actually observable in the 'real' world. A computer model is just a sophisticated version.

There are visual models, too. A map is a predictive model of the landscape, made so that a person can know what is over the next hill before they ever see it. There are models of atoms – many different models, in fact.

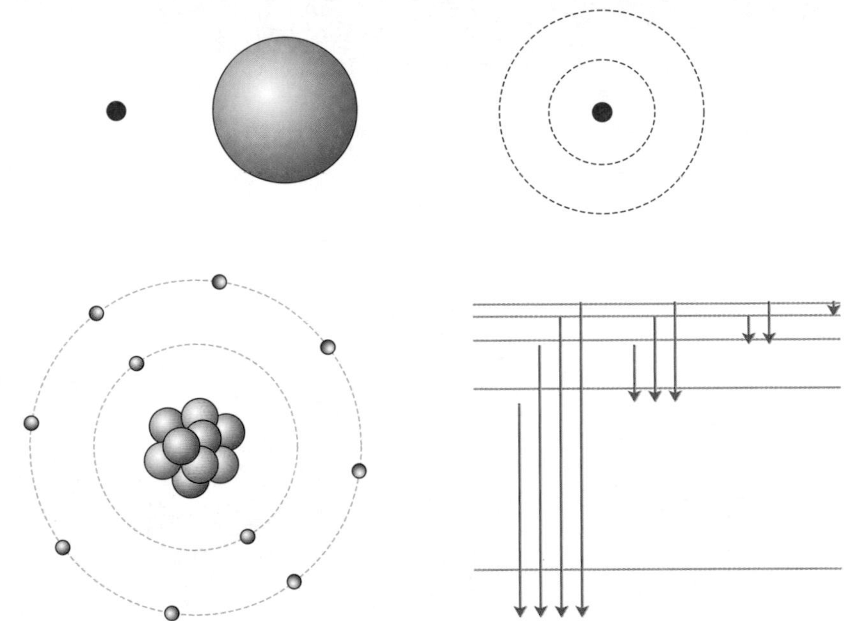

Different models of atoms used for making predictions of behaviour in different circumstances

Each of the above models predicts different aspects of the behaviour of atoms. A trained scientist will use the best model to match the required predictions.

A test of a model is not whether or not it is absolutely correct, but whether it gives good predictions. That means that models have to be used carefully – different models of the same real systems are useful in different circumstances, so that the skill of using the right model is an important one, and quite hard to learn.

A scientific theory is more than a model. An established scientific theory is the best available system of interpretation of observations. So a scientist can move from model to model freely, but will work within just one theory.

Many models use simplifications. Think of the light around you now. It is spreading from a source, or sources. It is reflecting from surfaces. Some of the light passes through the pupils of your eyes, and helps you know and learn about the world so that you can predict its behaviour. Most of the light misses your pupils. You are immersed in a sea of light that you will never see. But physicists interpret light using just simple straight lines drawn on paper – rays of light. It is almost absurdly simple, compared with the real sea of light, but it is very good at producing valuable predictions.

Newton's ideas about force and motion still have huge predictive ability, and so they remain on your exam specifications. But they make assumptions about space as a framework in which events take place and about time as absolute and universal.

These assumptions have been shown, by ideas of relativity, to be false. Newtonian ideas cannot explain behaviour of very fast motion, nor gravitational lensing of light from very distant sources. They do not provide the best possible source of explanations of all of our observations.

As scientific theory, they have failed, but as predictive models their importance lives on. It is just necessary to remember not to use them as predictive models in more extreme situations, such as when considering the motions of particles in accelerators or the travel of light around massive objects like black holes.

Sample questions and model answers

The definitions of gravitational and electric field strength are 'force per unit mass' and 'force per unit positive charge'.

All radial fields follow inverse square laws.

A negative charge experiences a force the same size as that experienced by a positive charge, but in the opposite direction.

Potential means 'the energy per unit mass/charge placed at that point'.

1

The gravitational field due to a point mass and the electric field due to a point positive charge show some similarities and differences.

One similarity is that both follow inverse square laws.

(a) (i) Write down expressions for the gravitational field strength due to a point mass, *m*, and the electric field strength due to a point positive charge, q. [2]

$g = Gm/r^2$ 1 mark

$E = q/4\pi\varepsilon_0 r^2$ 1 mark

(ii) What feature of these expressions shows that the fields follow an inverse square law? [1]

The r^2 term in the denominator shows that the field strengths vary as the inverse of the square of the distance from the mass or charge. 1 mark

(iii) State one other way in which the definitions of gravitational field strength and electric field strength are similar. [1]

They are both proportional to the size of the mass or charge. 1 mark

(iv) State one way in which the definitions of gravitational field strength and electric field strength are different. [2]

The definition of gravitational field strength does not specify the sign of the mass as mass can only have positive values. 1 mark

The definition of electric field strength specifies that it is the size of the force on a positive charge. 1 mark

(b) Gravitational potential and electric potential are properties of points within a field.

How is potential related to the potential energy of:

(i) a mass in a gravitational field? [1]

Potential energy = potential x mass. 1 mark

(ii) a charge within an electric field? [1]

Potential energy = potential x charge (including the sign of the charge). 1 mark

(c) Explain why the expressions for gravitational potential and electric potential have opposite signs. [4]

Gravitational forces can only be attractive (1 mark) so a point in a gravitational field always has a negative potential since energy would be needed to remove the mass to infinity (1 mark). The direction of an electric field is taken to be the direction of the force on a positive charge (1 mark). A positive charge in the field of another positive charge has a positive amount of energy since work would have to be done on the charge to move it from infinity to a point in the field (1 mark).

Sample questions and model answers (continued)

2

A motorist sees a bright light in the sky and she first thinks that it is the planet Venus. Then she notices that the movement across the sky is too rapid for a planet and realises that she is looking at a low-flying aircraft.

At its brightest, Venus reflects electromagnetic radiation from the Sun with a power of 2.8×10^{17} W. On this occasion, the distance between Venus and the Earth was 8.0×10^{10} m.

The power of the lights of a large aircraft is 1500 W.

Estimate the distance of the aircraft from the motorist for its lights to appear as bright as Venus. State the assumptions that it is necessary to make and explain whether your answer is likely to be too high or too low. [12]

The marks are awarded here for setting up a workable mathematical model – like the Earth, only half of Venus is receiving radiation from the Sun at any one time.

Assuming that Venus acts as a disc (1 mark) and reflects radiation over a hemisphere (1 mark):

You are not expected to know that the surface area of a sphere = $4\pi r^2$, this information is given in a data book.

Intensity of light from Venus at the Earth

$= P/2\pi r^2$ — 1 mark

$= 2.8 \times 10^{17}\ \text{W} \div (2 \times \pi \times (8.0 \times 10^{10}\ \text{m})^2)$ — 1 mark

$= 7.0 \times 10^{-6}\ \text{W m}^{-2}$ — 1 mark

At a distance d, for the aircraft lights to have the same intensity, assuming that they radiate light equally in all directions over a hemisphere — 1 mark

$P/2\pi r^2$ = observed intensity

$1500\ \text{W} \div (2 \times \pi \times d^2) = 7.0 \times 10^{-6}\ \text{W m}^{-2}$ — 1 mark

$d = \sqrt{(1500\ \text{W} \div (2 \times \pi \times 7.0 \times 10^{-6}\ \text{W m}^{-2}))}$ — 1 mark

It is important when making an estimate to round off the answer to a sensible number of significant figures.

$= 5.8 \times 10^3$ m or about 6 km — 1 mark

This is assuming that the light from Venus and the aircraft contains the same proportion of visible radiation. — 1 mark

The reasons for the answer are important here. No marks are awarded for a 50–50 guess.

If this is the case, then at a distance of 6 km the aircraft is likely to appear brighter than Venus, since its lights are focused into a beam, (1 mark) so I would expect my answer to be too low (1 mark).

Practice examination questions

1

A mass spectrometer is a device for comparing atomic masses. Positive ions pass through slits A and B before entering a region where there is both an electric and a magnetic field.

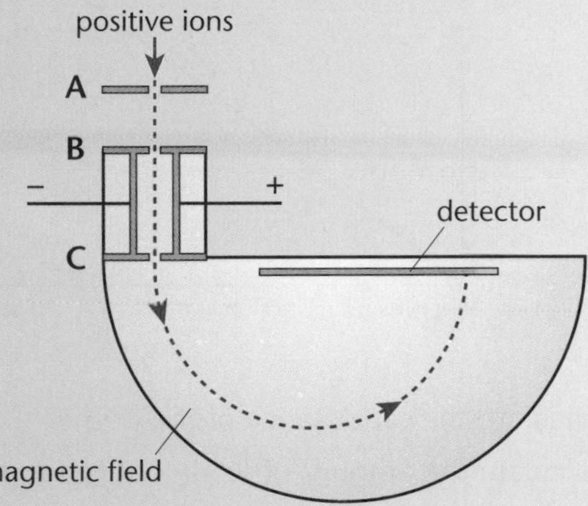

(a) Suggest why the ions must first pass through slits A and B. [1]

(b) State the direction of:
  (i) the electric force on the ions
  (ii) the magnetic force on the ions
  (iii) the magnetic field. [3]

(c) Show that an ion can only pass through slit C if its velocity is equal to the ratio of the field strengths. [2]

(d) (i) Explain why an ion follows a circular path after leaving slit C. [2]
  (ii) An ion of carbon-12 has a mass of 2.0×10^{-26} kg and carries a single positive charge of 1.6×10^{-19} C. It emerges from slit C with a speed of 2.4×10^{5} m s^{-1}.
  The magnetic field strength is 0.15 T.
  Calculate the radius of its circular path. [3]

(e) Radiocarbon dating is a method of estimating the age of objects made from materials that were once alive. In a living object the ratio of carbon-12 atoms to carbon-14 atoms is 10^{12} to one.
Suggest how a mass spectrometer could be used to estimate the age of wood from an old table. [3]

2

In a Van de Graaff generator, charge from a moving belt is transferred onto a spherical metal dome which is insulated from the ground.

The graph below shows how the potential of a dome of radius 0.12 m depends on the charge on it.

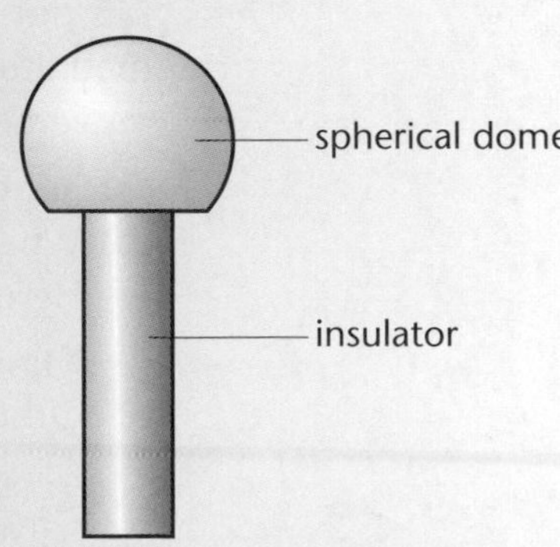

Practice examination questions (continued)

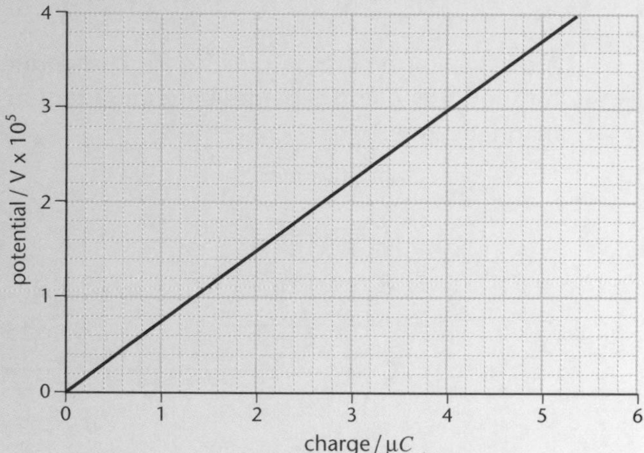

(a) Calculate the capacitance of the dome. [3]

(b) Calculate the amount of work that has to be done to place 4.0 μC of charge on the dome. [3]

The dome loses charge when the electric field strength at its surface reaches a value of 3.0×10^6 N C^{-1}.

(c) (i) Calculate the charge on the dome when the electric field strength at its surface has this value. [3]

$k = 1/4\pi\varepsilon_0 = 9.0 \times 10^9$ N m^2 C^{-2}

(ii) What is the potential of the dome when the electric field strength at its surface has this value? [1]

(d) (i) Calculate the force on an electron, charge = -1.6×10^{-19} C, near the surface of the dome when the electric field strength is 3.0×10^6 N C^{-1}. [3]

(ii) Suggest how the dome loses charge. [3]

3

Read the passage and answer the questions that follow.

Seismometers on the Moon's surface have detected very slight tremors in the body of the Moon. Some of these occur at irregular intervals and appear to originate at the Moon's surface. Others are regular and are repeated on a 14-day cycle. These regular Moonquakes originate deep below the Moon's surface and could be due to tidal stresses caused by the Moon's elliptical orbit.

It is not known whether the Earth undergoes similar tidal stresses, since the small magnitude of the tremors means that they would be indistinguishable from movement caused by other factors.

(a) Suggest one cause of the irregular tremors. [1]

(b) What 'other factors' could cause small Earth tremors? [2]

(c) Here is some data about the Moon and its orbit.

Mass of Moon = 7.4×10^{22} kg
Orbital period = 27.3 days
Closest distance between Moon and Earth = 3.56×10^8 m
Greatest distance between Moon and Earth = 4.06×10^8 m.

Estimate the difference between the greatest and smallest forces that the Earth exerts on the Moon. State clearly any assumptions and approximations that you make and what effect these are likely to have on your answer. [12]

Practice examination answers

1 Mechanics

1 (a) $\Delta E_p = mg\Delta h$ [1] = 40 kg × 10 m s^{-2} × 0.60 m [1] = 240 J [1].
(b) $\frac{1}{2}mv^2$ = 240 J [1] $v = \sqrt{(2 \times 240 \text{ J} \div 40 \text{ kg})}$ [1] = 3.5 m s^{-1} [1].
(c) The child is accelerating [1] upwards [1].
(d) $F = mv^2/r$ [1] = 40 kg × (3.5 m $s^{-1})^2$ ÷ 3.2 m [1] = 153 N [1].
(e) The downward gravitational pull [1] of the Earth [1].
(f) (i) 400 N [1] (ii) 553 N [1].

2 (a) Taking momentum to the right as positive, 0.30 kg m s^{-1} [1] – 0.16 kg m s^{-1} [1] = 0.14 kg m s^{-1} [1].
(b) 0.14 kg m s^{-1} ÷ 1.00 kg [1] = 0.14 m s^{-1} [1] to the right [1].
(c) Total kinetic energy before collision = 0.107 J [1].
Kinetic energy after collision = 9.8 × 10^{-3} J [1].
The collision is inelastic as there is less kinetic energy after the collision than before [1].

3 (a) $v = 2\pi r/T$ [1] = 2 × π × 4.24 × 10^7 m × (24 h × 60 min h^{-1} × 60 s min^{-1}) [1] = 3.08 × 10^3 m s^{-1} [1].
(b) Acceleration = v^2/r [1] = (3.08 × 10^3 m $s^{-1})^2$ ÷ 4.24 × 10^7 m [1] = 0.22 m s^{-2} [1] towards the centre of the Earth [1].
(c) The gravitational pull [1] of the Earth [1].

4 (a) 0.06 kg × 15 m s^{-1} [1] – (– 0.06 kg × 24 m s^{-1}) [1] = 2.34 kg m s^{-1} [1].
(b) Force = rate of change of momentum [1] = 2.34 kg m s^{-1} [1] ÷ 0.012 s [1] = 195 N [1].
(c) 195 N
(d) From kinetic energy of the ball and racket [1] to potential energy of the squashed ball and stretched strings [1] and back to kinetic energy of the ball [1].

5 (a) 600 N × cos 20° [1] = 564 N [1].
(b) Friction between the ground and the log [1]; 564 N [1].
(c) $W = F \times s$ [1] = 564 N × 250 m [1] = 1.41 × 10^5 J [1].
(d) It is transferred to the log and the ground [1] as heat [1].

6 (a) v^2/r [1] = (14 m $s^{-1})^2$ ÷ 100 m [1] = 1.96 m s^{-2} [1].
(b) $F = ma$ [1] = 800 kg × 1.96 m s^{-2} [1] = 1.57 × 10^3 N [1].
(c) (i) The frictional push [1] of the road on the tyres [1].
(ii) The frictional push [1] of the tyres on the road [1].
(d) Friction between the tyres and the road is reduced [1] and may not be sufficient for the centripetal force required [1].

7 (a) 5.0 × 10^5 kg × 60 m s^{-1} [1] = 3.0 × 10^7 kg m s^{-1} [1].
(b) 3.0 × 10^7 kg m s^{-1} ÷ 20 s [1] = 1.5 × 10^6 N [1].
(c) Resistive forces oppose the motion [1] so a greater forwards force is needed to produce this unbalanced force [1].
(d) The aircraft gains momentum in the forwards direction [1]. The exhaust gases gain an equal amount of momentum in the opposite direction [1].

8 (a) $\Delta E_p = mg\Delta h$ [1] = 3.2 × 10^5 kg × 10 m s^{-2} × 180 m [1] = 5.76 × 10^8 J [1].
(b) $\frac{1}{2}mv^2$ = 5.76 × 10^8 J [1] $v = \sqrt{(2 \times 5.76 \times 10^8 \text{ J} \div 3.2 \times 10^5 \text{ kg})}$ [1] = 60 m s^{-1} [1].
(c) ½ × 3.2 × 10^5 kg × (60 m $s^{-1})^2$ – ½ × 3.2 × 10^5 kg × (3 m $s^{-1})^2$ [1] = 5.7 × 10^8 J [1].

9 (a) The frequency an object vibrates at, when it is displaced from its normal position [1] and released [1].
(b) When the frequency of a forcing vibration [1] is equal to the natural frequency [1].
(c) $f = v/\lambda$ [1] = 340 m s^{-1} ÷ 0.15 m [1] = 2.27×10^3 Hz [1].

10 (a) a represents acceleration [1], k represents the spring constant [1], x represents the displacement [1] and m represents the value of the mass [1].
(b) (i) 0.13 m [1].
(ii) $f = 1/T$ [1] = 1 ÷ 1.6 s = 0.625 Hz [1].
(c) $k = 4\pi^2f^2m$ [1]
$= 4\pi^2 \times (0.625)^2 \times 0.5$ [1]
= 7.7 N m^{-1} [1].
(d) (i) $v_{max} = 2\pi fA$ = 0.51 m s^{-1} [1], E_k at this speed = ½ × 0.50 kg × (0.51 m $s^{-1})^2$ [1] = 6.5×10^{-2} J [1].
(ii) At zero displacement [1].

11 (a) A = 0.28 m [1], f = 0.67 Hz [1].
(b) $v_{max} = 2\pi fA$ = 1.18 m s^{-1} [1].
(c) $v = 2\pi f\sqrt{(A^2 - x^2)}$ [1] = 0.82 m s^{-1} [1]

2 Fields

1 (a) $F = kQ_1Q_2/r^2$ [1] = 9.0×10^9 N m^2 C^{-2} × $(1.6 \times 10^{-19}$ C$)^2$ ÷ $(5.2 \times 10^{-11}$ m$)^2$ [1] = 8.5×10^{-8} N [1].
(b) (i) $E = kQ/r^2$ [1] = 9.0×10^9 N m^2 C^{-2} × 1.6×10^{-19} C ÷ $(5.2 \times 10^{-11}$ m$)^2$ [1] = 5.3×10^{11} N C^{-1} [1].
(ii) From the proton towards the electron [1].
(c) $v = \sqrt{(r \times F \div m_e)}$ [1] = $\sqrt{(5.2 \times 10^{-11}}$ m × 8.5×10^{-8} N ÷ 9.1×10^{-31} kg) [1] = 2.2×10^6 m s^{-1} [1].

2 (a) (i) 1250 V [1]
(ii) $E = qV$ [1] = 10 × 1.6×10^{-19} C × 1250 V [1] = 2.0×10^{-15} J [1].
(b) 4.0×10^{-15} J [1].
(c) (i) Top plate positive and lower plate negative [1].
(ii) $F = qE = Vq/d = mg$ [1] $V = mgd/q$ = 5.0×10^{-5} kg × 10.0 m s^{-2} × 6.0×10^{-3} m ÷ 1.6×10^{-18} C [1] = 1.9×10^{12} V [1].

3 (a) (i) $g = GM_E/r^2$ [1] = 6.7×10^{-11} N m^2 kg^{-2} × 6.0×10^{24} kg ÷ $(3.8 \times 10^8$ m$)^2$ [1] = 2.8×10^{-3} N kg^{-1} [1].
(ii) 2.8×10^{-3} N kg^{-1} × 7.4×10^{21} kg [1] = 2.1×10^{19} N [1].
(iii) $T = 2\pi\sqrt{(rM_M/F)}$ [1] = $2\pi\sqrt{(3.8 \times 10^8}$ m × 7.4×10^{21} kg ÷ 2.1×10^{19} N) [1] = 2.3×10^6 s [1].
(b) (i) $F = GMm/r^2$ [1] = 6.7×10^{-11} N m^2 kg^{-2} × 7.4×10^{21} kg × 1.0 kg ÷ $(3.8 \times 10^8$ m$)^2$ [1] = 3.4×10^{-6} N [1].
(ii) The Sun has a greater mass than the Moon [1] but it is very much further away from the Earth [1] so its gravitational field strength at the surface of the Earth is less [1].
(iii) In position A the gravitational pulls of the Moon and the Sun are in the same direction [1], in position B they are in opposite directions, so the resultant force on the water is smaller [1].

4 (a) From N to S (radially outwards) [1].
(b) Out of the paper [1].
(c) The direction of the force on the electromagnet reverses [1] when the current reverses direction [1].
(d) $F = BIl = BI2\pi rN$ [1] $= 0.85\text{ T} \times 0.055\text{ A} \times 2 \times \pi \times 2.5 \times 10^{-2}\text{ m} \times 150$ [1] $= 1.1\text{ N}$ [1].

5 (a) (i) $E_k = eV$ [1] $= 1.6 \times 10^{-19}\text{ C} \times 2500\text{ V}$ [1] $= 4.0 \times 10^{-16}\text{ J}$ [1].
(ii) $v = \sqrt{(2E_k/m)}$ [1] $= \sqrt{(2 \times 4.0 \times 10^{-16}\text{ J} \div 9.1 \times 10^{-31}\text{ kg}}$ [1] $= 3.0 \times 10^{7}\text{ m s}^{-1}$ [1].
(b) (i) Down (from top to bottom of the paper) [1].
(ii) There is an unbalanced force on the electrons [1] which is always at right angles to the direction of motion [1].
(iii) $r = mv/Bq$ [1] $= 9.1 \times 10^{-31}\text{ kg} \times 3.0 \times 10^{7}\text{ m s}^{-1} \div (2.0 \times 10^{-3}\text{ T} \times 1.6 \times 10^{-19}\text{ C})$ [1] $= 8.5 \times 10^{-2}\text{ m}$

6 (a) (i) $\phi = BA$ [1] $= 0.15\text{ T} \times 8.0 \times 10^{-2}\text{ m} \times 5.0 \times 10^{-2}\text{ m}$ [1] $= 6.0 \times 10^{-4}\text{ Wb}$ [1].
(ii) 0 [1].
(b) (i) The reading is a maximum when the plane of the coil passes through the vertical position [1] as the rate of change of flux is greatest [1]. The reading decreases to zero when the plane of the coil is horizontal [1] as the rate of change of flux decreases to zero [1].
(ii) Generator/dynamo [1].
(c) Any three from: increase the speed of rotation, increase the number of turns on the coil, increase the area of the coil, increase the strength of the magnetic field [1 mark each].

3 Thermal and particle physics

1 (a) (i) A temperature difference of 10 K can be measured more precisely than one of 1 K [1]
(ii) There is less energy loss to the surroundings [1]
(iii) The energy supplied heats the block and replaces the energy lost to the surroundings [1] so this should give a higher value [1]
(iv) Insulate the block to reduce energy losses [1].
(b) $c = E/m\Delta\theta$ [1] $= 24\text{ W} \times 350\text{ s} \div (0.80\text{ kg} \times 14\text{ K})$ [1] $= 750\text{ J kg}^{-1}\text{ K}^{-1}$ [1].

2 (a) ΔU is the increase in internal energy [1]
Q is the heat supplied to the gas [1]
W is the work done on the gas [1].
(b) The temperature drops [1] as the heat removed from the gas is greater than the work done on it [1].

3 (a) Efficiency = useful power output ÷ power input = 2 kW ÷ 8 kW [1] = 0.25 [1].
(b) $(T_H - T_C) \div T_H = (1023\text{ K} - 423\text{ K}) \div 1023\text{ K}$ [1] $= 0.59$ [1].
(c) Some energy is wasted as heat [1].

4 (a) 25 + 273 = 298 K [1]
(b) pV = nRT

$n = \frac{pV}{RT}$ [1] = $\frac{2.02 \times 10^5 \times 2.2 \times 10^{-6}}{8.3 \times 298}$ [1] = 1.80×10^{-4} [1]

(c) (i) The total kinetic and potential energy of its particles [1]
(ii) Kinetic energy [1]
(iii) It doubles [1], internal energy of an ideal gas is proportional to temperature [1]
(iv) pV = nRT and since p, n and R all remain the same the volume doubles when the temperature does [1]. New volume is 4.40×10^{-6} m^3 [1].

5 (a) (i) $^{239}_{92}U$ [1]
(ii) the nucleus is unstable [1] it splits into two smaller nuclei [1] and a small number of free neutrons [1]
(iii) the free neutrons from any previous fission [1] can combine with more nuclei (of $^{239}_{92}U$) and induce fission [1]
(b) (i) Binding energy per nucleon increases [1]
Mass of particles decreases [1]
(ii) Kinetic energy [1] of the fission products [1]
(iii) (Primary) coolant carries energy away from the reaction [1]
This heats steam [1]
The steam turns turbines [1]
(c) (i) absorb neutrons [1]
(ii) slows fast neutrons down to thermal energies [1]

6 (a) 295 K [1].
(b) $n = \frac{pV}{RT}$ [1] = $\frac{(1.2 \times 10^5 \times 1.5 \times 10^{-4})}{8.3 \times 295}$ [1] = 7.4×10^{-3} [1].
(c) $\frac{p_1}{T_1} = \frac{p_2}{T_2}$

$T_2 = \frac{p_2 T_1}{p_1}$ [1]

$= \frac{(2 \times 10^5 \times 295)}{1.2 \times 10^5}$ [1]

= 492 K [1].

(d) (i) mean kinetic energy = $\frac{3}{2}$ kT [1]

$= \frac{(3 \times 1.38 \times 10^{-23} \times 295)}{2}$

$= 6.1 \times 10^{-21}$ J [1]

(ii) 590 K [1], temperature in Kelvin has doubled, mean kinetic energy is proportional to Kelvin temperature [1]

7 $\Delta E = ml + mc\Delta\theta$ [1]
$= 1.8 \times 2.25 \times 10^6 + 1.8 \times 4.2 \times 10^3 \times 15$ [1]
$= 4.16 \times 10^6$ J [1]

8 (a) $t = mc\Delta\theta/P$ [1] = 1.50 kg × 4.2 × 10^3 J kg^{-1} K^{-1} × 90 K ÷ 2.40 × 10^3 W [1] = 236 s [1].
(b) The element absorbs energy [1], the casing absorbs energy [1], energy is lost through conduction/convection/evaporation/radiation [1].

9 (a) Gamma [1], gamma can penetrate tissue but beta is absorbed by tissue [1].
(b) The beta radiation emitted by iodine-131 is absorbed and causes ionisation [1], resulting in the damage of cells/tissue [1].
(c) Iodine-123 has a shorter half-life [1] so less is needed to produce the same rate of decay [1].
(d) After 6 half-lives [1] = 78 hours [1].

10 (a) (i) Electrons are not repelled by the nucleus [1]. They have wave-like behaviour so are diffracted by a nucleus [1].
(ii) The higher the energy, the shorter the wavelength [1]. The wavelength needs to be short so that it is of the order of the size of a nucleus [1] for diffraction to occur [1].
(b) (i) $R = 1.20 \times 10^{-15}$ m $\times 4^{1/3}$ [1] = 1.90×10^{-15} m [1].
(ii) 3.81×10^{-15} m [1].
(c) The densities are the same [1]. The radius of the sulphur nucleus is twice that of the helium nucleus [1] so its mass should be $2^3 = 8$ times as great if it has the same density, which it is [1].

11 (a) This shows that most alpha particles do not go close to a region of charge [1], so most of the volume of an atom is empty space [1].
(b) There is a tiny concentration of charge in the atom [1] which must be the same charge as the alpha particles [1] to repel them [1]

12 (a) (i) hf is the energy of a photon [1], Φ is the work function, i.e. minimum photon energy that causes emission [1] and $(\frac{1}{2}mv^2)_{max}$ is the maximum amount of kinetic energy of an emitted electron.
(ii) Some electrons need more than the minimum energy to liberate them, so they have less than the maximum kinetic energy [1].
(b) (i) $E = hf = 2.02 \times 10^{15}$ Hz $\times 6.63 \times 10^{-34}$ J s [1] = 1.34×10^{-18} J [1].
(ii) 4.5×10^{-6} W m^{-2} $\times 2.5 \times 10^{-6}$ m^2 $\div 1.34 \times 10^{-18}$ J [1] = 8.40×10^6 [1].
(iii) $hf - (\frac{1}{2}mv^2)_{max}$ [1] = 1.34×10^{-18} J $- 4.05 \times 10^{-19}$ J = 9.35×10^{-19} J [1] $\div 1.60 \times 10^{-19}$ J eV^{-1} = 5.8 eV [1].

4 Further physics

1 (a) The intensity decreases [1].
(b) (i) Intensity = $10 \log (I/I_0)$ [1] = $10 \log (6.0 \times 10^{-5}$ W m$^{-2} \div 1.0 \times 10^{-12}$ W m$^{-2})$ [1] = 78 dB [1].
(ii) $r_0 = \sqrt{r_1^2(I/I_0)}$ [1] = $\sqrt{(6.0 \times 10^{-5} \text{ W m}^{-2} \div 1.0 \times 10^{-12} \text{ W m}^{-2})}$ [1] = 7.7×10^3 m [1].
(iii) 0 [1].
(c) The perceived loudness of a sound is greatest at a frequency of 3000 Hz [1]. It decreases [1] as the frequency becomes lower or higher [1].

2 (a) An A-scan [1] as this measures depth below the surface of the skin [1].
(b) $t = 2 \times s \div v$ [1] = 0.12 m ÷ 1500 m s^{-1} = 8.0×10^{-5} s [1].
(c) So that there is no interference [1] between the pulse and its reflection [1].
(d) There needs to be a time lag when the crystal is not emitting ultrasound [1] so that it can detect the reflection [1].
(e) $\lambda = 6.0 \times 10^{-2}$ m ÷ 200 = 3.0×10^{-4} m [1]; $f = v/\lambda$ = 1500 m s^{-1} ÷ 3.0×10^{-4} m [1] = 5.0×10^{6} Hz [1].
(f) The gel maximises the energy transmitted into the body [1] by reducing the amount that is reflected [1].
(g) (i) A CAT-scanner uses an X-ray beam [1] that rotates around the body [1] and produces a three-dimensional image of the body [1].
(ii) Advantages: the CAT-scan produces a three-dimensional image [1], tumours are more easily detected [1]. Disadvantage: X-rays can be more damaging to body cells and tissue [1].

3 (a) (i) Apparent magnitude is a measure of the brightness when viewed from the Earth [1].
Absolute magnitude is the apparent magnitude the star would have at a distance of 10 pc [1].
(ii) The perceived brightness of a star decreases with increasing distance [1]. Absolute magnitude takes account of this [1].
(b) $M = m - 5 \log (d/10)$ [1] = 1.3 [1].
(c) $d = 10 \log^{-1} (m-M)/5$ [1] = 240 pc [1].
(d) White dwarfs are smaller than red giants [1] so radiation is emitted from less surface area [1].

4 (a) An absorption spectrum is what remains of the full spectrum after passing through an element [1]. Elements absorb the same wavelengths as they emit [1] so the absorption spectrum has these wavelengths removed from the full spectrum, leaving dark lines [1].
(b) (i) The star is moving away from the Earth [1], causing the wavelengths received to be longer than those emitted [1].
(ii) $v = c\Delta\lambda/\lambda$ [1] = 3.00×10^{8} m s^{-1} × (0.277 ÷ 6.563) [1] = 1.27×10^{7} m s^{-1} [1].
(iii) $d = v/H$ [1] = 5.80×10^{24} m [1] to 6.20×10^{24} m [1].

5 A shows an open Universe [1], the Universe will continue to expand if there is insufficient mass for gravitational forces to stop this [1].
B shows a steady-state [1] where the expansion stops due to gravitational attraction of the mass in the Universe, for this to happen the Universe needs to have the critical mass [1].
C shows the Universe contracting to a 'Big Crunch' [1], this will only happen if the mass of the Universe is greater than the critical mass [1]

5 Synthesis and How Science Works

1 (a) To produce a parallel beam [1].
(b) (i) from right to left [1]
(ii) from left to right/anticlockwise [1]
(iii) into the paper [1].
(c) To move in a straight line, $Bqv = Eq$ [1] so $v = E/B$ [1].
(d) (i) The only force acting is the magnetic force [1], this is always at right angles to the direction of motion [1].
(ii) $r = mv/Bq$ [1] $= 2.0 \times 10^{-26}$ kg $\times 2.4 \times 10^{5}$ m s^{-1} $\div$ (0.15 T $\times 1.6 \times 10^{-19}$ C) [1] $= 0.20$ m [1].
(e) The ratio of carbon-14 atoms to carbon-12 atoms decreases after a tree is felled. Use a sample of carbon from the wood in the table in the spectrometer; the carbon-14 atoms follow a path with a greater radius of curvature than that of the carbon-12 atoms, as m increases then the radius of curvature increases [1]. The ratio of carbon-14 to carbon-12 can be determined by using counters [1].

2 (a) $C = Q/V$ [1] $= 4.8 \times 10^{-6}$ C $\div 3.6 \times 10^{5}$ V [1] $= 1.33 \times 10^{-11}$ F [1].
(b) $W = ½QV$ [1] $= ½ \times 4.0 \times 10^{-6}$ C $\times 3.0 \times 10^{5}$ V [1] $= 0.60$ J [1].
(c) (i) $E = kQ/r^2$, hence $Q = Er^2/k$ [1] =
3.0×10^{6} N C^{-1} $\times$ (0.12 m)^2 $\div 9.0 \times 10^{9}$ N m^2 C^{-2} [1] $= 4.8 \times 10^{-6}$ C [1].
(ii) $V = kQ/r = Er$ [1] $= 3.0 \times 10^{6}$ N V^{-1} $\times 0.12$ m [1] $= 3.6 \times 10^{5}$ V [1].
(d) (i) $F = Eq$ [1] $= 3.0 \times 10^{6}$ N C^{-1} $\times 1.6 \times 10^{-19}$ C [1] $= 4.8 \times 10^{-13}$ N [1].
(ii) The force on the outer electrons ionises the air [1], movement of these ions causes further ionisation [1], the dome discharges as it attracts ions of the opposite charge to it [1].

3 (a) Collisions with meteors [1].
(b) Any two from traffic, wind, tides, minor earthquakes [2].
(c) The mean speed of the Moon is calculated by assuming that the path is circular [1] with a mean radius of 3.81×10^{8} m [1].
mean speed = ($2 \times \pi \times 3.81 \times 10^{8}$ m) $\div$ ($27.3 \times 24 \times 60 \times 60$ s) [1]
$= 1.01 \times 10^{3}$ m s^{-1} [1].
The gravitational force between the Moon and the Sun is the centripetal force [1].
Force at closest distance = mv^2/r [1]
$= 7.4 \times 10^{22}$ kg $\times$ (1.01×10^{3} m s^{-1})^2 $\div$ (3.56×10^{8} m) [1]
$= 2.12 \times 10^{20}$ N [1]
Force at greatest distance $= 1.88 \times 10^{20}$ N [1].
The difference between these forces $= 2.4 \times 10^{19}$ N.

The actual difference is likely to be greater than this because of the assumption that the Moon travels at a constant speed [1]. In fact it speeds up as it gets closer to the Earth and slows down as it moves further away [1]. The effect of this would be to make the largest force larger and the smallest force smaller [1].

Notes

Notes

Notes